BIODIVERSITY OF FRESHWATER ECOSYSTEMS

Threats, Protection, and Management

BIODIVERSITY OF FRESHWATER ECOSYSTEMS

Threats, Protection, and Management

Edited by

Suhaib A. Bandh, PhD

Basharat Mushtaq, PhD

Mohammad Farooq Mir, PhD

Shahid Ahmad Wani

AAP | APPLE ACADEMIC PRESS

First edition published 2023

Apple Academic Press Inc.
1265 Goldenrod Circle, NE,
Palm Bay, FL 32905 USA

760 Laurentian Drive, Unit 19,
Burlington, ON L7N 0A4, CANADA

CRC Press
6000 Broken Sound Parkway NW,
Suite 300, Boca Raton, FL 33487-2742 USA

4 Park Square, Milton Park,
Abingdon, Oxon, OX14 4RN UK

Library and Archives Canada Cataloguing in Publication

Title: Biodiversity of freshwater ecosystems : threats, protection, and management / edited by Suhaib A. Bandh, PhD, Basharat Mushtaq, PhD, Mohammad Farooq Mir, PhD, Shahid Ahmad Wani.
Names: Bandh, Suhaib A., editor. | Mushtaq, Basharat, editor. | Farooq Mir, Mohammad, editor. | Wani, Shahid Ahmad, editor.
Description: First edition. | Includes bibliographical references and index.
Identifiers: Canadiana (print) 2022024426X | Canadiana (ebook) 20220244359 | ISBN 9781774910023 (hardcover) | ISBN 9781774910030 (softcover) | ISBN 9781003277125 (ebook)
Subjects: LCSH: Freshwater biodiversity. | LCSH: Aquatic biodiversity. | LCSH: Freshwater ecology.
Classification: LCC QH96.8.B53 B56 2023 | DDC 577.6—dc23

Library of Congress Cataloging-in-Publication Data

CIP data on file with US Library of Congress

ISBN: 978-1-77491-002-3 (hbk)
ISBN: 978-1-77491-003-0 (pbk)
ISBN: 978-1-00327-712-5 (ebk)

About the Editors

Suhaib A. Bandh, PhD

Suhaib A. Bandh, PhD, is working as an Assistant Professor in the Higher Education Department of Jammu and Kashmir Government. He teaches courses on environmental science and disaster management at the graduate level. He is President of the Academy of EcoScience in addition to being a life member of the Academy of Plant Sciences India, and National Environmental Science Academy. He is a recipient of many awards and has participated in a number of national and international conferences organized by different reputed scientific bodies in India and abroad. He has a number of scientific publications to his credit, published in some highly reputed and impacted journals, which attests to his scientific insight, fine experimental skills, and outstanding writing skills. Dr. Bandh has recently edited and authored a number of books, including *Freshwater Microbiology: Perspectives of Bacterial Dynamics in Lake Ecosystem, Environmental Management and Perspectives of Environmental Science* with leading publishing houses like Elsevier, Springer Nature, and Callisto Reference.

Basharat Mushtaq, PhD

Basharat Mushtaq, PhD, is working as Chief Chemist, 4 MLD Sewage Treatment Plant, Mehandi Kadal Ananatnag, J&K under Bionics Consortium Pvt Ltd., Bionics Advanced Filtration System, New Delhi. He earned his doctorate in Environmental Science (Aquatic Ecology) on the topic "Variations in macrozoobenthos communities in various basins of Dal Lake, Srinagar, Kashmir." He has a strong interest in aquatic ecology and studied the water and sediment chemistry and different aspects of macrozoobenthic communities. His overarching goal is to provide a general and predictive understanding of aquatic ecology and to promote the application of basic ecological science to water quality and conservation. A central focus of his research has been lake ecosystems and how human influences affect water quality and biological integrity. Further, he has worked as an Associate Functional Area Expert in an Environmental

Impact Assessment (EIA) project at the University of Kashmir, where he focused on air and soil quality and achieved expertise on these trust areas. He has published many research articles in reputed refereed international and national journals.

Mohammad Farooq Mir, PhD

Mohammad Farooq Mir, PhD, is working as a Principal in the Higher Education Department, Govt. of Jammu and Kashmir, India. Prof. Mir has also worked as Coordinator of the Hydrobiology Research Laboratory located at Sri Pratap College Campus of Cluster University Srinagar, India. He earned his doctorate in Zoology on "Macrozoobenthos fauna in different freshwater ecosystems of Kashmir." He has completed a number of major and minor research projects from different funding agencies of India. Presently, his focus of study is cold water fisheries and the trophic status of freshwater bodies of Kashmir. He has a strong interest in aquatic ecology and, as such, has guided many research students in this field. Prof. Mir has worked in different capacities in the Jammu and Kashmir Higher Education Department since its inception. He has published a number of research articles in reputed refereed international and national journals.

Shahid Ahmad Wani

Shahid Ahmad Wani, is working as Head P.G. Department of Environmental Science, S. P. College Campus, Cluster University Srinagar, India. He has cleared the National Eligibility Test (NET) in Environmental Science conducted by UGC, New Delhi, India. He is pursuing his PhD at Lovely Professional University, Chandigarh Punjab, in the area of aquatic ecology. He has a strong interest in aquatic ecology and, as such, has guided a number of postgraduate students in this area. Dr. Wani has also worked as in-charge controller of examination at Cluster University Srinagar. He has also worked in different capacities in the Jammu and Kashmir Higher Education Department. He has published many research articles in reputed refereed international and national journals.

Contents

Contributors

Uzma Afaq
Assistant Professor, Department of Biosciences, Integral University, Lucknow – 226026, Uttar Pradesh, India

Asmat Ara
Research Scholar, Division of Genetics and Plant Breeding, SKUAST-K, Jammu and Kashmir, India

Sudesh Batuwita
The Society for the Biodiversity Conservation, Kandy, Sri Lanka

Gresh Chander
Indian Council of Medical Research-Center for Advance Research, Shri Mata Vaishno Devi University, Katra, Jammu and Kashmir, India, E-mail: gresh.chander@smvdu.ac.in

Naseer Ahmad Dar
Department of Environmental Sciences, Government Degree College, Charar-i-Sharief, Jammu and Kashmir – 191112, India

Udeni Edirisinghe
Postgraduate Institute of Agriculture, University of Peradeniya, Sri Lanka

Zia Ur Rahman Farooqi
Institute of Soil and Environmental Sciences, University of Agriculture, Faisalabad – 38040, Pakistan, Mobile: +923336809609, E-mail: ziaa2600@gmail.com

Bashir Ahmad Ganaie
Center of Research for Development, University of Kashmir, Jammu and Kashmir, India

Aadil Gulzar
Department of Environmental Science, University of Kashmir, Srinagar, Jammu and Kashmir – 190006, India, E-mail: ehaan900@gmail.com

Nouman Gulzar
Institute of Soil and Environmental Sciences, University of Agriculture, Faisalabad – 38040, Pakistan

Ruquia Gulzar
Department of Botany, University of Kashmir, Srinagar, Jammu and Kashmir – 190006, India

Musheerul Hassan
Clybay Private Limited, Bangalore – 560114, Karnataka, India, E-mail: musheer123ni@gmail.com

Sumaya Hassan
Department of Biochemistry, University of Kashmir, Hazratbal, Srinagar – 190006, Jammu and Kashmir, India

Tabasum Hassan
Department of Botany, University of Kashmir, Srinagar, Jammu and Kashmir – 190006, India

Predrag Ilic
PSRI Institute for Protection and Ecology of the Republic of Srpska, Banja Luka, Vidovdanska 43, Banja Luka – 78000, Republic of Srpska, Bosnia and Herzegovina

Mahroofa Jan
Lecturer Department of Botany, Government Degree College, Ganderbal – 191201, Jammu and Kashmir, India

Nafeesa Farooq Khan
Department of Botany, Biological Invasion Lab, University of Kashmir, Hazratbal, Srinagar – 190006, Jammu and Kashmir, India, E-mail: khnnafisa@gmail.com

Moaz Khursheed
Institute of Soil and Environmental Sciences, University of Agriculture, Faisalabad – 38040, Pakistan

Ibraq Khurshid
Department of Zoology, Central University of Kashmir, Jammu and Kashmir – 191201, India

Gyanendra Kumar
Assistant Professor and Head, Department of Zoology, National P.G. College, Lucknow – 226017, Uttar Pradesh, India, Mobile: 7518763185, E-mail: gyanendra.mail@gmail.com

Rakesh Kumar
Indian Council of Medical Research-Center for Advance Research, School of Biotechnology, Shri Mata Vaishno Devi University, Katra, Jammu and Kashmir, India, E-mail: kumar.rakesh@smvdu.ac.in

Lone Rafiya Majeed
Vivekananda Global University, Jaipur, Rajasthan – 303012, India, E-mail: rafiyaevs@gmail.com

Uzma Naseer
Department of Zoology, Central University of Kashmir, Jammu and Kashmir – 191201, India

Bisma Nisar
Department of Environmental Science, University of Kashmir, Jammu and Kashmir – 190006, India

Heena Nisar Pahlavi
Department of Environmental Science, University of Kashmir, Jammu and Kashmir – 190006, India

Sami Ullah Qadir
Department of Environmental Sciences, Government Degree College, Kokernag, Kashmir – 192202, India, Phone: 9797831608; 7889310156; E-mail: samievs84@gmail.com

Muhammad Usama Raheem
Institute of Soil and Environmental Sciences, University of Agriculture, Faisalabad – 38040, Pakistan

Vaseem Raja
Department of Botany, Government Degree for Women, Pulwama, Jammu and Kashmir – 192301, India

Sumaira Rashid
Department of Environmental Science, University of Kashmir, Jammu and Kashmir – 190006, India

Zafar A. Reshi
Department of Botany, Biological Invasion Lab, University of Kashmir, Hazratbal,
Srinagar – 190006, Jammu and Kashmir, India

Manzoor Ahmad Shah
Department of Botany, University of Kashmir, Jammu and Kashmir, India

Ruchi Shah
Department of Biotechnology, Kashmir University, Jammu and Kashmir, India,
E-mail: scientistdobt@gmail.com

Sampath Udugampala
Wildlife Conservation Society-Galle, Hiyare Reservoir, Hiyare, Sri Lanka

Sonali Verma
Indian Council of Medical Research-Center for Advance Research, Shri Mata Vaishno Devi University,
Katra, Jammu and Kashmir, India, E-mail: Sonali.verma@smvdu.ac.in

Rifat Ara Wani
Center of Research for Development, University of Kashmir, Jammu and Kashmir, India,
E-mail: rifatevs@gmail.com

Sayar Yaseen Wani
Department of Environmental Sciences, Government Degree College, Shopian,
Jammu and Kashmir – 192303, India

Muhammad Saad Bin Zafar
Institute of Soil and Environmental Sciences, University of Agriculture, Faisalabad – 38040,
Pakistan

Abbreviations

BCF	bio-concentration factor
CBD	convention on biological diversity
CCS	carbon capture and sequestration
Cd	cadmium
COI	cytochrome oxidase I
Cr	chromium
EF	enrichment factor
ELOHA	ecological limits to hydrologic alteration
GHGs	greenhouse gases
GM	genetically modified
GSI	gonadosomatic index
HYR	higher yield rice
IFEBC	integrative freshwater ecology and biodiversity conservation
IMD	Indian Metrological Department
IPCC	intergovernmental panel on climate change
IUCN	International Union for Conservation of Nature
LAWDA	Lakes and Waterways Development Authority
LCIA	life cycle impact assessment
LHPs	hydropower plants
LPI	living planet index
LULCC	land use land cover change
MEA	millennium ecosystem assessment
NAS	National Academy of Sciences
NO_3^-	nitrate
PAH	polynuclear aromatic hydrocarbons
PCB	polychlorinated biphenyls
PGPB	plant growth-promoting bacteria
SDGs	sustainable development goals
SHPs	small hydropower plants
TFs	transcription factors

TRI	Tea Research Institute
UT	union territory
WHS	World Heritage Site

Preface

The water in lakes, reservoirs, rivers, a large variety of wetlands, or other water bodies outside the oceans is freshwater. The interrelationship of freshwater species and their natural and cultural environments are known as freshwater ecology. Freshwater ecosystems provide various plant and animal species, whose composition depends on food supply, oxygen, temperature, and sunlight. Freshwater ecosystems are essential biodiversity centers, though less vast than the oceans. Freshwater habitats include a large variety of plants, animals, and microbes. The smallest of these are the microscopic phytoplankton and zooplankton that form the food chains based in the freshwater ecosystems. Some invertebrates, including worms and insects, also inhabit freshwater. Amphibians, including frogs, live on soil and water among freshwater vertebrates, while fish are solely a water residents. Many species of birds also live in or near a freshwater environment, such as kingfishers and ducks.

There are several ecosystems of diverse complexities within a wide range of freshwater ecosystem types, which sustain a diverse range of permanent and seasonal species that are adaptable to living in an often variable and extreme climate. The importance of freshwater habitats is high for many reasons. They offer humans essential ecosystem services and form a crucial component of the Earth with comparatively higher species resources than marine and terrestrial ecosystems. In terms of freshwater mega-biodiversity, India is one of the world's largest diverse countries and ranks eighth. Freshwater biodiversity represents a significant natural resource in terms of science, education, and socio-economics. It gives human communities a range of valued services and products.

High altitude lakes and the Valley Lakes in India are picturesque water bodies of the Indian Himalayas. These water bodies (streams, reservoirs, pools, and wetlands) are of particular ecological interest as they provide essential food, forage, fish, wildfowls, medicinal plants, manure, hydroelectricity, irrigation, navigation, and drinking water to the local community. Freshwater ecosystems show trophic gradients ranging from oligotrophy to mesotrophy to eutrophy as a result of increased human activity. Due to differences in flow rates, dissolved oxygen, physical substrates, altitude, and the riparian zones that offer nutrition, shade, and refuge are heterogeneous.

The diversity of freshwater ecosystems potentially serves as a model for fundamental research in evolutionary biology and ecosystem studies. Freshwater biodiversity is a much underestimated part of global biodiversity. Freshwater species are often used to monitor ecosystem health improvements, representing the nature of the water bodies. In particular, at the upper multispecies level and in the organization, the freshwater habitats are especially vulnerable to natural and anthropogenic drivers and other stressors. The most used, depended upon, and abused by humans for survival are the biological diversity of freshwater ecosystems. As the population increases exponentially, human beings are using freshwater ecosystems so severely that a range of negative impacts is becoming visible on the water bodies. Habitat destruction and breakdown, unplanned dredging and de-weeding processes, immediate disposal of sewer systems, introduction of invading invasive organisms, water diversions, environmental contamination, and climate change are the leading causes of habitat loss in freshwaters. These impacts have led to significant decreases in the numbers and productivity of many freshwater species and decreased biodiversity in freshwater.

This book consists of a unique set of primary research and review papers covering several aquatic biodiversity studies. It involves evaluating the plant, macroinvertebrate, macrophyte, benthic, and fish diversity in freshwater ecosystems. It provides a data platform on freshwater biodiversity distribution, status, and patterns. It makes comprehensive information currently accessible on the biodiversity of freshwater across various databases. This book aims to promote the enhancement and strengthening of freshwater protection and its unique biodiversity for scientists, policymakers, scholars, researchers, NGOs, and the public.

CHAPTER 1

Freshwater Biodiversity: Importance, Threats, and Management

GYANENDRA KUMAR[1] and UZMA AFAQ[2]

[1]*Department of Zoology, National P.G. College, Lucknow – 226017, Uttar Pradesh, India, Mobile: 7518763185, E-mail: gyanendra.mail@gmail.com*

[2]*Department of Biosciences, Integral University, Lucknow – 226026, Uttar Pradesh, India*

ABSTRACT

Freshwater biodiversity exhibits a wide variety of plants and animals. All of them are synchronously acting and contributing to different ecosystems linked with each other. Natural habitats within freshwater provide them with a place suitable to embellish their population and build harmonious communities. Every organism in an organization is essential to play a significant role at different trophic levels. Humans are directly dependent on freshwater for their daily needs and survival. Their daily needs for livelihood, sports, fishing, ultra-structured engineering, channelization, dam formation, etc., are fragmenting and devastating the natural habitat of freshwater organisms. As a result, most of the species are at risk of extinction and need particular conservation concerns. Recent data collected and analyzed by IUCN, postulate that most aquatic plants, microorganisms, molluscs, fishes, and amphibians are fighting critically for their survival. So, this is the time to think, decide, plan,

Biodiversity of Freshwater Ecosystems: Threats, Protection, and Management. Suhaib A. Bandh, PhD, Basharat Mushtaq, PhD, Mohammad Farooq Mir, PhD, & Shahid A. Wani (Eds.)

and implement the conservational practices. Although the conservational strategies are being planned and executed for several decades and centuries, there is still some need to work efficiently and effectively. Management of them is critically essential to maintain biodiversity and sustain the Earth. Assessing the quality and traits by utilizing next-generation DNA sequencing plays a vital role in evaluating the cause of any change in animal physiology and behavior caused by stressors or mutations. Species richness can be maintained and improvised by a wide variety of parameters discussed in this chapter.

1.1 INTRODUCTION

Biodiversity exhibits all of the diverse living beings present on Earth, including the plants, animals, bacteria, fungi, or other living microorganism put together. There is a countless and rich biodiversity present on Earth, with most of them still discovered. Many of them are standing at the edge of extinction because of the malfunctioned human practices. Biodiversity act as a tool to define the variety of living being at genetic, species, and ecosystem level. Genetic diversity defines the diverse genetic makeup and phenotypic existence of the organisms. Species diversity defines the diversified species of different populations harboring an area. Ecological diversity shows the variations in the plant and animal species living together, their habitats, biological communities and their interactions through different ecological communications and energy cycles.

The Earth is divided superficially into some realms or geospheres like atmosphere, lithosphere, hydrosphere, and biosphere. The atmosphere acts as a scaffold of gases surrounding the Earth. The Lithosphere represents the rigid outer shell; Hydrosphere holds all the water girdled on or in the Earth; biosphere encompasses the space in which all the living beings are present. All of these geospheres act as habitats for a variety of organisms. The biosphere includes biodiversity divisible into aerial, terrestrial, and aquatic zones. Freshwater habitat is divided mainly into lotic (running) and lentic (standing) water bodies. While lotic water bodies act as geologically stable habitats for the species, the less stable lentic water bodies need their organisms to be more dispersive for their maintenance and survival.

Water is a mixture of dissolved chemical gradients and components essential for the survival of aquatic organisms. The freshwater composition is directly or indirectly influenced by ecological conditions, crust layers of the geosphere, biotic, and abiotic interactions. Increasing foliage debris in freshwater habitats can result in eutrophication and lowered concentration of dissolved oxygen which may cause a mass extinction of inhabitants. Freshwater bodies act as a unique environment for biodiversity. Due to various atmospheric, ecological, and evolutionary events like evaporation, precipitation, and deposition, they develop stable and variable biomes (Dijkstra et al., 2014).

Freshwater ecosystems that comprise less than 1% of the Earth; represent the most varied and vigorous system over the continents (Harrison et al., 2016). They serve to ensure water purification, energy flow, oxygen demand, carbon restoration, etc., for sustainable development (WWAP, 2015; Carrizo et al., 2017). Despite these critical functions, freshwater ecosystems are considered as most threatened globally (WWF, 2016). The rate of depletion of freshwater species compared to the marine and terrestrial species is so high that one out of three species from the freshwater ecosystem is threatened (IUCN, 2016; WWF, 2016, 2018).

The expanding human populations are exploiting all the available niche and resources drastically, due to which the ecosystem and biodiversity are in a significant threat. These are the reasons that are making us witness the sixth mass extinction (Ceballos et al., 2017; Tickner et al., 2020). Freshwater bodies like rivers, lakes, ponds, ditches, large pools, streams, deltas, swamps, springs, etc., act as an excellent biodiversity source. The available data reveals that the freshwater ecosystems are under threat. Being less than 1% of total Earth, they are home to 10% of organisms (IUCN, 2019; Fricke et al., 2019). The human population serves to ensure the continuous supply of the necessities needed for their survival. Regardless of the services provided to nature and humankind, the freshwater ecosystems have been extensively exploited at the expense of ecological benefits, species, population size, interactions among different communities and ecosystems (Ruhi et al., 2018). That is why the depletion and loss of freshwater ecosystems, their members, communities, and populations is greater than terrestrial and forest ecosystem (Gardner and Finlayson, 2018; Grooten and Almond, 2018).

In reduced freshwater bodies and ecosystems, a species' capability to develop, flourish, and increase its population is also affected drastically. Alterations in water composition and flow rate are dependent upon the zoogeography of the terrestrial areas because the phenotypic traits and behavioral responses of a species get affected. All the benefits that man receives from these freshwater bodies and its inhabitants can affect man's health, vigor, and economy. Many environmental factors have been recognized as regulators of any species in a given habitat. These factors act as promoters or stressors for any species for its richness in any ecosystem. Freshwater biodiversity conservation requires the comprehensive perception of pattern, distribution, identification, and sampling of species in response to the cause because they are getting flourished or reduced. Identifying the unknown facts behind the depletion of biodiversity is very important to implement appropriate management practices. It is well known that the management and conservation of the freshwater ecosystem are meagerly attempted and supported compared to the terrestrial ecosystem. Freshwater biodiversity hotspots are examples in this concern where conservational programs, funds, and public awareness have been raised much.

1.2 THREATS

Freshwater supports a wide variety of inhabitants. Biodiversity in freshwater ecosystems is more diverse in comparison to others. However, it is considered one of the most threatened habitats for living beings. A considerable number of known species rely on freshwater as their natural habitat for survival. More than 55% of all known fishes are confined to freshwater ecosystems. Most of the invertebrates, amphibians, fishes, microorganisms, plants, etc., inhabiting freshwater bodies are facing lots of threats (Reid et al., 2019). Humans act as super predators of most fishes and one of the leading causes of the decline in their sustainable population development (Darimont et al., 2015).

About 85% of wetlands have been lost from 1,700 to 2,000. At present, wetland ecosystems are at the edge of extinction more than any terrestrial or forest ecosystem (Diaz et al., 2019). Due to the direct contact with the inland water system, freshwater biodiversity is important at any local level to support basic food and livelihood sources for humans. Regardless of the benefits, they provide to nature and man; they will be extinct faster. About one-third of freshwater species are facing extinction just because of habitat destruction, the introduction of exotic species, pollution, uncontrolled over-harvesting, water extraction, drainage, modification of hydrological systems, fluctuation, and alterations in habitat composition, water habitat division, channelized zones, etc. (Hill et al., 2016).

Although many conservational programs with legislation are being implemented the world over, there is still a need to achieve all the goals planned to execute and get the required results successfully (Bower et al., 2014; Reis et al., 2017). Confined distribution of lotic freshwater bodies, limited expansion towards adjoining terrestrial ecosystems, threatened migratory species without conservational strategies, pollutant effluents, avoiding or disregarding the conservational laws and legislations, limited manpower followed by insufficient funding are some of the major points which demand extra attention to be worked upon for the better conservational strategies of freshwater biodiversity and ecosystem (Adams et al., 2015; Finlayson et al., 2018; Acreman et al., 2020). Some of the threats for freshwater biodiversity of significant recognition are listed below (Figure 1.1):

1. **Changing Climate:** It has considerable impacts on freshwater biodiversity. Temperature impacts different ecosystems differently. Freshwater flow is one of the critical parameters, which is primarily affected by temperature because of which sedimentary deposition accumulates very fast, thus disturbing the migration of aquatic organisms. Precipitation and evaporation caused by increasing temperature affect the water level and habitat composition. Small lentic freshwater bodies get dried off. Because of the change in species community, the natural food web and flow of energy get affected. Reduced ice cover in lakes, ponds or ditches affects the fishes' winter survival that affects the natural food-web compositions. Increased temperature as the primary stressor increases the oxygen demand in aquatic organisms and alters a species' population characteristics, and the restriction among individuals to be present in a limited habitat is increased. Many of the aquatic organisms, especially the salmonid fishes and stenothermic species, are unexpectedly lost.

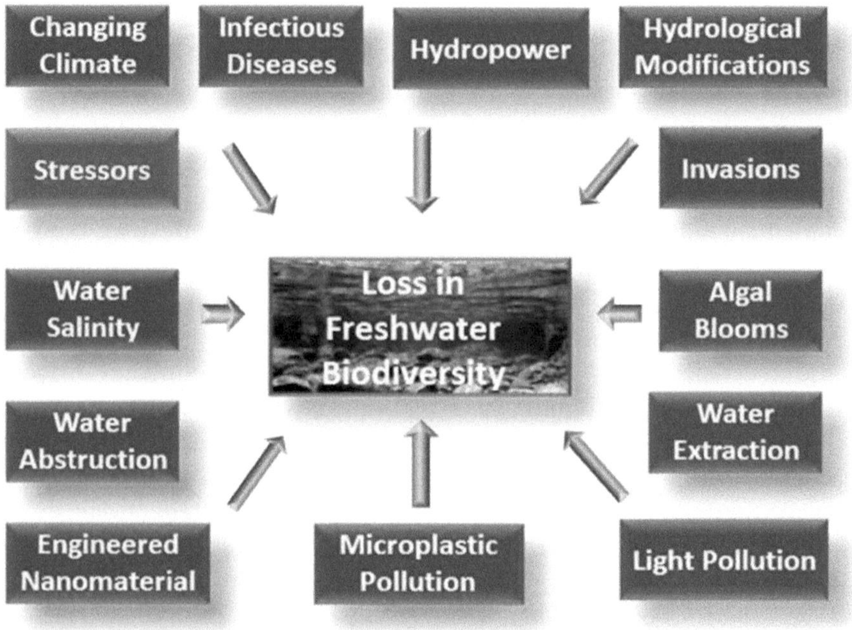

FIGURE 1.1 Different parameters acting as major threats to global freshwater biodiversity.

2. **Infectious Diseases:** Freshwater habitats are an interacting medium between aquatic organisms and a variety of pathogens. Environmental fluctuation plays a vital role in the transmission, host finding and survival of pathogens. Freshwater habitats harbor various pathogens of many infectious diseases for aquatic organisms like whirling disease (in *Salmon*), chytridiomycosis (in amphibians), plague (in crayfish), etc. Introduced non-native aquatic species pose a high risk of pathogen transmission and a significant threat to biodiversity. This introduction could result from human interference with freshwater habitats, especially in the aquaculture industry, where the probability of pathogen or disease transmission is much intensified from cultured fishes to sympatric species.
3. **Hydropower:** Expansion and development of renewable energy resource infrastructure have (Fizko and Martina, 2017) emerged as one of the significant threats to freshwater ecosystems and

biodiversity (Simpson, 2019). Hydropower affects freshwater biodiversity by altering the natural habitats, water quality or composition, flow alteration, habitat fragmentation and land use pattern. Life cycle impact assessment (LCIA) and ecological limits to hydrologic alteration (ELOHA) are some of the methods which can be utilized to assess the global impact on freshwater biodiversity (Gracey and Verones, 2016).

4. **Hydrological Modifications:** This resulted from anthropogenic activities (such as water diversion, urbanization, sealing, drainage, etc.), negatively influence freshwater biodiversity. They cause habitat alteration or fragmentation, obstruction in the flow, thermal variation due to the variable rate of flow, sedimentary deposition, flood wave generation, etc. (Cantonati et al., 2020). It also affects average annual discharge, stabilization of periodic release and accumulated silt or sedimentary depositions (Vella and Vella, 2020).
5. **E-Commerce and Invasions:** E-commerce, trade, and biological invasions are posing attempts towards freshwater biodiversity threats. The introduction of exotic or invasive species for e-commerce and business is negatively affecting the natural freshwater ecosystem. The introduction of invasive species is a threat to native communities when they intercommunicate with them or escape from there (Humair et al., 2015). For example, Turtles of genera, *Chelydra*, and *Macrochelys* native to America are considered highly invasive species in China, where they have been found affecting the natives (Liu et al., 2020).
6. **Harmful Algal Blooms:** The frequency and geographic exposure of algal blooms are increasing globally. Although they are a part of the viable aquatic biodiversity, they become harmful when their population increases or blooms above a level where the toxins produced by them become significantly evident to harm other aquatic organisms and even man. For example, a cyanobacterial algal bloom in the year 2014 devastated Lake Erie, Ohio's natural composition, biodiversity, and ecosystem (Gatz, 2020).
7. **Microplastic Pollution:** Plastic is a material used most commonly in different forms all over the world. However, these are harmful to the environment because of their proper management plans (Windsor et al., 2019). Microplastic is a part of plastic (<5 mm),

which results from degraded plastic over 450–1,000 years. These microplastics enter any natural ecosystem from a wide variety of sources and industrial processes. Microplastics are readily available to neustonic, pelagic, and benthonic freshwater species, and they alter food webs at many trophic levels and niches (Scherer et al., 2018).

8. **Light Pollution:** It is caused by artificial lightening by man. It adversely affects the natural ecosystem by altering natural light: dark cycle and unnecessary illumination in the surroundings (Zapata et al., 2019). When the level of light pollution increases in freshwater habitats by anthropogenic activities, it masks the natural light and modifies the behavior, physiology, and chronobiology of aquatic organisms (Liu et al., 2020).
9. **Sedimentation:** Soil erosion, transportation, deposition, agricultural practices, riverbed incision, aquatic habitat degradation, effluent discharge, industrial release, and intrusion, etc., have increased the issue of sedimentation globally into the freshwater. Sedimentation increases water turbidity, which consequently affects aquatic plants and animals. Sediment causes blocked interstitial spaces for micro-and macro aquatic organisms (Graf et al., 2016). Fine sediments are considered an essential factor in affecting several fishes' population dynamics (Doretto et al., 2020).
10. **Water Abstraction and Dam Construction:** It affects the freshwater ecosystem by altering the water flow, sedimentary deposition, change in temperature, drainage system, the formation of isolated wetlands, etc. All of these, in turn, affect the habitat range, migratory routes, spawning areas, mating, and reproductive behavior of most of the aquatic organisms (Pittock et al., 2015). Dam with a point of use treatment infrastructures is advantageous to hold or release and accumulate or filter the described proportion of freshwater to ensure the reliable composition of high-quality water (Abell et al., 2018).
11. **Ground and Surface Water Extraction:** These activities misfunction the distribution of all the water reservoirs above or below the surface. It is radically modified with eruptive effects on water quality, quantity, composition, and biota in most cases. Water extraction from these bodies causes a reduction in species

number, population, and community structure, which affects the energy flow in the food web. Change in water level causes the migration of the macro-species to suitable sites because of the structure of both the habitat and ecosystems changes drastically (Brkic et al., 2019).

12. **Engineered Nanomaterials:** Nanomaterials, when released directly into the environment, cause eco-toxicological effects. Nanomaterials are typically made from metals or metal oxides, or carbon. They also act as vectoral agents through the adsorption of co-occurring contaminants and pollutants in freshwater. The ultra-fine nanomaterial particles resulted from combustion processes also add toxic effects in this regard (Rist and Hartmann, 2018).
13. **Water Salinity:** Freshwater organisms need balanced ion concentration inside their body cells and bodily fluids. The salinity of the surrounding water plays a significant role in this regard. Freshwater ecosystems are being salinized by anthropogenic activities, causing additive and antagonistic effects on the freshwater inhabitants. Freshwater species show intolerance after exposure to certain salinity threshold levels and becomes toxic after a significant period. However, interactions between ions and the animal body can vary from species to species (Canedo-Arguelles, 2018).
14. **Stressors:** Depletion or extension of any species is primarily affected by several stressors present around. Some of the stressors like high population density, social instability, predation, parasitism, pathogens, resource richness or availability, captivity, sports hunting confinement, translocation, and transportation, etc., play a leading role in maintaining or declining the freshwater biodiversity. Most of the stressors are associated with anthropogenic activities (Jackson et al., 2016).

All of the above-discussed threats, either in single or in combination, may propagate population decline, reduced genetic variability, and can drastically result in species extinction (He et al., 2017).

1.3 MANAGEMENT

The threatened aquatic biodiversity affects the services provided by any freshwater ecosystem. The ecosystem and the large proportion of fresh drinking water available to humans and others are also drastically altered. More than 750 million people cannot access drinking water, which is getting worse over time. By 2050, water demand will increase by about 400% (WWAP, 2015). A large proportion of the global ecosystem suffers from a threat to water security, especially the freshwater organisms in different communities and habitats under significant threat.

Several efforts and practices have been made for the prevention of loss of freshwater biodiversity. Convention on biological diversity (CBD) did set decadal targets (CBD, 2010), but it was a short-term plan named as 'Strategic Plan for Biodiversity 2011–2020.' It was a project to be implemented and completed in a decade and was followed by all country's world over. This project provided 20 targets named 'Aichi Targets' (Leadley et al., 2014). However, a limited capacity framework was a hindrance in achieving these targets successfully. This was improvised to achieve the Aichi Targets and another new implementation of sustainable development goals (SDGs) up to 2030 (Osborn et al., 2015). United Nations adopted SDGs in September 2015, which were important in freshwater conservation, management, restoration, availability, accessibility, sanitation, and sustainable use to enhance the freshwater biodiversity and its different ecosystems. The SDGs-6, i.e., clean water and sanitation, allows, and focuses on the importance of clean water for anyone. It ensures availability, management, and sanitation of water. It explored the extent of wetlands and initiated management of aquatic ecosystems like wetlands, ponds, lakes, and, rivers, etc. The SDGs-12 supports to meet sustainable exploration and management of nature and natural resources. Its prime importance includes minimizing or stopping the release of effluents and pollutants directly into the water. The SDGs-15, i.e., life on land, postulated how freshwater biodiversity is vital for humans' sustenance and livelihood (Turak et al., 2016). It ensures and commits conservation, management, and sustainable use of freshwater ecosystems (Harrison et al., 2016).

In-situ and *ex-situ* conservation of biodiversity helps conserving species according to their habitat. Various national parks, wildlife sanctuaries and biosphere reserves comprise categorized core and buffer zones confined especially for natural water bodies or grasslands. The CBD set 20 targets. In its target 11, 17% of global inland freshwater areas have been conserved. From 1992 at a number of about 575, 96 river reserves and 2,314 wetlands have been managed to increase (Ramsar, 2018). Inland water is always found stable and consistently dynamic, acting as an individual or connecting units with other water bodies. They provide countless interfaces to terrestrial ecosystem also. The vibrant stability of freshwater bodies allows biodiversity to get richly diversified. Their specifications include mobility, stability, outsource obstructions, communication, interaction, fluctuation, and adaptations. All of these are known to contribute significantly to biodiversity. These bear scope to assess, evaluate, quantify, and apply the necessities at species, community or population level based on adaptations towards fluctuations. Many novel conservational tools like environmental flows, environmental DNA, removing the dams, legislative habitat protection, relocation of species, etc., may contribute much better (Reid et al., 2019).

Conservational legislations for freshwater ecosystems and biodiversity are included in two categories. The first one, i.e., 'Pollution and Water Resources, aims to refine the freshwater quality and composition. The second one, i.e., 'Nature Conservation, ' aims to protect the threatened habitats and species in freshwater ecosystems (Hassall et al., 2012). These two categories are now considered the framework in shaping the legislation to incorporate conservational tactics at water resources and natural composition. These were later included in national legislations among 28 European Union Member states (Hassall et al., 2012).

Before implementing conservational tactics, we should first know the present status of the individual species in different ecosystems. Sampling and evaluating the genetic makeup at the allelic range is essential to quantify a species' individual or community level. In this regard, genetic or DNA isolation protocols can significantly assess organisms' genetic variation (Boothroyd et al., 2016). Sampling the freshwater species at the community level in a given habitat or ecosystem is essential in evaluating the taxonomic or zoogeographic range and status of any freshwater species. IUCN Red List is contributing a good deal in this regard. Besides this, the living planet index (LPI) is also helpful to assess threatened species. These clarify the present status, variety, distribution, specific allocation, endemic zones, conservational surplus, etc., for any given species. Change in a species' traits is also one of the essential measures in evaluating a species' characteristics in terms of alterations in specific, generic or population level (Turak et al., 2016). It helps to assess the accumulation of a species in different freshwater habitats according to any fluctuation caused by the surrounding environment, because of which there should always be a tendency to change the life history traits of a freshwater species.

Some of the essential measures which should be taken into account for the management and conservation of freshwater biodiversity are given below (Figure 1.2):

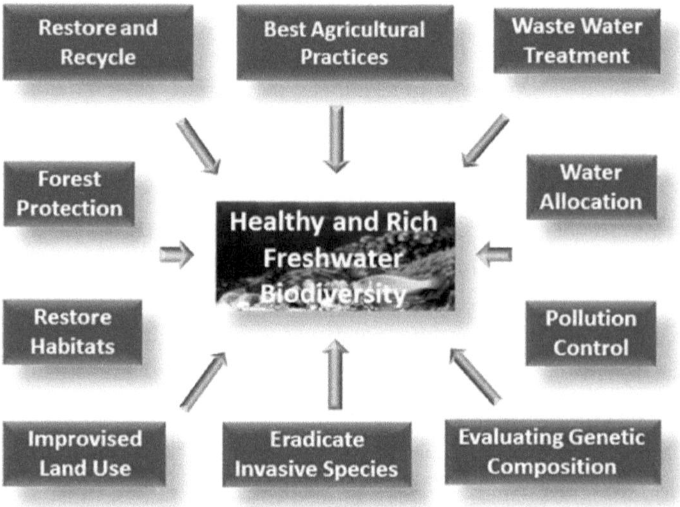

FIGURE 1.2 Different management practices which can improvise and enhance the freshwater biodiversity.

1. **Restoration Practices:** Restoration of natural composition and dynamics of freshwater bodies is essential to maintain sustainable growth and development in diverse aquatic organisms. Ecological integrity is needed to be managed and conserved accordingly (Hwang et al., 2014). Freshwater bodies and biota which are at the risk of extinction need special attention for conservation and restoration. Aquatic ecosystems, particularly the lotic freshwater bodies, demand extensive improvements. Lotic water bodies should have some scope to get connected whenever needed to recover stored water depletion.
2. **Forest Protection:** Forests render services to sustain freshwater biodiversity. They provide land area, textured soil profile, and inter-connecting channels for lotic and lentic freshwater bodies. They tend to pack the soil tightly at their place to stop soil erosion, which is considered one of the vital cause in changing the quality and composition of water quality and composition (Hanna et al.,

2020). They also tend to contribute to maintaining the natural eco-habitat of freshwater bodies.

3. **Best Agricultural Management Practices:** Agriculture includes crops utilizing the freshwater from lotic or lentic bodies. Planting hedgerows, keeping variation in type of species, mulching, and reducing the use of pesticides are some of the tools that can help encourage biodiversity (Tayleur et al., 2017). Aquatic insects and pollinators, which rely on freshwater, help maintain the balance in different freshwater ecosystems.
4. **Water Allocation:** This to fulfill demand and environmental management is essential to sustain and conserve freshwater biodiversity. Consumptive water demand is increasing day by day which is causing overexploitation of water resources. Many trade-offs always exist between the use and benefits of freshwater resources. Therefore, the fulfillment of proper demand and water allocation management are correlated with each other, which define their crucial role in maintaining freshwater biodiversity (Zamani et al., 2019).
5. **Waste Water Treatment:** It is essential to fulfilling our daily demand and utilizing the available freshwater resources equitably and sustainably. We use the water taken and processed from freshwater bodies. A considerable amount of freshwater is discarded and drained off daily from human uses. Daily discard of this water makes us rely on available freshwater resources in a non-independent way. Global water withdrawal, water discharge, contamination, sewage, or sludge is a worldwide threat in this regard (Jaramillo and Restrepo, 2017). Sewage treatment, reed bed filtration, nutrient filtration of wetlands, municipal wastewater management, industrial small scale wastewater management and even the individual home level wastewater management is essential to save freshwater biodiversity.
6. **Pollution Control:** Pollution and especially water pollution, is posing an emerging and growing issue for the sustenance of freshwater ecosystems (Mishra and Kumar, 2020). Human settlements, municipal waste, agricultural, and industrial waste release untreated polluted water directly into freshwater bodies. They are dumping millions of tons of pollutants and sediments in aquatic

bodies every year. This critically affects aquatic life at the level of their survival, growth, captivity, and migration (Baseem, 2020).

7. **Protect and Restore Natural Habitats:** Natural habitats provide their inhabitants with a safe set up to live, grow, mature, reproduce, and increase their population. The habitat is integrity adjoining the environment, climate, nutrients, and interactions with co-occurring organisms for a species. The genetic makeup and traits exhibited are the phenomenon which is associated with modifications or adaptions according to the change in these integrities (Thalinger et al., 2021). Therefore, they should be kept on prime concern to save and maintain freshwater biodiversity.
8. **Eradicate Invasive Non-Native Species:** Non-native species are always a severe threat to native freshwater biodiversity and ecosystems. The presence of non-native species poses many pressures and stresses against native species (Cuthbert et al., 2021). It is mandatory to assess the overlapping zones between native and non-native species distribution to identify and implement particular conservational strategies.
9. **Improvised Land Use:** More than a third of the global land area has been modified, and half of the surface freshwater has been utilized for anthropogenic activities. Improper use of landmasses causes a reduction in available habitat for freshwater species. Their limitation exceeds in confined zones. The ecosystems in these restricted zones get overlapped, and energy flow among different trophic levels gets disturbed (Linke et al., 2019). This leads to the mass devastation of a variety of many freshwater species. Many freshwater organisms like fishes, molluscs, and amphibians are at risk of extinction (Singh et al., 2020).
10. **Evaluating the Change in Genetic Composition:** Genetic diversity among a species' individuals makes it different from other species. It is critical to identify and assess the species richness affected by any stressor or environmental factor. Unfortunately, this point has been meagerly worked upon in the case of freshwater ecosystems. Application of next-generation DNA-sequencing technologies in freshwater biodiversity assessment is required to safeguard and conserve freshwater species richness (Turak et al., 2016).

1.4 CONCLUSIONS

Species diversity and area covered by freshwater is richer in comparison to terrestrial and marine ecosystems. Despite this, it is facing the risk of extinction essentially. Many conservation biologists have proposed and implemented environmental flow, legislation, biodiversity conservation laws, developing the fishways, and removing nano-engineered aquatic infrastructures and dams. Although these options got some significant results, critical success is still required. The human population is increasing fast and is causing extra pressure on water and land area demand. To fulfill their needs, they are exploiting the natural freshwater resources drastically without any future concern. As the top part of the ecosystem present at the superior trophic level, we must get aware of the importance of freshwater biodiversity and its conservation. Despite implementing several laws and legislations, any conservational strategy would fail if we don't contribute our best in this regard. Therefore, we have to come ahead not only at the level of society or population but also as individuals. Tactics to recycle and restore the wastewater should be implemented from house to industry level. Searching for the alternative for the protection of freshwater fish capturing should be done to maintain their population. Effective and sustainable land use planning should be done to safeguard the aquatic organisms' natural freshwater habitats. Environmental DNA assessment tool can also augment the actual data procurement for implementing the conservational strategies better.

KEYWORDS

- **conservation**
- **ecosystem threats**
- **freshwater biodiversity**
- **life cycle impact assessment**
- **living planet index**
- **sustainable development goals**

REFERENCES

Abell, R., Vigerstol, K., Higgins, J., Kang, S., Karres, N., Lehner, B., Sridhar, A., & Chapin, E., (2019). Freshwater biodiversity conservation through source water protection: Quantifying the potential and addressing the challenges. *Aquatic Conservation: Marine and Freshwater Ecosystem, 29*, 1022–1038.

Acreman, M., Hughes, K. A., Arthington, A. H., Tickner, D., & Duenas, M. A., (2020). Protected areas and freshwater biodiversity: a novel systematic review distils eight lessons for effective conservation. *Conservation Letters, 13*, e12684. https://doi.org/10.1111/conl.12684.

Adams, V. M., Setterfield, S. A., Douglas, M. M., Kennard, M. J., & Ferdinands, K., (2015). Measuring benefits of protected area management: Trends across realms and research gaps for freshwater systems. *Philosophical Transactions of the Royal Society B., 370*(1681). https://doi.org/10.1098/rstb.2014.0274.

Baseem, S. M., (2020). Water pollution and aquatic biodiversity. *Biodiversity International Journal, 4*(1), 10–16.

Bower, S. D., Lennox, R. J., & Cooke, S. J., (2014). Is there a role for freshwater protected areas in the conservation of migratory fish? *Inland Waters, 5*, 1–6. https://doi.org/10.5268/IW-5.1.779.

Brkic, Z., Kuhta, M., Larva, O., et al., (2019). Groundwater and connected ecosystems: an overview of groundwater body status assessment in Croatia. *Environmental Sciences Europe, 31*, 75.

Canedo-Arguelles, M., Kefford, B., & Schafer, R., (2018). Salt in freshwaters: causes, effects and prospects - introduction to the theme issue. *Philosophical Transactions of the Royal Society of London: Series B, Biological Sciences, 374*(1764), 20180002. https://doi.org/10.1098/rstb.2018.0002.

Cantonati, M., Poikane, S., Pringle, C. M., Stevens, L. E., Turak, E., Heino, J., Richardson, J. S., et al., (2020). Characteristics, main impacts, and stewardship of natural and artificial freshwater environments: consequences for biodiversity conservation. *Water, 12*(1), 260. https://doi.org/10.3390/w12010260.

Carrizo, S. F., Jahnig, S. C., Bremerich, V., Freyhof, J., Harrison, I., He, F., Langhans, S. D., et al., (2017). Freshwater megafauna: Flagships for freshwater biodiversity under threat. *BioScience, 67*(10), 919–927.

CBD, (2010). *Aichi Biodiversity Targets*. Convention on biological diversity, Montreal, Canada Available from http://www.cbd.int/sp/targets/(accessed on 13 December 2021).

Ceballos, G., Ehrlich, P. R., & Dirzo, R., (2017). Biological annihilation via the ongoing sixth mass extinction signaled by vertebrate population losses and declines. *Proceedings of the National Academy of Sciences, 114,* E6089–E6096.

Cuthbert, R. N., Pattison, Z., Taylor, N. G., Verbrugge, L., Diagne, C., Ahmed, D. A., Leroy, B., et al., (2021). Global economic costs of aquatic invasive alien species. *Science of Total Environment*. doi: https://doi.org/10.1016/j.scitotenv.2021.145238.

Darimont, C. T., Fox, C. H., Bryan, H. M., & Reimchen, T. E., (2015). The unique ecology of human predators. *Science, 349*, 858–860.

Díaz, S., Settele, J., Brondízio, E., Ngo, H., Guèze, M., Agard, J., & Zayas, C., (2019). The global assessment report on biodiversity and ecosystem services. *Intergovernmental*

Science-Policy Platform on Biodiversity and Ecosystem Services. Nairobi, Kenya: United Nations Environment Program.

Dijkstra, K. D. B., Monaghan, M. T., & Pauls, S. U., (2014). Freshwater biodiversity and insect diversification. *Annual Review of Entomology, 59*, 143–163.

Doretto, A., Bona, F., Falasco, E., Morandini, D., Piano, E., & Fenoglio, S., (2020). Stay with the flow: How macroinvertebrate communities recover during the rewetting phase in Alpine streams affected by an exceptional drought. *River Research and Applications, 36*(1), 91–101.

Fijko, R., & Martina, Z., (2017). Impact of small hydropower plant on the biodiversity of the selected area. In: *10th International Conference "Environmental Engineering: Water Engineering*." 10.3846/enviro.2017.077.

Finlayson, C. M., Arthington, A. H., & Pittock, J., (2018). *Freshwater Ecosystems in Protected Areas*. Oxford, UK: Routledge.

Freshwater Blog, (2020). *An Emergency Recovery Plan for Freshwater Biodiversity*. Source: https://freshwaterblog.net/2020/02/27/an-emergency-recovery-plan-for-global-freshwater-biodiversity (accessed on 13 December 2021).

Fricke, R., Eschmeyer, W., & Van, D. L. R., (2019). *Catalog of Fishes: Genera, Species, References*. California Academy of Sciences. http://researcharchive.calacademy.org/research/ichthyology/catalog/SpeciesByFamily.asp (accessed on 13 December 2021).

Gardner, R., & Finlayson, C., (2018). *Global Wetland Outlook: State of the World's Wetlands and Their Services to People*. Ramsar Convention Secretariat.

Gatz, L., (2020). *Freshwater Harmful Algal Blooms: Causes, Challenges, and Policy Considerations*. Congressional Research Service. https://crsreports.congress.gov/product/pdf/R/R44871 (accessed on 13 December 2021).

Gracey, E. O., & Verones, F., (2016). Impacts from hydropower production on biodiversity in an LCA framework- review and recommendations. *The International Journal of Life Cycle Assessment, 21*, 412–428.

Graf, W., Patrick, L., Iris, H., & Christoph, H., (2016). Ecological degradation of a meandering river by local channelization effects: a case study in an Austrian lowland river. *Hydrobiologia, 772*, 145–160.

Grooten, M., & Almond, R., (2018). *Living Planet Report 2018: Aiming Higher.* World Wildlife Fund.

Hanna, D. E. L., Raudsepp-Hearne, C., & Bennett, E. M., (2020). Effects of land use, cover, and protection on stream and riparian ecosystem services and biodiversity. *Conservation Biology, 34*, 244–255.

Harrison, I. J., Green, P. A., Farrel, T. A., Juffe-Bignoli, D., Saenz, L., & Vorosmarty, C. J., (2016). Protected areas and freshwater provisioning: a global assessment of freshwater provision, threats and management strategies to support human water security. *Aquatic Conservation: Marine and Freshwater Ecosystems, 26*(1), 103–120.

Hassall, C., Hill, M., Gledhill, D., & Biggs, J., (2016). The ecology and management of urban pondscapes. In: Francis, R., Millington, J. D. A., & Chadwick, M. A., (eds.), *Urban Landscape Ecology: Science: 536 Policy and Practice*. Routledge: Abingdon, UK.

Hassall, C., Hollinshead, J., & Hull, A., (2012). Temporal dynamics of aquatic communities and implications for pond conservation. *Biodiversity Conservation, 21*, 829–852.

He, F., Zarfl, C., Bremerich, V., Henshaw, A., Darwall, W., Tockner, K., & Jähnig, S. C., (2017). Disappearing giants: A review of threats to freshwater megafauna. *Wires Water, 4*, e1208.

Hill, M., Ryves, D. B., White, J., & Wood, P. J., (2016). Macroinvertebrate diversity in urban and rural ponds: Implications for freshwater biodiversity conservation. *Biological Conservation, 201*, 50–59.

Humair, F., Humair, L., Kuhn, F., & Kueffer, C., (2015). E-commerce trade in invasive plants. *Conservation Biology, 29,* 1658–1665.

Hwang, S. J., Lee, S. W., & Yoo, B., (2014). Ecological conservation and the restoration of freshwater environments in Korea. *Paddy Water Environ., 12*(1), 01–05.

IUCN (International Union for Conservation of Nature), (2019). T*he IUCN Red List of Threatened Species, Version 2019-1*. IUCN. www.iucnredlist.org.

IUCN, (2016). *About the Freshwater Biodiversity Unit.* http://www.iucn.org/about/work/programmes/species/our_work/about_freshwater/(accessed on 13 December 2021).

Jackson, M. C., Loewen, C. J., Vinebrooke, R. D., & Chimimba, C. T., (2016). Net effects of multiple stressors in freshwater ecosystems: a meta-analysis. *Global Change Biology, 22*(1), 180–189. doi: 10.1111/gcb.13028. Epub 2015 Oct 23. PMID: 26149723.

Jaramillo, M. F., & Restrepo, I., (2017). Wastewater reuse in agriculture: a review about its limitations and benefits. *Sustainability, 9*, 1734. https://doi.org/10.3390/su9101734.

Leadley, P. W., Krug, C. B., Alkemade, R., Pereira, H. M., Sumaila, U. R., Walpole, M., Marques, A., et al., (2014). Progress towards the Aichi biodiversity targets: an assessment of biodiversity trends, policy scenarios and key actions. *Secretariat of the Convention on Biological Diversity* (p. 500). Montreal, Canada. Technical Series 78.

Linke, S., Hermoso, V., & Januchowski-Hartley, S., (2019). Toward process-based conservation prioritizations for freshwater ecosystems. *Aquatic Conservation: Marine Freshwater Ecosystem, 29,* 1149–1160.

Liu, S., Newman, C., Buesching, C., Macdonald, D., Zhang, Y., Zhang, K., & Zhou, Z., (2020). E-commerce promotes trade in invasive turtles in China. *Oryx,* 1–4. doi: 10.1017/S0030605319001030.

Liu, Z., Lv, Y., Ding, R., Chen, X., & Pu, G., (2020). Light pollution changes the toxicological effects of cadmium on microbial community structure and function associated with leaf litter decomposition. *International Journal of Molecular Sciences, 21*, 422.

Mishra, S., & Kumar, A., (2020). Estimation of physicochemical characteristics and associated metal contamination risk in river Narmada, India. *Environmental Engineering Research, 26*, 190521. https://doi. org/10.4491/eer.2019.521.

Osborn, D., Cutter, A., & Ullah, F., (2015). *Universal Sustainable Development Goals: Understanding the Transformational Challenge for Developed Countries.* Report of a Study by Stakeholder Forum Available at https://kestavakehitys.fi/documents/2167391/2330167/Universal+Sustainable+Development+Goals.pdf (accessed on 13 December 2021).

Pittock, J., Finlayson, M., Arthington, A. H., Roux, D., Matthews, J. H., Biggs, H., Harrison, I., Blom, E., Flitcroft, R., Froend, R., et al., (2015). Managing freshwater, river, wetland and estuarine protected areas. In: Worboys, G. L., Lockwood, M., Kothari, A., Feary, S., & Pulsford, I., (eds.), *Protected Area Governance and Management* (pp. 569–608). ANU Press: Canberra.

Prokop, P., & Randler, C., (2018). Biological predispositions and individual differences in human attitudes toward animals. In: Alves, R. R. N., & Albuquerque, U. P., (eds.), *Ethnozoology: Animals in Our Lives* (pp. 447–466). Academic Press, London, United Kingdom.

Ramsar Convention on Wetlands, (2018). Report of the secretary general pursuant to article 8.2 concerning the list of wetlands of international importance. *13th Meeting of the Conference of the Contracting Parties.* Dubai, United Arab Emirates.

Reid, A. J., Calson, A. K., Creed, I. F., Eliason, E. J., Gell, P. A., Johnson, P. T. J., Kidd, K. A., et al., (2019). Emerging threats and persistant conservation challenges for freshwater biodiversity. *Biological Reviews, 95*, 849–873.

Reis, V., Hermoso, V., Hamilton, S. K., Ward, D., Fluet-Chouinard, E., Lehner, B., & Linke, S., (2017). A global assessment of inland wetland conservation status. *Bioscience, 67*(6), 523–533. https://doi.org/10.1093/biosci/bix045.

Rist, S., & Hartmann, N. B., (2018). Aquatic ecotoxicity of microplastics and nanoplastics: Lessons learned from engineered nanomaterials. Freshwater microplastics. In: Wagner, M., & Lambert, S., (eds.), *The Handbook of Environmental Chemistry* (Vol. 58). Springer, Cham. https://doi.org/10.1007/978-3-319-61615-5_2.

Ruhi, A., Messager, M. L., & Olden, J. D., (2018). Tracking the pulse of the Earth's fresh waters. *Nature Sustainability, 1*, 198–203.

Scherer, C., Weber, A., Lambert, S., & Wagner, M., (2018). Interactions of microplastics with freshwater biota. Freshwater microplastics. In: Wagner, M., & Lambert, S., (eds.), *The Handbook of Environmental Chemistry* (Vol. 58). Springer, Cham. https://doi.org/10.1007/978-3-319-61615-5_8.

Simpson, D., (2019). *Does Hydropower Harm Biodiversity?* Environmental Impact. From climate change to biodiversity loss- documenting human impacts on the environment. doi: https://www.cabi.org/environmentalimpact/news/65969.

Singh, P., Kumar, A., & Mishra, S., (2020). Performance evaluation of conservation plan for freshwater lakes in India through a scoring methodology. *Environment, Development and Sustainability*. doi: https://doi.org/10.1007/s10668-020-00744-8.

Tayleur, C., Balmford, A., Buchanan, G. M., Butchart, S. H. M., Ducharme, H., Green, R. E., & Phalan, B., (2017). Global coverage of agricultural sustainability standards, and their role in conserving biodiversity. *Conservation Letters, 10*(5), 610–618.

Thalinger, B., Rieder, A., Teuffenbach, A., Putz, Y., Schwerte, T., Wanzenbock, J., & Traogott, M., (2021). The effect of activity, energy use, and species identity on environmental DNA shedding of freshwater fish. *Frontiers in Ecology and Evolution.* doi: https://doi.org/10.3389/fevo.2021.623718.

Tickner, et al., (2020). Bending the curve of global freshwater biodiversity loss: an emergency recovery plan. *BioScience, 70*(4), 330–342.

Turak, E., Harrison, I., Dudgeon, D., Abell, R., Bush, A., Darwall, W., Finlayson, C. M., et al., (2016). Essential biodiversity variables for measuring change in global freshwater biodiversity. *Biological Conservation.* doi: 10.1016/j.biocon.2016.09.005.

Vella, A., & Vella, N., (2020). First population genetic structure analysis of the freshwater crab *Potamon fluviatile* (*Brachyura*: *Potamidae*) reveals fragmentation at small geographical scale. *Genetics of Aquatic Organisms, 4*(1), 49–59.

Windsor, F. M., Tilley, R. M., Tyler, C. R., & Ormerod, S. J., (2019). Evidence of microplastic entry routes into freshwater food webs. *Science of the Total Environment, 646*, 69–74.

WWAP (United Nations World Water Assessment Program), (2015). *The United Nations World Water Development Report 2015: Water for a Sustainable World.* UNESCO, Paris.

WWF, (2016). *Living Planet Report 2016: Risk and Resilience in a New Era.* WWF International, Gland, Switzerland.

WWF, (2018). In: Grooten, M., & Almond, R. E. A., (eds.), *Living Planet Report 2018: Aiming Higher.* WWF International, Gland, Switzerland.

Zamani, S. H., Rezapour, S., Fovargue, R., Moreno, H., & Neeson, T. M., (2019). Strategic allocation of water conservation incentives to balance environmental flows and societal outcomes. *Ecological Engineering, 127*, 160–169.

Zapata, M. J., Sullivan, S. M. P., & Gray, S. M., (2019). Artificial lighting at night in estuaries- implications from individuals to ecosystems. *Estuaries and Coasts, 42*, 309–330.

CHAPTER 2

Freshwater Biodiversity: Advances and Challenges

SUMAYA HASSAN,[1] MAHROOFA JAN,[2] and ASMAT ARA[3]

[1]*Department of Biochemistry, University of Kashmir, Hazratbal, Srinagar – 190006, Jammu and Kashmir, India*

[2]*Department of Botany, Government Degree College, Ganderbal – 191201, Jammu and Kashmir, India*

[3]*Division of Genetics and Plant Breeding, SKUAST-K, Jammu and Kashmir, India*

ABSTRACT

Freshwater habitats occupy less than 1% of the Earth's area, yet they support nearly 10% of all the reported species. Freshwaters provide a platform for human activities that contribute to extensive habitat degradation, fisheries overexploitation, water extraction, pollution, and foreign species invasion. Such effects have caused critical declines in the variety and abundance of many freshwater species, making them far more vulnerable than their marine or terrestrial counterparts. Freshwater ecosystems are systematically altered on progressively large scales since the introduction of irrigation. However, these ecosystems' historic degradation has been augmented periodically by surges of economic and technological influences. Existing extinction rates, habitat disruption, and evolving challenges depict that these ecosystems encounter greater loads than any other ecosystem around. Such threats are anticipated to intensify in forthcoming years as freshwater

Biodiversity of Freshwater Ecosystems: Threats, Protection, and Management. Suhaib A. Bandh, PhD, Basharat Mushtaq, PhD, Mohammad Farooq Mir, PhD, & Shahid A. Wani (Eds.)

resources' overuse continues to provide human necessities. Conservative and vital steps will be helpful at this scale only if based on supporting the processes that strengthen freshwater ecosystems, biodiversity, and the diverse threats to their flow modifications and connectivity disruption.

2.1 INTRODUCTION

Freshwater is a renewable but certainly a limited natural resource that provides us with exceptional services. The quality and quantity of freshwaters determine biodiversity, ecosystem productivity, and human well-being indirectly through biogeochemical processes and ecological dynamics at all levels, be it local, regional, or global. Freshwater ecosystems and their respective riparian areas are amongst the most diverse on Earth biologically. They have immeasurable health, economic, scientific, cultural, and educational benefits to all the globe's living communities (Lynch et al., 2016). Still, our influence on water bodies like rivers, lakes, streams, wetlands, groundwater, etc., is dramatically shrinking the biodiversity and depriving the critical natural resources and services. The decline in freshwater biodiversity is rapid on every continent and in each major river basins of Earth, and the degradation is taking place more rapidly than in terrestrial ecosystems. At present, nearly one-third of all Earth's freshwater releases are utilized for human agricultural practices, industrial set up or urban organizations. Global evolution is reaching critical limits of what the biosphere can sustain (Steffen et al., 2015). Human activities modify the distribution and flows of surface, subsurface, and atmospheric waters at regional scales, declining aquatic ecosystems' flexibility (Rodell et al., 2018). So far, researchers have recognized eight global and overlying environmental degradation trends with all of them as the immediate concern for the human future (Heino et al., 2009; Vorosmarty et al., 2010; Ripple et al., 2017, 2019). Freshwater ecosystems are central to only five of these trends: decreasing freshwater accessibility, loss of forest, diminishing biodiversity, climate change and expanding human population. Freshwater is among the most diverse habitations of Earth (Reid et al., 2019). Its biodiversity is progressively threatened by human actions, including habitat amendment, water pollution, exotic species invasion, overfishing, diversion of rivers, their fragmentation and flow alterations, enlargement of agricultural and urban lands, rising sea levels, climate change and modified precipitation patterns (Dudgeon, 2019; Grill et al., 2019; IPBES, 2019). The disintegration of river corridors also shrinks the population size and gene flow of water species, increasing species extinction risk (Cohen et al., 2016; Dias et al., 2017). Because of the unprecedented socioeconomic developments and demographic makeovers since the past five decades (Crist et al., 2017), human, and livestock population and economic growth are marking an ever-mounting freshwater footprint. The current freshwater biodiversity loss rate stands three times more than forests' loss (Gardner and Finlayson, 2018). The freshwater vertebrate population has dropped at more than two times the rate of their terrestrial or oceanic counterparts (Grooten and Almond, 2018). As reflected by IUCN Red List, 27% of freshwater dependent species are susceptible to extinction. Among these, 62% of turtle, 47% of gastropods, 42% of mammals, 33% of amphibians, 30% of decapod crustaceans, 28% of fishes and 20% of bird species are in danger (Figure 2.1; IUCN, 2019).

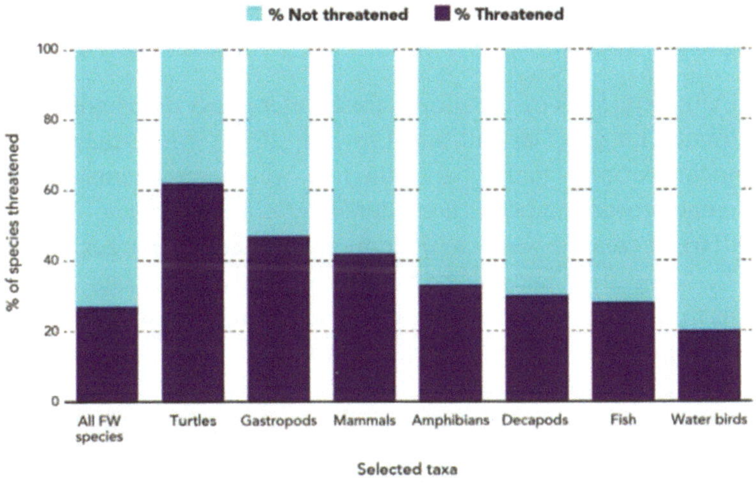

FIGURE 2.1 The proportion of freshwater taxa susceptible to extinction.
Source: IUCN (2019).

2.2 THREATS TO FRESHWATER BIODIVERSITY

Experts' opinion and research literature has helped identify few persistent and emerging dangers to freshwater biodiversity. These pressures differ in their geographic ranges, sternness, and degree of understanding.

2.2.1 HABITAT DEGRADATION

Habitat degradation undoubtedly represents a prominent and ubiquitous reason for population regressions and risk to biodiversity in freshwater ecosystems (WWF, 2018), which is expected to amplify as new threats arise. Habitat degradation through flow modification because of dam constructions without considering the ecological outcome is another persistent risk to freshwater biodiversity globally (Dudgeon et al., 2006; Zarfl et al., 2015; Winemiller et al., 2016). Freshwater ecosystems are at a possible threat of suffering at the hands of dam-induced modification inflow (e.g., disturbed fish migration, reduced release and reduced river-floodplain association) (Juracek, 2015). Overexploitation of living creatures for consumption is an additional critical element of freshwater biodiversity damage. Disease and invasive species are other substantial causes of biodiversity decline (Dudgeon et al., 2006; WWF, 2018).

2.2.2 CLIMATE CHANGE

Approximately 50% of the global freshwater ecosystems are intimidated by climate change (Darwall and Freyhof, 2015). More than 70% of 31 ecological processes that enhance the freshwater ecosystems' functioning, from genes to populations, are altered by climate change (Scheffers et al., 2016). Increased water temperatures, altered discharges and interactions between these and other stressors are constant threats from climate change to the freshwater ecosystem. Escalating freshwater temperatures alters species distribution, disease outbreaks, phenology, and survival (Parmesan, 2006; Krabbenhoft, Platania, and Turner, 2014; Bassar et al., 2016; Hermoso, 2017). Climate change exacerbates these issues, and in fact, freshwater habitats are already disturbed in regions relatively unaffected by human activity so far. Higher water temperatures are causing changes in the distribution of freshwater species, further exacerbating water quality issues, especially in those systems in which nutrient loads are relatively higher. In several rivers and wetlands, precipitation variations can dramatically disrupt flow regimes' critical ecological properties (Döll and Bunn, 2014). Many of these environmental developments will hurt freshwater environments' essential goods and services (e.g., drinking water, fisheries) and make water protection task more complicated.

2.2.3 INVASION

Invasive species pose a significant threat to the biodiversity of freshwater ecosystems, and approaches to the invasion of species can develop further in the future. The long-standing primary mechanisms are biological invasions by global trade and the related movement of live organisms, especially the recent emanation of global e-commerce linked to the purchase of new invasive species via the internet (Walters et al., 2006; Humair et al., 2015).

2.2.4 HARMFUL ALGAL BLOOMS

Freshwater algae, which provide nutrients and energy for associated aquatic food webs, occupy a fundamental trophic position. Periodically, the selection of algal species by environmental or ecological forces allows the accumulation of biomass. This conditional accumulated biomass of algae makes them ecologically prevalent and are termed as harmful algal blooms. Global changes have increased the algal blooms causing hydrological intensification, climate warming, eutrophication, and brownification (Kritzberg and Ekström, 2012; Elliott, 2012; Downing, 2014; Huisman et al., 2018).

2.2.5 POLLUTION

Surface waters are mainly polluted by point-source discharges such as agriculture, aquaculture, oil, and gas production, urban runoff, and mining. These can adversely affect freshwater biodiversity indirectly by affecting the habitat or directly through toxicity. Antimicrobial compounds, such as antibiotics and the additives present in personal care products, are present in the wastewater from municipalities and runoff water from heavily fertilized agricultural lands. The chemicals designed to kill harmful microbes in humans would also harm natural microorganisms' communities (Barra Caracciolo, Topp, and Grenni, 2015). Some recent studies substantiate the effects of other emerging pollutants on algae communities (Bacsi et al., 2016; Richmond et al., 2016). Moreover, these antimicrobial compounds may also impact the aquatic ecosystem's functioning (Nietch et al., 2013).

The threats mentioned above to freshwater biodiversity do not act independently; instead, they interact with one another, leading to disastrous results. For example, higher temperatures can increase toxin's toxicity (Wang et al., 2019), while greater human water abstraction in coordination with climate change can decrease water bodies' ability to dilute pollutants. Flow regulation and pollution change the habitat conditions in ways favorable to the invasive species. The invasive species usually have more excellent resistance than the native populations, further facilitated by climate change (Strayer, 2010). The ongoing global construction of dams for reservoirs (Zarfl et al., 2015) is expected to lead to a 40% increase in reservoir water storage by 2030, directly impacting freshwater biodiversity. Furthermore, water reservoirs act as the breeding ground for invasive species to propagate (Johnson et al., 2008).

2.3 ADVANCES IN FRESHWATER BIODIVERSITY

Since 2005, information on the distribution, diversity, and conservation status of freshwater ecosystems has dramatically increased. In 2008, the first description of taxa dependent on freshwater habitats spanning 126,000 species in eight large zoogeographic regions was completed (Balian et al., 2008). In the same year, 426 freshwater ecoregions, primarily focused on freshwater fish, were identified by a higher-resolution global bio-regionalization (Abell et al., 2008), showing global trends of species abundance and endemism for amphibians, fish, turtles, crocodiles, mammals, and birds as well as some evidence on the level of threat (FEOW, 2013). The IUCN Red List of Endangered Species is the most valuable source of frequently updated species distribution and conservation status information. While significant progress has been made in the last decade in assessing freshwater species, considerable gaps remain in the taxonomic and geographic domains, particularly among invertebrates (Carrizo et al., 2013). The latest compilation of global trends of aspects like endemism, diversity, and conservation status of freshwater biodiversity show high regional threats to freshwater species (Collen et al., 2014). The living planet index (LPI) also shows much higher levels of decrease in freshwater biodiversity relative to its terrestrial and marine counterparts at the level of population trends (McLellan et al., 2014). The origins of these reductions have been synthesized comprehensively. (Dudgeon et al., 2006; Reid et al., 2019), however, there is no global framework to direct policy responses following the crisis's magnitude and urgency. The steps that have been taken so far to conserve freshwater habitats are insufficient (Harrison et al., 2018).

To classify freshwater assemblages or populations, species characteristics including anatomy, morphology, structure, ecology, behavior, habitat utilization, reproduction, and life history have been commonly used. The reaction of various taxa to different pressures should be predictable, depending on the correct combination of characteristics, e.g., stenothermic, and rheophilic taxa are likely to be "losers" under most climate change predictions (Chessman, 2009). Moreover, the effect of changes in an ecosystem's functioning on the composition of functional characteristics should be relatable (Durance et al., 2016). However, there is still a lack of detailed evidence on many freshwater taxa characteristics in most parts of the world. There is a lack of facts on many common species' basic ecology, especially in the tropics. There is also a lack of knowledge about how species characteristics respond to climate change and other human impacts (Carr et al., 2014). The survival of freshwater biodiversity is dependent on the geographical configuration of appropriate habitats in the landscape and the impediments to movement, present naturally or due to the human interruption. Developments in remote sensing technology result in fine, regular, and more reliable data on systemic characteristics associated with species persistence (MacKay et al., 2009). More extensive global databases are now available for natural and artificial obstacles, while coverage limitations exist (Lehner et al., 2011; Lehner, 2013). Detailed maps of sub-catchments and the river networks associated with them are now open to the world (Lehner and Grill, 2013), thus providing a global platform for systematic data collection in a coordinated manner and assess changes in freshwater biodiversity.

Since the detailed analysis by Revenga et al. (2005) and the millennium ecosystem assessment (MEA) of 2005, remarkable progress has been made in evaluating the status of global freshwater biodiversity. In this direction, novel concepts and approaches have been developed for assessing freshwater biodiversity and its significant components. There have been significant advances in the technology associated with the acquisition, storage, transfer, and management of relevant data and a toolkit for assessing and implementing environmental flow allocations (Poff et al., 2010).

2.4 CHALLENGES FOR THE FUTURE OF FRESHWATER BIODIVERSITY

The freshwater available on this planet can be argued to be abundant to sustain human society. However, it is not uniformly distributed across space and time, causing the unavailability of the same for required use by humans. This is further aggravated due to the global problem of pollution and overuse of water in certain sections of society, resulting in 80% of the global population being exposed to water security threat (Vörösmarty et al., 2010). This threat is being mitigated by considerable investments in water infrastructure (Vörösmarty et al., 2010). However, such investments have taken a toll on the natural environment by damaging freshwater ecosystems. Water quality deterioration, overfishing, biological invasions and habitat modifications are considered significant concerns in the present world (Dudgeon et al., 2006). These concerns continue to aggravate (Vörösmarty et al., 2013). It is expected that during the upcoming decade, such threats shall be faced by most of the freshwater ecosystems (Settele et al., 2014).

Water management is generally considered a regional or local level problem. However, it can be argued that there is a global perspective on this challenge. Containment of potable water is caused by water movement due to international trade and the changing and variable climate (Vörösmarty et al., 2015). Millennium Development Goals' idea can be considered the beginning of a global commitment to a sustainable environment. Still, when it came to freshwater, it focused only on sanitation and safe drinking water. The goal of controlling the loss of biodiversity was limited to marine and terrestrial ecosystems. Freshwater biodiversity was neglected from this goal, although 10% of all described animal species sustain on lakes and rivers, including 40% of fish diversity and one-third of all vertebrate species (Balian et al., 2008).

Many steps can be taken to mitigate the risks posed on freshwater ecosystems in terms of quality and quantity, both at local and regional levels. Due to economic and environmental reasons, improvement in water use efficiency in agriculture is very critical. It is the responsibility of freshwater scientists to guide the policy on the design, operation, and location of dams while ensuring aquatic ecosystems' protection. Tools to improve systematic planning (Hermoso et al., 2015) must be employed to help policymakers identify cost-effective outcomes with social and environmental acceptance. There should be a better understanding of the importance of flow-ecology relationships (Bunn and Arthington, 2002). It must be a priority to develop practical techniques and tools to identify environmental flow requirements (Poff et al., 2010). Due to poorly managed and degraded riparian lands, there is a need to solve diffuse pollution (Allan, 2005).

Freshwater science is in such a stage of evolution that there is an imminent need for global action and analysis (Vörösmarty et al., 2015). Security of biodiversity and freshwater ecosystem can be indirectly enhanced by the programs to provide potable water, food, and energy to humans. Barrier effects of dams must be reduced to ensure preservation and restoration of connectivity. Efforts to control pollution at its sources must prioritize diversifying investments to mitigate point sources of pollution and eliminate its diffused forms. Social and economic aspects of these problems must be taken care of in a systematic manner. This demands a shift in governance models based on the outcomes of efforts made in water-stressed regions of the world (Grafton et al., 2012).

The "water literacy" of common masses is generally neglected. Effective communication must be established with the broader community about the causes, sources, results, and potential solutions to these problems. Awareness programs must be conducted to dispel unfounded myths in conservative populations that serve as barriers for positive change in this direction. There must be an enhancement in methods used for quantification and analysis of the management and governance programs.

2.5 CONSERVATION AND MANAGEMENT TOOLS

A question posed to us all is: What could be done to alleviate the calamity of freshwater biodiversity loss or ward off further species decline? The certainty of promoting awareness among people is the first and principal initiative to preserve the incredible richness of freshwater biodiversity. Information should be provided regarding the countless threats and subsequent results. It is equally important to educate people about the veiled or unheeded biodiversity of freshwaters. A central feature of freshwater commons' disaster is that people must restrict their actions to preserve the communal good. Decelerating and reversing the intense losses of freshwater species is still conceivable. The most efficient conservation management synchronizes activities at local to regional; and national to international platforms of organizations (Ceballos et al., 2015). Irrespective of the overall gloomy picture of freshwater biodiversity, there are multiple prospects for effective conservative plans and management. Specific evolving threats can be controlled with new tools and techniques (Jackson et al., 2016). Considering the rapidity and extent of downfall in freshwater biodiversity, parallels can be drawn for post-disaster recovery managements.

- Water management for power generation, reducing flood risks or stocking and supply water for industrial, agricultural, or domestic utility changes the quantity and water flows. While doing so, the physical accessibility of freshwater habitats, ambient circumstances, habitat connectivity, ecosystem processes, sediment flow, etc., is directly altered. Such alterations, in turn, influence functional associations of hydrological regimes with the life histories of the inhabiting species (Bunn and Arthington, 2002). Therefore, it contributes significantly to the losses occurring in freshwater biodiversity. Sustaining or reestablishing ecologically important hydrological regimes advances biodiversity products (Olden et al., 2014). A refined methodological toolbox is available for developing environmental flow setups and recommendations in an extensive range of water resource management settings, from slightly altered to profoundly managed freshwater systems (Acreman et al., 2014, Poff et al., 2017). Several environmental flow valuation tools study desired socioeconomic and cultural purposes together with biodiversity goals, supporting the incorporation of approvals into river basin management plans, water distribution regimes and designing water infrastructure.
- Pollution influences on freshwater biodiversity can be intense and may reflect direct harmfulness or disruption to ecosystem practices. Water pollution is attributed to nutrients coming from sewage or animal wastes; salinization by seawater infiltration or over-irrigation; synthetic herbicides, pesticides; heavy metals, persistent biological pollutants, unsafe constituents from agriculture lands and industries; plastics; pharmaceuticals and their human metabolites; sediments mobilized by agricultural practices, forestry, and mining procedures and heat released from industrial and power sectors (Reid et al., 2019). Policy and management options include

improved wastewater treatment or reuse, regulation of polluting industries, market instruments that reflect downstream pollution costs, improved agricultural practices, and nature-based solutions such as floodplain wetland restoration or riparian buffer zones (WWAP 2017). Better farm management, often combined with market mechanisms, can reduce pollution loads while maintaining agricultural yields.

- A range of interventions can diminish the impacts of the previous degradation on freshwater biodiversity and reduce upcoming risks. These include community conservation of habitats for flagship or culturally essential species, formally protected area designations; land-use planning (often associated with markets for ecosystem services); and habitat restoration programs (UN Water, 2018).
- The misuse of living organisms and mineral substrates distresses freshwater biodiversity directly by eliminating individuals and their habitats and indirectly through modifications to freshwater ecosystem developments. A range of freshwater taxa are overexploited, and policy frameworks to guide such harvests are often insufficient, and enforcement is also poor, making sustainable management difficult (Cooke et al., 2016).
- Freshwater habitats are particularly susceptible to invasion by alien species (Strayer, 2010). Their impact on freshwater biodiversity ranges from behavioral modifications of innate species to complete reorganization of food webs. Preventing the introduction of non-native species is the best approach to limit biodiversity loss.
- The flow of nutrients and sediment through water ecosystems is essential for the regulation of biodiversity. Various species rely on convenient connectivity between upstream and downstream reaches of rivers or between river channels and floodplain areas for relocation and reproduction (McIntire et al., 2016). Construction of dams distinctly alters the flow patterns to which biota is used, and a reservoir substitutes a riverine area above the dam. The movement of animals is obstructed, especially migratory species, and material is entrained. Dams fragment upstream to downstream connectivity and affect lateral, vertical, and temporal connectivity due to alteration in water flow. Engineered barriers and other flood management structures separate rivers from their floodplains. To

prevent freshwater biodiversity loss, we need to Safeguard and restore freshwater connectivity.

The prevailing degree of habitat degradation, species extinction and evolving challenges confirm that freshwater ecosystem stresses are much more significant than any other ecosystem. With the growing exploitation of freshwater resources to meet human demand, these threats are estimated to deepen in the future. This puts more responsibilities on conservation scientists working on freshwater ecosystems to provide evidence for actions to protect, stop the decline or reestablish the world's water bodies. Conservation and restoration actions will only be operational at this scale provided these are grounded on the understanding of the practices that strengthen freshwater ecosystems, their biodiversity, and numerous risks to them. Conservation strategies are most effective when executed with full acknowledgement of the open fragility of ecosystems. Damania et al. (2019) has wisely stated that "An ounce of conservation prevention is worth a pound of technological cures."

2.6 CONCLUSIONS

Freshwater contains a unique biodiversity inside it, that is a repertoire for human use and a part of the diverse mix of ecosystems. The vast and fast decline seen in freshwaters is because of their comparatively small extent and discrete internal relations and their close relations to surrounding land areas. Freshwater ecosystems are remarkably influenced by local fishes' overharvesting, the over abstraction of water, unchecked pollution, and river fragmentation. The successful management of these risks is more complicated when river catchments cross political or administrative boundaries (WWF, 2016). Freshwater bodies are global hotspots of biodiversity but at the same time represent hotspots of endangerment as well. The environmental elements bringing about the preliminary pattern of richness are accountable for the susceptibility of biodiversity to anthropogenic risks. The current range of threats to freshwaters is diverse, and with increasing human populations and water demands, the intensity of these threats increases. Climate change is further going to amplify these effects.

Despite these significant challenges and the critical value of freshwater ecosystems in biodiversity and livelihoods, freshwater ecosystems deteriorate at alarming rates. Thus, there is always a major conflict between the utilization of freshwater by humans and the conservation of its ecosystems. The conflict is expected to increase, given that the accessibility of freshwater is anticipated to decline in many areas in the coming times. Freshwater biodiversity is not given due attention regarding conservation literature. Overexploitation, habitat degradation, and species invasion, all stressors recognized by Dudgeon et al. (2006), remain persistent and worldwide threats to freshwater biodiversity with potent socioeconomic detrimental effects on human welfare. A planet with shrunken freshwaters will deprive many facets of human interest, and we are further putting our essential life-support systems at stake. Therefore, the hour has come to act.

KEYWORDS

- **biodiversity**
- **conservation**
- **degradation**
- **ecosystem**
- **freshwater**
- **global evolution**

REFERENCES

Abell, R., Thieme, M. L., Revenga, C., Bryer, M., Kottelat, M., Bogutskaya, N., Coad, B., et al., (2008). Freshwater ecoregions of the world: a new map of biogeographic units for freshwater biodiversity conservation. *BioScience, 58*, 403–414.

Acreman, M., Arthington, A. H., Colloff, M. J., Couch, C., Crossman, N. D., Dyer, F., Overton, I., et al., (2014). Environmental flows for natural, hybrid, and novel riverine ecosystems in a changing world. *Frontiers in Ecology and the Environment, 12*, 466–473.

Allan, J. D., Abell, R., Hogan, Z., Revenga, C., Taylor, B. W., Welcomme, R. L., & Winemiller, K., (2005). Overfishing of inland waters. *BioScience, 55*,1041–1051.

Bacsi, I., B-B´eres, V., K´okai, Z., Gonda, S., Nov´ak, Z., Nagy, S. A., & Vasas, G., (2016). Effects of non-steroidal anti-inflammatory drugs on cyanobacteria and algae in laboratory strains and in natural algal assemblages. *Environmental Pollution, 212*, 508–518.

Balian, E. V., Leveque, C., Segers, H., & Martens, K., (2008). The freshwater animal diversity Assessment: an overview of the results. *Hydrobiologia, 595*, 627–637.

Barra, C. A., Topp, E., & Grenni, P., (2015). Pharmaceuticals in the environment: biodegradation and effects on natural microbial communities: A review. *Journal of Pharmaceutical and Biomedical Analysis, 106*, 25–36.

Bassar, R. D., Letcher, B. H., Nislow, K. H., & Whiteley, A. R., (2016). Changes in seasonal climate outpace compensatory density-dependence in eastern brook trout. *Global Change Biology, 22*, 577–593.

Bunn, S. E., & Arthington, A. H., (2002). Basic principles and ecological consequences of altered flow regimes for aquatic biodiversity. *Environmental Management, 30*, 492–507.

Burford, M. A., Beardall, J., Willis, A., Orr, P. T., Magalhaes, V. F., Rangel, L. M., Azevedo, S. M. F. O. E., & Neilan, B. A., (2016). Understanding the winning strategies used by the bloom-forming cyanobacterium *Cylindrospermopsis raciborskii. Harmful Algae, 54*, 44–53.

Carr, J. A., Hughes, A. F., & Foden, W. B., (2014). *A Climate Change Vulnerability Assessment of West African Species*. UNEP-WCMC Technical Report Available from http://www.unep-wcmc.org (accessed on 13 December 2021).

Carrizo, S. F., Smith, K. G., & Darwall, W. R. T., (2013). Progress towards a global assessment of the status of freshwater fishes (Pisces) for the IUCN red list: application to conservation programs in zoos and aquariums. *International Zoo Yearbook, 47*, 46–64.

Ceballos, G., Ehrlich, P. R., Barnosky, A. D., Garcı´a, A., Pringle, R. M., & Palmer, T. M., (2015). Accelerated modern human-induced species losses: Entering the sixth mass extinction. *Science Advances, 1*, e1400253.

Chessman, B. C., (2009). Climatic changes and 13-year trends in stream macroinvertebrate assemblages in New South Wales, Australia. *Glob. Chang. Biol., 15*, 2791–2802.

Cohen, A. S., Gergurich, E. L., Kraemer, B. M., McGlue, M. M., McIntyre, P. B., Russell, J. M., Simmons, J. D., & Swarzenski, P. W., (2016). Climate warming reduces fish production and benthic habitat in Lake Tanganyika, one of the most biodiverse freshwater ecosystems. *Proceedings of the National Academy of Sciences of the United States of America, 113*, 9563–9568.

Collen, B., Whitton, F., Dyer, E. E., Baillie, J. E., Cumberlidge, N., Darwall, W. R., Pollock, C., et al., (2014). Global patterns of freshwater species diversity, threat and endemism. *Glob. Ecol. Biogeogr., 23*, 40–51.

Cooke, S. J., Allison, E. H., Beard, T. D., Arlinghaus, R., Arthington, A. H., Bartley, D. M., Cowx, I. G., et al., (2016). On the sustainability of inland fisheries: Finding a future for the forgotten. *Ambio, 45*, 753–764.

Crist, E., Mora, C., & Engelman, R., (2017). The interaction of human population, food production, and biodiversity protection. *Science, 356*, 260–264.

Damania, R., Desbureaux, S., Rodella, A. S., & Russ, J., (2019). *Quality Unknown: The Invisible Water Crisis*. World Bank Publications.

Darwall, W. R., & Freyhof, J., (2015). Lost fishes, who is counting? The extent of the threat to freshwater fish biodiversity. In: Closs, G. P., Krkosek, M., & Olden, J. D., (eds.), *Conservation of Freshwater Fishes* (pp. 1–35). Cambridge University Press, Cambridge.

Dias, M. S., Tedesco, P. A., Hugueny, B., Je´ze´quel, C., Beauchard, O., Brosse, S., & Oberdorff, T., (2017). Anthropogenic stressors and riverine fish extinctions. *Ecological Indicators, 79*, 37–46.

Döll, P., & Bunn, S. E., (2014). Cross-chapter box on the impact of climate change on freshwater ecosystems due to altered river flow regimes. In: Field, C. B., et al., (eds.), *Climate Change Impacts, Adaptation, and Vulnerability* (pp. 143–146). Part A: Global and Sectoral Aspects. Contribution of Working Group II to the Fifth Assessment Report of the Intergovernmental Panel on Climate Change. Cambridge University Press.

Downing, J., (2014). Limnology and oceanography: two estranged twins reuniting by global change. *Inland Waters, 4*, 215–232.
Dudgeon, D., (2019). Multiple threats imperil freshwater biodiversity in the Anthropocene. *Current Biology, 29*, 960–967.
Dudgeon, D., Arthington, A. H., Gessner, M. O., Kawabata, Z. I., Knowler, D. J., Lévêque, C., & Sullivan, C. A., (2006). Freshwater biodiversity: importance, threats, status and conservation challenges. *Biological Reviews, 81*, 163–182.
Dudgeon, D., Arthington, A. H., Gessner, M. O., Kawabata, Z. I., Knowler, D. J., L'ev^eque, C., Naiman, R. J., et al., (2006). Freshwater biodiversity: importance, threats, status and conservation challenges. *Biological Reviews, 81*, 163–182.
Durance, I., Bruford, M. W., Chalmers, R., Chappell, N. A., Christie, M., Cosby, B. J., Noble, D., et al., (2016). Chapter three-the challenges of linking ecosystem services to biodiversity: lessons from a large-scale freshwater study. *Adv. Ecol. Res., 54*, 87–134.
Elliott, J. A., (2012). Is the future blue-green? A review of the current model predictions of how climate change could affect pelagic freshwater cyanobacteria. *Water Research, 46*, 1364–1371.
FEOW, (2013). *Freshwater Ecoregions of the World*. (Available from: http://www.feow.org (accessed on 13 December 2021).
Gardner, R., & Finlayson, C., (2018). *Global Wetland Outlook: State of the World's Wetlands and Their Services to People*. Ramsar Convention Secretariat.
Grafton, R. Q., Pittock, J., Davis, R., Williams, J., Fu, G., Warburton, M., et al., (2012). Global insights into water resources, climate change and governance. *Nat. Clim. Chang., 3*, 315–321.
Grill, G., Lehner, B., Thieme, M., Geenen, B., Tickner, D., Antonelli, F., Babu, S., Borrelli, P., et al., (2019). Mapping the world's free flowing rivers. *Nature, 569*, 215.
Grooten, M., & Almond, R., (2018). *Living Planet Report 2018: Aiming Higher*. World Wildlife Fund.
Harrison, I., Abell, R., Darwall, W., Thieme, M. L., Tickner, D., & Timboe, I., (2018). The freshwater biodiversity crisis. *Science, 362*, 1369.
Heino, J., Virkkala, R., & Toivonen, H., (2009). Climate change and freshwater biodiversity: Detected patterns, future trends and adaptations in northern regions. *Biological Reviews, 84*, 39–54.
Hermoso, V., (2017). Freshwater ecosystems could become the biggest losers of the Paris agreement. *Global Change Biology, 23*, 3433–3436.
Hermoso, V., Clavero, M., Blanco-Garrido, F., & Prenda, J., (2010). Assessing the ecological status in species-poor systems: a fish-based index for Mediterranean Rivers (Guadiana River, SW Spain). *Ecol. Indic., 10*, 1152–1161.
Huisman, J., Codd, G. A., Paerl, H. W., Ibelings, B. W., Verspagen, J. M. H., & Visser, P. M., (2018). Cyanobacterial blooms. *Nature Reviews Microbiology, 16*, 471–483.
Humair, F., Humair, L., Kuhn, F., & Kueffer, C., (2015). E-commerce trade in invasive plants. *Conservation Biology, 29*, 1658–1665.
IPBES, (2019). In: Dı´az, S., Settele, J., Brondı´zio, E. S., Ngo, H. T., Gue'ze, M., Agard, J., Arneth, A., Balvanera, P., et al., (eds.), *Summary for Policymakers of the Global Assessment Report on Biodiversity and Ecosystem Services of the Intergovernmental Science-Policy Platform on Biodiversity and Ecosystem Services* (p. 56). IPBES secretariat, Bonn.

IUCN International Union for Conservation of Nature, (2019). *The IUCN Red List of Threatened Species, Version 2019–1.*

Jackson, M., Loewen, C. J. G., Vinebrooke, R. D., & Chimimba, C. T., (2016a). Net effects of multiple stressors in freshwater ecosystems: a meta-analysis. *Global Change Biology, 22*, 180–189.

Johnson, P. T., Olden, J. D., & Vander, Z. M. J., (2008). Dam invaders: impoundments facilitate biological invasions in freshwaters. *Front. Environ. Sci., 6*, 357–363.

Juracek, K. E., (2015). The aging of America's reservoirs: in-reservoir and downstream physical changes and habitat implications. *Journal of the American Water Resources Association, 51*, 168–184.

Krabbenhoft, T. J., Platania, S. P., & Turner, T. F., (2014). Interannual variation in reproductive phenology in a riverine fish assemblage: implications for predicting the effects of climate change and altered flow regimes. *Freshwater Biology, 59*, 1744–1754.

Kritzberg, E. S., & Ekström, S. M., (2012). Increasing iron concentrations in surface waters - a factor behind brownification? *Biogeosciences, 9*, 1465–1478.

Lehner, B., & Grill, G., (2013). Global river hydrography and network routing: baseline data and new approaches to study the world's large river systems. *Hydrol. Process., 27*, 2171–2186.

Lehner, B., (2013). *Global Waterfall Database*. Technical Documentation, Version 1.0. Available at www.hydrosheds.org (accessed on 13 December 2021).

Lehner, B., Liermann, C. R., Revenga, C., Vörösmarty, C., Fekete, B., Crouzet, P., Döll, P., et al., (2011). High-resolution mapping of the world's reservoirs and dams for sustainable river-flow management. *Front. Ecol. Environ., 9*, 494–502.

Lynch, A. J., Cowx, I. G., Fluet-Chouinard, E., Glaser, S. M., Phang, S. C., Beard, T. D., Bower, S. D., et al., (2017). Inland fisheries – invisible but integral to the UN Sustainable Development Agenda for ending poverty by 2030. *Global Environmental Change, 47*, 167–173.

MacKay, H., Finlayson, C. M., Fernandez-Prieto, D., Davidson, N., Pritchard, D., & Rebelo, L. M., (2009). The role of earth observation (EO) technologies in supporting implementation of the Ramsar convention on wetlands. *J. Environ. Monit., 90*, 2234–2242.

McIntyre, P. B., Reidy, L. C., Childress, E., Hamann, E. J., Hogan, J. D., Januchowski-Hartley, S. R., Koning, A. A., et al., (2016). *Conservation of Migratory Fishes in Freshwater Ecosystems*. Cambridge University Press.

McLellan, R., Iyengar, L., Jeffries, B., & Oerlemans, N., (2014). *Living Planet Report 2014: Species and Spaces, People and Places*. WWF, Gland, Switzerland.

Nietch, C. T., Quinlan, E. L., Lazorchak, J. M., Impellitteri, C. A., Raikow, D., & Walters, D., (2013). Effects of a chronic lower range of triclosan exposure on a stream mesocosm community. *Environmental Toxicology and Chemistry, 32*, 2874–2887.

Olden, J. D., Konrad, C. P., Melis, T. S., Kennard, M. J., Freeman, M. C., Mims, M. C., Bray, E. N., et al., (2014). Are large-scale flow experiments informing the science and management of freshwater ecosystems? *Frontiers in Ecology and the Environment, 12*, 176–185.

Paerl, H. W., & Huisman, J., (2009). Climate change: a catalyst for global expansion of harmful cyanobacterial blooms. *Environmental Microbiology Reports, 1*, 27–37.

Parmesan, C., (2006). Ecological and evolutionary responses to recent climate change. *Annual Review of Ecology, Evolution, and Systematics, 37*, 637–669.
Poff, N. L., Richter, B., Arthington, A. H., Bunn, S. E., Naiman, R. J., Kendy, E., Acreman, M., et al., (2010). The ecological limits of hydrologic alteration (ELOHA): a new framework for developing regional environmental flow standards. *Freshw. Biol., 55*, 147–170.
Poff, N. L., Tharme, R. E., & Arthington, A. H., (2017). Evolution of environmental flows assessment science, principles, and methodologies. In: Horne, A. C., Webb, J. A., Stewardson, M. J., Richter, B., & Acreman, M., (eds.), *Water for the Environment* (pp. 203–236). Elsevier.
Reid, A. J., Carlson, A. K., Creed, I. F., Eliason, E. J., Gell, P. A., Johnson, P. T., Kidd, K. A., et al., (2019). Emerging threats and persistent conservation challenges for freshwater biodiversity. *Biological Reviews, 94*, 849–873.
Revenga, C., Campbell, I., Abell, R., De Villiers, P., & Bryer, M., (2005). Prospects for monitoring freshwater ecosystems towards the 2010 targets. Phil. *Trans. R. Soc. London. Ser. B Biol. Sci., 360*, 397–413.
Richmond, E. K., Rosi-Marshall, E. J., Lee, S. S., Thompson, R. M., & Grace, M. R., (2016). Antidepressants in stream ecosystems: influence of selective serotonin re-uptake inhibitors (SSRIs) on algal production and insect emergence. *Freshwater Science, 35*, 845–855.
Ripple, W. J., Wolf, C., Newsome, T. M., Barnard, P., & Moomaw, W. R., (2019). World Scientists' warning of a climate emergency. *BioScience*.
Ripple, W. J., Wolf, C., Newsome, T. M., Galetti, M., Alamgir, M., Crist, E., Mahmoud, M. I., Laurance, W. F., et al., (2017). World scientists' warning to humanity: A second notice. *BioScience, 67*, 1026–1028.
Rodell, M., Famiglietti, J. S., Wiese, D. N., Reager, J. T., Beaudoing, H. K., Landerer, F. W., & Lo, M. H., (2018). Emerging trends in global freshwater availability. *Nature, 557*, 651–659.
Scheffers, B. R., De Meester, L., Bridge, T. C., Hoffmann, A. A., Pandolfi, J. M., Corlett, R. T., & Watson, J. E., (2016). The broad footprint of climate change from genes to biomes to people. *Science, 354*(6313).
Steffen, W., Richardson, K., Rockstro¨m, J., Cornell, S. E., Fetzer, I., Bennett, E. M., Biggs, R., De Carpenter, S. R., et al., (2015). Planetary boundaries: Guiding human development on a changing planet. *Science, 347*, 1259855.
Strayer, D. L., & Dudgeon, D., (2010). Freshwater biodiversity conservation: recent progress and future challenges. *Journal of the North American Benthological Society, 29*, 344–358.
Strayer, D. L., (2010). Alien species in fresh waters: ecological effects, interactions with other stressors, and prospects for the future. *Freshwater Biol., 55*, 152–174.
UN Water, (2018). *UN World Water Development Report, Nature-Based Solutions for Water.* United Nations.
Vorosmarty, C. J., Hoekstra, A. Y., Bunn, S. E., Conway, D., & Gupta, J., (2015). Fresh water goes global. *Science, 349*, 478, 479.
Vorosmarty, C. J., McIntyre, P. B., Gessner, M. O., Dudgeon, D., Prusevich, A., Green, P., & Davies, P. M., (2010). Global threats to human water security and river biodiversity. *Nature, 467*, 555–561.

Vörösmarty, C. J., Pahl-Wostl, C., Bunn, S. E., & Lawford, R., (2013). Global water, the Anthropocene and the transformation of a science. *Current Opinion in Environmental Sustainability, 5*, 539–550.

Walters, L. J., Brown, K. R., Stam, W. T., & Olsen, J. L., (2006). E-commerce and Caulerpa: unregulated dispersal of invasive species. *Frontiers in Ecology and the Environment, 4*, 75–79.

Wang, P., Lombi, E., Zhao, F. J., & Kopittke, P. M., (2016). Nanotechnology: a new opportunity in plant sciences. *Trends in Plant Science, 21*, 699–712.

Wang, Z., Liu, G. C. S., Burton, G. A., & Leung, K. M. Y., (2019). Thermal extremes can intensify chemical toxicity to freshwater organisms and hence exacerbate their impact to the biological community. *Chemosphere, 224*, 256–264.

Wen, Y., Schoups, G., & Van De, G. N., (2017). Organic pollution of rivers: combined threats of urbanization, livestock farming and global climate change. *Sci. Rep., 7*, 43289.

Winemiller, K. O., McIntyre, P. B., Castello, L., Fluet-Chouinard, E., Giarrizzo, T., Nam, S., Baird, I. G., Darwall, W., Lujan, N. K., Harrison, I., Stiassny, M. L. J., Silvano, R. A. M., Fitzgerald, D. B., Pelicice, F. M., Agostinho, A. A., et al., (2016). Balancing hydropower and biodiversity in the Amazon, Congo, and Mekong. *Science, 351*, 128–129.

WWAP, U., (2017). *Wastewater: The Untapped Resource, the United Nations World Water Development Report*. UNESCO.

WWF, (2016). *Living Planet: Report 2016: Risk and Resilience in a New Era*. Gland: World Wide Fund for Nature.

WWF, (2018). In: Grooten, M., & Almond, R. E. A., (eds.), *Living Planet Report 2018: Aiming Higher.* WWF International, Gland, Switzerland.

Zarfl, C., Lumsdon, A. E., Berlekamp, J., Tydecks, L., & Tockner, K., (2015). A global boom in hydropower dam construction. *Aquat. Sci., 77*, 161–170.

CHAPTER 3

Diversity and Management of Freshwater Macrophytes in the Kashmir Himalayas

NAFEESA FAROOQ KHAN[1] and ZAFAR A. RESHI[1]

[1]Department of Botany, Biological Invasion Lab, University of Kashmir, Hazratbal, Srinagar – 190006, Jammu and Kashmir, India,
E-mail: khnnafisa@gmail.com

ABSTRACT

It is a well-known fact that Jammu and Kashmir has been an area rich in unexplored biodiversity and is one of the Indian hotspots with high numbers of endemic species, including species from an extraordinarily wide range of aquascapes. Like lakes, accompanied by wetlands and rivers, freshwater ecosystems constitute the largest part of the aquatic ecosystems recorded in the union territory (UT). A rich diversity of macrophytes is found in these freshwater ecosystems. The biotic assemblage of macrophytes forms a group of macroscopic plant species that are evolutionarily diverse. They comprise vascular aquatic plants, bryophytes, macroalgae, and charophytes, with high adaptation levels at the morphological, physiological, and biochemical levels. The chapter examines existing evidence on the biological properties and roles of aquatic macrophytes that are essential components of the aquatic ecosystems, including the wetland ecosystems. Nevertheless, in many freshwater ecosystems, anthropogenic

Biodiversity of Freshwater Ecosystems: Threats, Protection, and Management. Suhaib A. Bandh, PhD, Basharat Mushtaq, PhD, Mohammad Farooq Mir, PhD, & Shahid A. Wani (Eds.)

activities have likely caused surface water contamination, and eutrophication turning into a concomitant decrease in macrophytic diversity. Subsequently, proper protection of freshwater and wetland habitats, including surveillance and regulation, is a prerequisite for significant biodiversity restoration and management.

3.1 INTRODUCTION

The Himalaya is among diversity hotspots bestowed with Asia's eight large rivers and is referred to as a "third pole" (in terms of ice and glacial ice). Himalayan natural riches support about one-fifth of humankind. Many researchers have often discussed total floristic wealth in the Indian Himalayas' freshwaters, their vegetation distribution, and managing strategies. As water is the cornerstone of all life on earth, environmental, and economic processes are inextricably connected to aquatic environments. Flora and fauna present in them hold an ample supply of goods and services that remain essential to our economic development and other health aspects. Hence, such habitats are also regarded as the most productive systems of planet earth. The term "Aquatic" encompasses ecosystem conditions where water is commonly abundant and a powerful physiochemical force. Such habitats include streams, wetlands, seas, oceans, springs, rivers, estuaries, ponds, lakes, and pools, influencing living organisms' growth and sustenance. Generally, Water systems are widely classifiable into two significant categories: freshwater and marine, differing in several ways. The present chapter shall offer a baseline for assessment studies in the diversity and management of aquatic macrophytes. It shall also set-up a stage for future research in the field.

3.2 PHYSIO-GEOGRAPHIC ASPECT OF KASHMIR HIMALAYAN REGION

In the union territory (UT) of J&K, Kashmir has rich repositories of freshwater lakes, rivers, and wetlands, with a significant fraction of the human population directly dependent on these freshwater ecosystems. Jammu and Kashmir occupy a peculiar location within the Himalayas due to its magnificent exquisiteness, enriching the culture and natural peculiarity.

Geographically, Kashmir is towards the north-western direction in the Himalayas with a latitude of 32.28–37.06 degrees and a longitude of 72.53–80.32 degrees. The region boasts 3,651 water bodies (Panigrahy et al., 2010), covering an estimated area of 3,915 km^2, and including commonly reported high-low altitudinal lake(s), river(s), wetland(s), reserve(s), and dam(s) (Dar and Khuroo, 2020). Major surveys are taken up by researchers and the scientific community in the region from various fields accrued enormous endemic biodiversity documentation. This westernmost Indian Himalayan region comprises around half of the biodiversity contained in this major mountain chain of the Indian portion. Jammu and Kashmir harbor few great rivers sourcing from high mountain glacier(s) with their origin in Pakistan and India serving diverse regions of both the countries (Dar and Khuroo, 2020). In the current scenario, despite considerable scientific fervor and societal significance, the region's serene lakes and wetlands are dwindling as a consequence of fast and unforeseen urbanization, destruction, and degradation of habitats. Many aquascapes, such as rivers, lakes, and wetlands, adjacent to human occupancy, encounter manifold routes of deterioration, primarily by anthropogenic actions (Dar and Khuroo, 2020; Mushtaq and Nee, 2016). Main factors which negatively influence aquatic systems include colonization of waters by fast-growing weeds and other plants (Romshoo and Rashid, 2014), encroachment (Mushtaq and Pandey, 2014), ferocious application of fertilizers and pesticides, cyclic alterations in precipitation pattern, lesser streamflow (Lal et al., 2001; Romshoo et al., 2015), floods (Romshoo, Mohammd, and Irfan, 2018), excessive nutrient load, and weather.

3.3 DIVERSITY OF FRESHWATER MACROPHYTES

The word "aquatic macrophyte" refers to some water inhabiting photosynthetic species (Dar and Khuroo, 2020) large enough to be visible to the unaided eye. It includes macroalgae, bryophytes, pteridophytes, and spermatophytes. They can grow permanently or regularly on or over the water's surface (Chambers, Patricia et al., 2008). Aquatic macrophytes have a well-established character of guiding structure, function, and the resilience of the ecosystem. Aquatic macrophytes are found in seven plant divisions, consisting of some 41 plant and 103 families of cyanobacteria, chlorophytes, rhodophytes, xanthophytes, bryophytes, pteridophytes, and spermatophytes (Chambers, Patricia et al., 2008). Aquatic Flora varies significantly on a spatial scale with various abiotic and biotic ecological players (Lacoul and Bill, 2006). The main external factors that affect the vegetational macrophytic distribution and abundance include water currents, flow rate, substratum distinctiveness, brightness (mainly adjusted as per shading and water relative clarity), and features of water (hardness, ionic content, pH, nutrients) (Szoszkiewicz et al., 2017).

Macrophytes (synonymous with hydrophytes) serve as the primary producers among freshwater ecosystems and participate in sustaining breakdown (Swan and Kominoski, 2012; Paccagnella, Bianchini, and Da Cunha-Santino, 2020). Degradation of aquatic plants corresponds to carbon kinetics in an aquatic environment via heterotrophic organic carbon metabolization, carbon accumulation, and nutrient oscillations derived from plant debris (Zhang, Lü, Shao, and He, 2014; Han et al., 2019; Paccagnella, Bianchini, and Da Cunha-Santino, 2020). Due to the admixture of these factors, macrophyte-dominated environments are clean. They have an excellent water quality, which, along with color, the composition of dissolved organic material, nutrient content, and oxygen demand, are changed by the leachate generated by residues (He et al., 2018; Liu et al., 2020). Also, they avoid sediment resuspension and act as significant competitors of phytoplankton for nutrients and light. They assimilate N (Racchetti et al., 2017) from pore and bottom water and even add to the total amount of "Oxic" sediments via radial oxygen loss (ROL). They improve the Nitrogen mineralization (organic), NH_4 oxidation, transient nitrate (NO_3^-) accumulation in pore water, and subsequently coupled nitrogen fixation and denitrification (Bartoli et al., 2020; Petersen and Jensen, 1997; Vila-Costa et al., 2016). As such, macrophytes play a crucial role in carbon accumulation and processing and serve ecological purposes for other invertebrates and maintain rivers and streams' stability. The diverse flora present in the freshwater systems represents the environmental and ecosystem conceptual intricacy. Macrophytes add to the total fitness and diversity of balanced aquatic ecosystems by serving as quality markers and facilitating water nutrient cycling. Macrophytes are also a central part of aqua food webs and perform a host of ecological functions in these ecosystems. They improve the spatial heterogeneity of the aquatic ecosystems, including shelter, food supplies, and freshwater fauna breeding sites (Chambers, Patricia et al., 2008; Paccagnella et al., 2020; Camargo, 2018). The other primary functions include synthesis and storage of organic compounds and oxygen release, absorption, and accumulation of chemical elements, water filtration and detoxification of pollutants, the release of biologically active compounds involved in interspecies communications, and impact on the hydrological regimes. A wide array of macrophytic species are also used in various human activities, including bio-indication of water quality, phytoremediation of contaminated water bodies, and wastewater treatment. However, social activities leading to surface water pollution, eutrophication, and global warming have led to a concomitant decrease in macrophyte diversity in many freshwater ecosystems and marine environments. Diverse growth forms and plasticity characterize aquatic macrophytes in physiological and metabolic processes under the influence of changes in environmental conditions.

Most macrophytic species are cosmopolitan, while individuals of similar species are considered to replace each other in the aquatic habitats of various parts of the globe. A few macrophytes are exclusively aquatic; to complete their lifecycle, they need to be immersed in water, whereas other organisms can develop and reproduce in occasionally waterlogged habitats, adapted to water level changes, or occupy so-called ephemeral bodies of water such as floodplains, intermittent lakes, wetlands, etc.

Around 33 orders and 88 families are estimated to represent aquatic tracheophytes, totalling around 2,614 species in 412 plant genera (dominated by angiosperms-Magnoliopsida). Few reported angiosperms that emerge under the surface of water include *Ceratophyllum demersum* and *Lemna trisulca*. An assortment of macrophytic vegetation consists of emergent species whose parts, in submerged and floating plants, rise above the water surface, with unique features in each ecological community's morphological and physiological processes. Based on the classic and the contemporary research, two main classes of aquatic macrophytes are generally distinguished: macrophytes (embedded/attached in bottom sediments) and free floaters. Former has two subgroups; i) Emergent macrophytes (include upright emergents) such as members of the genera *Phragmites, Typha, Sagittaria,* and *Butomus*, etc., ii) Creeping emergents (flourishing in shallow littoral waters often developing their leaves and flowers above the water surface). Such plants are ideal for surviving in water-saturated habitats such as wetlands, marshes, swamps, and flooded areas. Their roots and rhizomes are also tailored for continuous anaerobic sediments, like *Myriophyllum aquaticum, Nasturtium aquaticum, Hydrocotyle* spp., etc.

Macrophytes with floating-leaves, such as *Nymphaea* spp., *Nuphar lutea, and Potamogeton natans, use their creeping rhizomes or roots to anchor* sediments and embody floating or aerial reproductive system, but can be submerged as in the case of *Ceratophyllum* spp. The class of macrophytes growing on the water surface includes groups like *Eichhornia crassipes,* members of *Lemnoideae* like *Spirodella polyrhysa* (posing well-developed submerged roots), *Lemna* spp. (have short roots excluding species *Lemna trisulca* which has no roots), *Wolffia* spp. and *Wolffiella* spp. Among others are ferns: *Azolla* spp., *Salvinia* spp. Alternatively, owing to their relationship to environmental influences, aquatic macrophytes are split into multiple ecological groups: Hydatophytes (subsurface/submerged), pleistophytes (floating leaved) as well as helophytes (aerial but water evolving species). Besides, some species are amphibious that grow and reproduce both in aquatic and terrestrial habitats. Hence, a wide variety of the adaptive mechanisms formed by macrophytes at the morphological, physiological, and biochemical levels encourages them to occupy different types of aquatic habitats such as freshwaters, brackish water, and even marine habitat. Their well-being and longevity are impaired by changes in neighboring land and waterways, marked by frequent, seasonal, or periodic inundations. Variables such as rainfall, salinity, and UV-B shifts, generally correlated with climate change, also impact various wetland plant populations differently, suggesting distinct trends and occupancy rates than those found for other biological classes (Kaul and Zutshi, 1967). Consequently, the careful maintenance of freshwaters, including wetland environments, especially their monitoring and enforcement, is a requirement for the restoration of aquatic macrophytic biodiversity.

Talking about Jammu and Kashmir, the thicknesses of macrophytic evolution such as *Phragmites australis* and *Typha angustata* has been studies by some workers to form the most significant breeding grounds for waterfowls (Kaul and Zutshi, 1967). For example, Kaul and Zutshi (1967) recorded around 67 aquatic plant species, while Kak (1990) identified 58 species from the region's wetlands. Among known species, *Sparganium erectum* and *Typha angustata* (emergent spp.) occupy the littorals primarily. In contrast, *Nymphoides peltata* grow in comparatively shallow and open waters and are dominant rooted floating plant species. Reportedly, *Ceratophyllum*, and *Myriophyllum* tend to dominate the completely submerged forms while the free-floating *Lemna* and *Salvinia* expand luxuriantly among the side-channels.

The first report of species diversity has been documented since the 12th-century, as recounted in the Rajatarangni (2020), the oldest record available about Kashmir. William Moorcroft, a Kashmir marshland, and lake collector (Pandit et al., 2001), has given the primary documentation of water plants from the region (Dar and Khuroo, 2020). Royle (1833–1840) also collected aquatic plants from this part of the world and included his '*Botanical Illustrations of the Himalayan Mountains and the Cashmere Flora* (Dar and Khuroo, 2020). Diverse plants from rivers and lakes were documented from time to time, but Kaul and Zutshi (1967) provided the first detailed account of the aquatic vegetation by naming 117 macrophyte species. Stewart (Pandit et al., 2005) performed the region's most extensive floristic studies, which documented over 2,500 vascular plant species, together with 159 aquatic Kashmir Himalayan plant species (Dar and Khuroo, 2020). In line with this, Kak (1990) identified 196 Kashmir Himalayan aquatic angiosperm species, 82 genera in 44 families (Dar and Khuroo, 2020). However, Pandit et al. (2005) recorded only 117 aquatic macrophyte species from the north-western Himalayas. Later surprisingly, Pandit et al. (2005) recorded only 38 species from various Kashmir Himalayan freshwater bodies. Such diverse and dispersed knowledge for the region's aquatic flora is the key reason why its aquatic flora's precise estimation is always tricky. A read-through recent available scientific literature, herbarium research, and comprehensive field exploration showed that many species previously recorded from the area were either misidentified (Khuroo et al., 2007) or may have been eliminated due to natural and human-led changes occurring in aquatic bodies over time (Dar and Khuroo, 2020).

Furthermore, some alien aquatic plant species previously unknown from the region have now appeared in the water bodies, including lower and higher plants, few of which have also been naturalized and spread rapidly (Dar and Khuroo, 2020; Ahad et al., 2012; Ganie, Reshi, and Wafai, 2016; Dar et al., 2012; Saraswat and Rai, 2018). It is a standard for misidentification and inadequate species documentation, one of the significant challenges for proper management practices (Dar and Khuroo, 2020; Ahad et al., 2012). Therefore, the primary prerequisite for successful management is accurate taxonomic taxa recognition. In a way, precise identification is critical in identifying rare, threatened, and endangered plant species (Dar and Khuroo, 2020; Kardol et al., 2010). This is mainly due to the inconspicuous structural organization of vegetative and reproductive parts of the plant (Xin et al., 2020). However, aquatic plants are difficult to classify due to their discreet floral characters and convergent vegetative morphology.

3.4 AQUATIC MACROPHYTES AND THEIR ROLE

Water quality is deteriorating because of contaminants' prolog and nutrients above the maximum levels of aquatic habitats (Lu, 2010). Many chemicals, such as pesticides and fertilizers added to agricultural ecosystems, have made their way into nature, posing innumerable ecological threats. Scientists have found out different species of plants that can absorb, decay, and turn the dangerous and poisonous pollutants into defunct or non-toxic states. Given the impacts of burgeoning anthropogenic pressures and climate change on various ecosystems observed via plant dominance patterns and community composition (Goswami et al., 2019), macrophytes' nutrient sequestration's effectiveness is of vital importance. Acting as sinks of nutrients, phytosociology, and sequestering potentials of Kashmir Himalayan macrophytes has been elucidated in previous studies.

Multiple anthropogenic behaviors such as population explosion, industrial expansion, and urbanization adversely affect freshwater ecosystems directly or indirectly; such as residues in household drainage, waste products of agriculture and factories, pose a severe risk to aquatic habitats. These factors cause a bulk decline in the population of the essential aquatic organisms (Bhat, Hakeem, and Al-Saud, n.d.). Chemical contaminants such as pesticides, detergents, antimicrobials, solvents, and cleaning fluids, hormones, and sterols, and inorganic pollutants like heavy metals, minerals, radionuclides, etc., have also created a disparity in the standard output of the aquatic ecosystems (Suárez et al., 2005; Szoszkiewicz et al., 2006). The macrophyte population's response to anthropogenic stresses is usually scaled through changes in species' composition, range of species, and abundance of species (Szoszkiewicz et al., 2006; Baattrup-Pedersen et al., 2008; Ceschin et al., 2010; Dodkins et al., 2012; Benitez-Mora and Camargo, 2014; Abati et al., 2016; Zhou et al., 2017; Alencar et al., 2020). Quite often, macrophytes' competence to remove organic and inorganic contaminants has been assessed (Li et al., 2018; Saraswat and Rai,

2018; Sharma, Singh, and Manchanda, 2015; Kamarudzaman, Aziz, and Jalil, 2011; Kamiyango, Sajidu, and Masamba, 2011). In several studies conducted by researchers and scientists, the prevalence of Nitrogen and Phosphorus in water bodies were reported too far above maximum allowable limits (Kantawanichkul, Kladprasert, and Brix, 2009). Excess of these nutrients has often led to a negative effect on the water quality and organism activity in aquatic environments. Besides that, phosphate is also known as a 'limiting nutrient.' It contributes to unsustainable aquatic plant growth in water (Klapper, 1991). The critical root of 'eutrophication' in an aquatic habitat is an aggregation of species and nutrient load (especially N and P), undermining the environmental quality and efficiency of aquatic environments (Xin et al., 2020; Correll, 1998). Eutrophication of water sources is a vital cause of the excessive growth rate of algae and cyanobacterial blooms with deteriorating natural flora, fish mortality, unclear water, and diminishing "biodiversity index" (Pretty et al., 2003; Conley et al., 2009; Allan and Castillo, 2007; Dibble and Thomaz, 2009). Above and beyond, their role in uptake nutrients, macrophytes minimize river and stream bed flooding, encourage settling of solid suspensions, water uptake, water oxygenation through the process of photosynthesis, increase spatial heterogeneity and diversity of ecosystems, as well as provide refuge and nutritional supplies (Figure 3.1) (Chambers, Patricia et al., 2008; Ceschin et al., 2015; Sood et al., 2012; Ahmad, Bhat, and Buang, 2018).

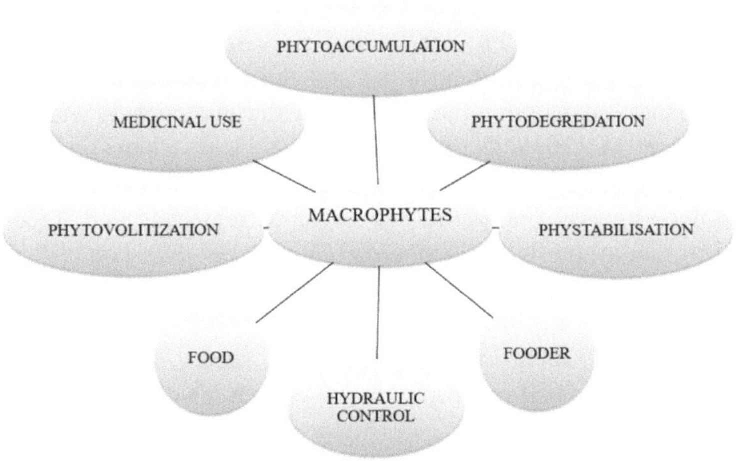

FIGURE 3.1 Important applications of macrophytes.

3.5 HEAVY METALS STRESS RESPONSE OF MACROPHYTES

Elements such as heavy metals placed a significant burden on aquatic biota's usual operations and sometimes have been the reason for denudation of body structures in aquatic Flora and fauna (Szoszkiewicz et al., 2006; Kushwaha et al., 2015). Factually, heavy metals are among the most lethal substances because of their rigorous effects on plants, animals, and human beings (Szoszkiewicz et al., 2006; Etesami, 2018; Goswami et al., 2017; Yu, Li, and Liu, 2017). Amplified anthropocentrism and hyper agricultural activities, to meet the mammoth population's demands, further strengthen the environmental degradation pace. Anthropocentric activities like metal ion processing, paint manufacturing, application, and dependence on synthetic fertilizers and pesticides, leather processing, and construction of electronic gadgets promote spoilage of aquatic environs by releasing a large wastewater volume a huge amount of toxic elements (Szoszkiewicz et al., 2006; Abbas, Ahmed, and Tariq, 2013; Dou et al., 2017). When released into the aquatic environment, these compounds typically appear to mix with other matter found in the water bodies, especially sediments, due to their inherent tendency of sedimentation under the force of gravity. Generally, toxic metals: lead (Pb), cadmium (Cd), mercury (Hg), arsenic (As), and chromium (Cr) (Szoszkiewicz et al., 2006) pose critical threats to humans even in low concentrations (Bayen, 2012). The concentration of such pollutants marked by cellular and genetic toxicity and is, therefore, high risk to life forms (Alahuhta et al., 2018; Ramachandra et al., 2018). Cadmium (Cd), lead (Pb), chromium (Cr) and copper (Cu) are metal pollutants that affect wetland biota and the physiochemical equilibrium of the aquatic environment and endanger the diversity of wetland plants (Alahuhta et al., 2018; Nahlik et al., 2019; Lee, Chen, and Schroeder, 2003; Zhang et al., 2012). Moreover, state, and city waste disposal involve partly decomposed (inorganic and organic) materials and trace metals, including Cadmium (Cd), Chromium (Cr), Nickel (Ni), Lead (Pb), Copper (Cu), Zinc (Zn), Manganese (Mn) and Iron (Fe). Absorption of elements like P, K, Mg, Ca, Cr, Mn, Fe, Co, Zn, Pb, Cu, and Cd, by *Azolla Apinata, Pistia stratiotes, Solvinia molesta, Trapa Natans, Persicaria maculosa* and *Nelumbo nucifera* macrophytes have been also reported (Verma and Suthar, 2018; Petrović et al., 2016; Lu et al., 2011; Mukesh, Dwivedi, and Rubee, 2011; Sahu et al., 2020; Miretzky, Andrea, and Alicia, 2004). Distinguishably, chemical metal element contamination poses a serious threat to all forms of aquatic life in comparison to organic contamination (Szoszkiewicz et al., 2006; Hamidian, Atashgahi, and Khorasani, 2014), which caste less harmful consequences on biota (Szoszkiewicz et al., 2006; Arul et al., 2016); but these are dealt with high biomagnification properties of tolerant aquatic plants (Szoszkiewicz et al., 2006; Abbas, Ahmed, and Tariq, 2013; Bind, Goswami, and Prakash, 2018; Singh et al., 2019; Grant et al., 2016; Harguinteguy, Cirelli, and Pignata, 2014). Thus, unlike organic pollutants, metal toxins are tough to eliminate from ecosystems by itself (Alahuhta et al., 2018). Generally, metal toxins get into the sediments or water. They are easily taken up by aquatic plants' roots, and get into the food chain, constituting a danger to living creatures because of biomagnification and intensification (Khurana and Pritpal, 2012). Consequently, sediments are considered the most well-known warehouse of toxic substances (Szoszkiewicz et al., 2006; Kushwaha, 2020). Largely, effluent discharge and wastewater discharge into the aquascapes add to detrimental impacts on all life forms (Szoszkiewicz et al., 2006; Baattrup-Pedersen et al., 2008; Scholz, 2010; Brix and Arias, 2005) and aquatic environs (Mitsch et al., 2005). In the present scenario, wetland habitats have been built to mitigate eutrophication issues of aquascapes (Phillips, Willby, and Moss, 2016; Vymazal, 2008). Evidence of oxygen release from the aerial part of a plant to root via the plant's aerenchyma tissue in wetland plants helps them enhance the quality of water (Alahuhta et al., 2018; Samecka-Cymerman and Kempers, 2002). More recently constructed wetlands, and engineered structures, have been designed and developed to replicate the natural wetland system. In engineered wetlands, all kinds of aquatic plants viz. free-floating, submerged, and even emergent types are used to manage wastewater associated with the substrate (soil, sand, gravel) and microbes for treating wastewater (Bates and Lowther, 1952; Samecka-Cymerman, Kolon, and Kempers, 2005). It involves various physical, chemical, and biological processes to eradicate different toxins such as dissolved and undissolved particulate organic matter, nitrogen, phosphorus, metals, and pathogenic organisms from wastewater to upgrade the water quality (Bates and Lowther, 1952). Plants like *Potamogeton pectinatus* L. and *Typha domingensis* are amongst the strongest entrants as hyper-accumulators in the bioremediation of metal elements in polluted water. The latest research assesses the effect of water quality, amount of heavy metals and a comparative phytoremediation, metal accumulation capacity in *Typha angustifolia* and *Echhornia crassipus.* Apart from phytoremediation (plants filtering heavy metals from the aquatic environs) (Szoszkiewicz et al., 2006; Vymazal, 2011; Alaa and Elsayed, 2015; Filho et al., 2015 Bolpagni et al., 2007; Boerema et al., 2014), macrophytes are also accepted food source for aquatic and higher life forms (Szoszkiewicz et al., 2006; Boerema et al., 2014; Jordan, Stoffer, and Nestlerode, 2011; Castaldelli et al., 2015; Lin et al., 2002; Shin et al., 2007). The whole aquatic macrophytic community shows a species explicit tolerance capacity towards holding heavy metals and nutrients, so species' combined effect may reduce the toxicity efficiently rather than using a single species (Naveed and Alka, 2017). Due to water-borne infections and health risks, water bodies composed of turbid heterogeneous liquid with residual contaminants pose significant challenges. Recently studies on climate change have led to a new outlook on carbon capture and sequestration (CCS) by natural habitats, especially concerning aquascapes (Cole et al., 2007). Doubtlessly, 2.1% of the global freshwater lake water cover, estimated to contribute around 260 g Carbon/m^2/yr (Tranvik et al., 2009). Therefore, it becomes obvious that the major hot spots for carbon storage, transformation, and recycling are lakes (Williamson et al., 2009; Downing and Leibold, 2010; Downing et al., 2006). Still, that carbon is primarily harvested through photosynthesis by prodigious macrophytes of these habitats. Even then, lakes remain ignored or unrecognized as main carbon gateways and storage inventories. This may be because of the comparatively limited 4.2 million Km^2 global surface area (Medina-Sánchez et al., 2010; Camargo, 2018).

Another essential and developing concern lies over graywater or sullage in aquatic environs, culminating in adverse implications on community and esthetic areas (Bates and Lowther, 1952). For such water treatment, conventional strategies such as aeration ponds, mechanically aerated lagoons, activated sludge systems, anaerobic sludge blanket systems and rotating biological contact systems are usually employed. Still, these require extra land, high capital and operating costs, qualified technical staff and long term preparation and installation (Bates and Lowther, 1952). Mostly, graywater is dumped into irrigation systems, parks, open fields, or natural waters in developed countries without any prior treatment. Compared to Blackwater and solid waste disposal (Bates and Lowther, 1952), it has relatively less priority in the municipal sanitation management systems (Bates and Lowther, 1952). Wastewater is usually drained either into domestic septic tanks or instantly into sewage systems in cities in developed countries, which eventually may leak and get in direct contact with groundwater, thus polluting it. Consequently, drainage is a vital challenge with contaminated surfaces and groundwater (Bates and Lowther, 1952). However, wastewater treatment plants are also insufficient to meet the need for densely populated cleanliness. As a result, untreated household wastewater that contributes 60% of pollutants can contaminate natural waters (Dar and Khuroo, 2020). The rapid urbanization and population growth and the ecosystem's unfriendly behavior of humans make it much worse. An alternate low-cost wastewater treatment infrastructure is also much needed, as conventional water treatment technologies remain expensive for developing countries. This includes expenditure, power supplies, replaceable parts, qualified technicians, operating, and repairing costs. Constructed wastewater wetlands seem to be a feasible way to treat domestic cum industrial water discharges in developing countries (Bates and Lowther, 1952). The technique works around most tropics due to trouble-free operation and low operation cost (Mushtaq and Nee, 2016). Through these effortless and cost-effective wastewater treatment strategies, employing regional water plants and sediment materials, aquatic environs can be filtered from threats of pollutants. Few of the macrophytes that have been documented to have more significant potential for phytoaccumulation are *Eichhornia crassipes* (water hyacinth) (Silva et al., 2019; Passos et al., 2019), *Lemna minor* (common duckweed) (Melo et al., 2019) and *Azolla spp.* (Kushwaha et al., 2015).

3.6 RESPONSE OF MACROPHYTES TO OTHER CHEMICAL CONTAMINANTS

Large-mobility and persistence of herbicides like diuron, hexazinone, and sulfometuron-methyl pose a greater risk of pollution of natural water bodies, including groundwater (Li et al., 2018; Qu et al., 2017; Ribeiro et al., 2019). Solutions to polluted areas, though, needs to be sought. Some macrophytes such as *Pistia stratiotes* and *Eichhornia crassipes* help remove clomazone (an herbicide residue in water), remediating both soil and water resources (Li, Gu, Yang, and Zhang, 2018; Ribeiro et al., 2019 Wang et al., 2019). Similarly, Qu et al. (2017) illustrated a couple of macrophytes, *Potamogeton crispus* and *Myriophyllum spicatum*, eliminate atrazine contaminants. Still, the presence of macrophytes, *Pistia stratiotes* and *Eichhornia crassipes* have hindered the degradation of hexazinone in waters containing residues of herbicide (Li, Gu, Yang, and

Zhang, 2018; Dhir, Sharmila, and Saradhi, 2009). Therefore, pursuing strategies for remediation of water resources is essential to ensure plants and animals' protection. Bioremediation, particularly phytoremediation, can benefit to thoroughly clean both soil and water supplies in this regard (Li et al., 2018; Ribeiro et al., 2019; Wang et al., 2019). Amounting to a mix of agricultural waste, toxic waste, overflow water and natural processes, most freshwater in India is contaminated. This water quality conversion is also because of unaddressed urban and partially handled persistent wastes and other biological influences. Thus, macrophytes can be phytoremediators of herbicides in aquatic systems and can show potential ways of minimizing deterioration of water quality. Recent studies have also tested phytoremediators for their ability to uptake pesticides in their tissues in addition to the absorption potential of the herbicide. The optimal remediation process requires eliminating the pesticide from the substrate and changing in form over some time into the least harmful substance (Li et al., 2018; Mohamed, 2017; Rezania et al., 2016). Various analysis articles have covered the role of aquatic macrophytes in phytoremediation (Sharma, Singh, and Manchanda, 2015). Some of these have focused on monitoring harmful blooms of cyanobacteria and the degradation of toxins (Vanhoudt et al., 2018), including few other reports of heavy metal and radionuclide reduction (Rahman and Hasegawa, 2011; Choi et al., 2002; Brummer et al., 2017) to arsenic removal (Cecal et al., 2002).

Unlike conventional methods, phytoremediation is an eco-friendly, low-cost, and low-maintenance renewable technique that can efficiently handle heavy metals and other toxins from polluted waters. The bioremediation of other critical elements such as S, Cl, Rb, Sr, Ba, Ti, Cr, Mo, and macrophyte polyphenols is yet to be investigated. Polyphenols have prominence in plant development, regulation, and structure and differ widely across various plant groups (Martin et al., 2017). Phenolic compounds from different macrophytes have reported interacting with several insect herbivores' growth and survival (Abubakr, Nahvi, and Kundangar, 2011).

3.7 NEED FOR MANAGEMENT AND SIGNIFICANCE OF MACROPHYTES

Freshwaters provide ample livelihood opportunities for the inhabitants living around them. For instance, aquatic forage constitutes a high-quality alternative to terrestrial plants. Some aquatic resources have been assessed to provide biomass intended for biogas production (Duke, 2019). Aquatic macrophytes serve quality control goals of water through heavy metal extraction, thus reducing their persistence in water (Lavnert, 1981). Most neighboring inhabitants/indigenous cultures have been dependent on lakes and rivers for centuries. Similarly, aquatic plants of Kashmir valley have been a major source of money-making gain in the shape of exploitable assets. Due to their medicinal use in wild habitats, local populations have been harvesting them from time to time to meet their needs (Hofstra et al., 2020). Many such aquatic plants are gradually promising to be a valued repository of substantive chemicals and extracts of folk medicine merit and have been used as household remedies, essentially in parts of Himalayas. For instance, *Nymphaea alba* is anthelmintic, diuretic, and febrifuge; also, its extract is used in the treatment of burns, fevers, ulcers, snake bites and swelling; *Trapa natans* therapeutically finds its use in dyspepsia, hemorrhages, diarrhea dysentery, astringent, cooling, diuretic, and tonic (Hofstra et al., 2020). Much commonly growing, *Nelumbo nucifera* is used as astringent, cardiotonic, febrifuge, and hypotensive with roots and rhizome as a vegetable source (Hofstra et al., 2020). Other plants like *Phragmites communis* (analgesic, antispasmodic, and hypotensive) (Hofstra et al., 2020; Zhang et al., 2019) and *Typha angustata* (nose bleeds, hematemesis, hematuria, uterine bleeding, dysmenorrheal, abscesses astringent, diuretic, tonic, febrifuge, and diuretic) (Thiébaut, 2008) have been much widely used as local traditional medicine (Hofstra et al., 2020). The emergence of excessive numbers of macrophytes has recently been the focus of India's ethnic-medico botanical survey. Regardless of their exceedingly significant ecosystem functions, several ecological issues reflect these plant assemblages (Hofstra et al., 2020). Even though in nature, we see aquatic macrophytes as "self-regulating" that typically do not require interference, decision-makers don't share that belief (Paccagnella et al., 2020). Therefore, a probable volume of conflict arises among stakeholders. They have collapsed from time to time under normal conditions without intense human intervention (Duke, 2019). All in all, relative location on a wetness gradient, surface temperature, water flow, oxygen, and nutrient content often impact the living species of freshwater environments, including rooted-floating leaf type, free-floating leaf type, submerged aquatic plant communities. However, a good mix of native herbivores, such as snails in water bodies, helps to regulate the development of aquatic macrophytes (Duke, 2019; Zhang et al., 2019). While some require other ways to manage unsustainable aquatic macrophytic growth, such as physical extraction (weeding), but all of them have substantial disadvantages. At large, the number of freshwater macrophytes is being used as a clean water predictor (Cuassolo, Villanueva, and Modenutti, 2020; Gimenes et al., 2020). It has been reported that when these plant populations become thick, they add to issues which are of ecological concern, for instance: (i) intensifying temperature of the water as a result of rising global temperature (global warming) (Paccagnella et al., 2020; Puche et al., 2020); (ii) Introducing of invasive plant species, changing the mechanism of the ecosystem (Paccagnella et al., 2020; Calero and Rodrigo, 2019); and (iii) increased concentration of nutrients and pollution of freshwater with metals (Paccagnella et al., 2020; Rojo, Puche, and Rodrigo, 2019). These effects of macrophytes on ecosystems have long term implications for the ecosystem services. Furthermore, insufficient disposal facilities and insufficient urban development preparation have exacerbated the problem and poisoned numerous rivers, reservoirs, ponds, lakes, and other water bodies. Certainly, water-borne pathogens and health issues from water sources composed of turbid heterogeneous material with residual contaminants pose significant challenges. It is concluded that environmental changes actively or indirectly compromise all ecosystems, including freshwater environments. The current vascular flora (native and endemic species) with 4/5 of its flora, bearing in mind both trophic and non-trophic interactions between its members, ranging from vulnerable to critically endangered species. Though it is not a convenient problem to solve the mesh of non-trophic relations with the widely researched trophic level interactions, yet, few attempts must be made towards this (Pysek and Richardson, 2010). To consider how environments are organized and how they react to shocks (like global climate change), a network approach is critical. Charophytes within macrophytes have shown to be extremely susceptible to changes in climate change-related factors such as soaring high temperature, eutrophication (nutrient increase), salinity, and UV-radiation (Pysek and Richardson, 2010; Rejmankova, 2011; Pimentel, Zuniga, and Morrison, 2005). Native hydrophytes aid as key components of balanced aquatic and semi-aquatic environments (Chambers, Patricia et al., 2008), while invasive macrophytes seriously harm the ecological balance by replacing native organisms, changing environments, and disturbing food chains

(Khan et al., 2004; Khan and Shah, 2010; Álvarez, Rojo, and Angeler, 2005). Besides, non-native aquatic Flora disrupts native biodiversity leaving a significant impact on ecology. Khan et al. (155) has detailed the initiatives of aquatic floral conservation and restoration. Competitive elimination and relocation preserve the life-form gradient of vegetation. Khan and Shah (156) have also examined improvements in productivity and the accumulation of nutrients. Mostly freshwater environments dominated by shallow macrophytes are sensitive to climate change (157, 158). Commonly, emergent species are far less reliant on environmental factors because they are readily spread by wind and are thus more frequently dispersed (Tombolini et al., 2014; Schneider et al., 2015). As such, they are less affected by climate change. Furthermore, floating, and submerged organisms are more vulnerable to factors like water quality and condition. They are also more dependent on physical encounters with aquatic ecosystems; additionally, they often rely on some other species as dispersal vehicles (Schneider et al., 2015, 2018). Consequently, these plants flourish because of their great physiological resistance to certain environmental stresses at the aquatic environmental gradient peaks. Other experiments have shown a significant impact on up-stream and down-stream aquatic Flora around dams (Ceschin et al., 2015; Zhou et al., 2017; Alencar et al., 2020; Hayati and Proctor, 1991). Preceding studies considering environmental features of lakes with the spatial location have shown to contribute to changes in macrophytic life structure and, accordingly, the extent of colonization of species and beta diversity trends (Janse et al., 2015). In the case of Jammu and Kashmir, Hokersar wetland has seen a drastic change in the pattern of vegetation and perhaps some commercially valuable aquatic plant species, like *Nelumbo nucifera, Euryale ferox* and *Acorus calamus*, and have almost vanished (Hofstra et al., 2020). From an early stage, this change in the aquatic macrophytic population's structure is due to changing environmental factors in the wetland's nutrient status, which is in line with the studies conducted by Hayati and Proctor (1991). The increasing trend in the macrophytes' nutritional composition from spring onwards correlates to the profitable growth of plants within wetlands, which drops and decomposes with the advent of late fall to winters when the lowest nutrient concentration levels are also observed. Since the lock-up productivity is positively linked to plant biomass, it might significantly assist in decreasing nutrient load and preventing eutrophication. In comparison, an inevitable reduction in biomass is a critical tandem for endangered species to drive into separate conservation biology strategies/approaches; this allows us to address our aquatic freshwater plant conservation objectives.

3.8 CONCLUSION

Zonation of plant species and related organismal assemblages characterize most of the lakes around the world. The Emergent type plants stand in water along the lake edges, with a larger portion floating up in the water. In the shallow water regions of the lake and along lake margins, small, free-floating plants occur. While macrophytes have deep social-ecological significance, there are minimal interventions in the recent past to comprehend their diversity and distribution within the region's lakes. Aquatic macrophytes represent a fascinating diversity that must be recorded and correctly described. The first and forefront prerequisite for efficient management of aquatic environments is to classify constituent forms of aquatic plants correctly. Their inconspicuous floral features, convergent vegetative morphology and prevalent hybridization are hard to recognize. Overall, J&K has rich high-diversity aquascapes. Upcoming research would produce a detailed database of environmental information on the occurrence of invasive alien organisms, growing or reducing the incidence of organisms and aiding locations for the degradation of marine habitats by species. This experience can build an efficient water weed control network management plan and help devise correct management strategies. Hopefully, it will be instrumental in aquatic environment science, structural, and sustainable organization.

KEYWORDS

- **aquascapes**
- **biodiversity**
- **eutrophication**
- **hotspot**
- **macrophytes**
- ***Myriophyllum spicatum***

REFERENCES

Abati, S., Minciardi, M. R., Ciadamidaro, S., Fattorini, S., & Ceschin, S., (2016). Response of macrophyte communities to flow regulation in mountain streams. *Environmental Monitoring and Assessment, 188*(7), 414.

Abbas, H. S., Ahmed, A. M., & Tariq, J. A., (2013). Multicomponent biosorption of heavy metals using fluidized bed of algal biomass. *J. Eng., 19*(4), 469–484.

Abubakr, A., Nahvi, A., & Kundangar, M. R. D., (2011). Ecological studies and uses of valued aquatic plants in Kashmir wetlands. *Nature, Environment and Pollution Technology, 10*(1), 115–118.

Ahad, B., Reshi, Z. A., Ganie, A. H., & Yousuf, A. R., (2012). Azolla cristata in Kashmir Himalaya. *Am. Fern. J., 102*(3), 224–227.

Ahmad, A., Bhat, A. H., & Buang, A., (2018). Biosorption of transition metals by freely suspended and Ca-alginate immobilized with *Chlorella vulgaris*: kinetic and equilibrium modeling. *J. Clean Prod., 171*, 1361–1375.

Alaa, M. Y., & Elsayed, M. A. N., (2015). Heavy metals and nutritional composition of some naturally growing aquatic macrophytes of Northern Egyptian Lakes. *J. Biodivers. Environ. Sci., 6*(3), 16–23.

Alahuhta, J., & Heino, J., (2013). Spatial extent, regional specificity and metacommunity structuring in lake macrophytes. *Journal of Biogeography, 40*(8), 1572–1582.

Alahuhta, J., Lindholm, M., Bove, C. P., Chappuis, E., Clayton, J., De Winton, M., & Hoyer, M. V., (2018). Global patterns in the metacommunity structuring of lake macrophytes: regional variations and driving factors. *Oecologia, 188*(4), 1167–1182.

Alencar, B. T. B., Ribeiro, V. H. V., Cabral, C. M., Dos, S. N. M. C., Ferreira, E. A., Francino, D. M. T., & De Freitas, S. M., (2020). Use of macrophytes to reduce the contamination of water resources by pesticides. *Ecological Indicators, 109*, 105785.

Allan, J. D., & Castillo, M. M., (2007). *Stream Ecology: Structure and Function of Running Waters*. Dordrecht: Springer.

Álvarez-Cobelas M., Rojo C., & Angeler D. G., (2005). Mediterranean limnology: current status, gaps and the future. *J. Limnology, 64*, 13–29.

Arul, M. N., Alemu, A. K., Goswami, L., Pakshirajan, K., & Pugazhenthi, G., (2016). Waste litchi peels for Cr (VI) removal from synthetic wastewater in batch and continuous systems: sorbent characterization, regeneration and reuse study. *J. Environ. Eng., 142*(9), C4016001.

Baattrup-Pedersen, A., Springe, G., Riis, T., Larsen, S. E., Sand-Jensen, K., & Larsen, L. M. K., (2008). The search for reference conditions for stream vegetation in northern Europe. *Freshwater Biology, 53*, 1890–1901.

Bartoli, M., Benelli, S., Lauro, M., Magri, M., Vybernaite-Lubiene, I., & Petkuviene, J., (2020). Variable oxygen levels lead to variable stoichiometry of benthic nutrient fluxes in a hypertrophic estuary. *Estuaries and Coasts,* 1–15.

Bashir, I., Bhat, R. A., & Mir, S. A., (2020). Applications of macrophytes as environmentally sound technique for cleaning of contaminated ecosystems. In: *Bioremediation and Biotechnology* (Vol. 3 pp. 269–290). Springer.

Bates, R. S. P., & Lowther, E. H. N., (1952). The history of bird photography in India. *J. Bombay Natural Society, 50*(2).

Bayen, S., (2012). Occurrence, bioavailability and toxic effects of trace metals and organic contaminants in mangrove ecosystems: a review. *Environ. Int., 48*, 84–101.

Benitez-Mora, A., & Camargo, J. A., (2014). Ecological responses of aquatic macrophytes and benthic macroinvertebrates to dams in the Henares River Basin (Central Spain). *Hydrobiologia, 728*, 167–178.

Bhat, R. A., Hakeem, K. R., & Al-Saud, N. B. S. (2021). *Bioremediation and Biotechnology, 3*.

Bind, A., Goswami, L., & Prakash, V., (2018). Comparative analysis of floating and submerged macrophytes for heavy metal (copper, chromium, arsenic and lead) removal: sorbent preparation, characterization, regeneration and cost estimation. *Geol. Ecol. Landsc., 2*(2), 61–72.

Boerema, A., Schoelynck, J., Bal, K., Vrebos, D., Jacobs, S., Staes, J., & Meire, P., (2014). Economic valuation of ecosystem services, a case study for aquatic vegetation removal in the nete catchment (Belgium). *Ecosyst. Serv., 7*, 46–56.

Bolpagni, R., Pierobon, E., Longhi, D., Nizzoli, D., Bartoli, M., Tomaselli, M., & Viaroli, P., (2007). Diurnal exchanges of CO_2 and CH_4 across the water-atmosphere interface in a water chestnut meadow *(Trapa natans* L.*). Aquat. Bot., 87*, 43–48.

Brix, H., & Arias, C. A., (2005). The use of vertical flow constructed wetlands for on-site treatment of domestic wastewater: New Danish guidelines. *Ecological Engineering, 25*(5), 491–500.

Brummer, V., Roth, S., Röhl, M., & Herbes, C., (2017). Nature conservation against all? Aquatic macrophyte de-weeding–cut or conserve? A stakeholder analysis. *Landscape Online, 54*, 1–14.

Calero, S., & Rodrigo, M. A., (2019). Reproductive phenology of submerged macrophytes: A tracker of year-to-year environmental variations. *Journal of Vegetation Science, 30*(6), 1217–1227.

Camargo, J. A., (2018). Responses of aquatic macrophytes to anthropogenic pressures: comparison between macrophyte metrics and indices. *Environmental Monitoring and Assessment, 190*(3), 173.

Castaldelli, G., Soana, E., Racchetti, E., Vincenzi, F., Fano, E. A., & Bartoli, M., (2015). Vegetated canals mitigate nitrogen surplus in agricultural watersheds. *Agric. Ecosyst. Environ., 212*, 253–262.

Cecal, A., Popa, K., Caraus, I., & Craciun, I., (2002). Uranium and thorium uptake on hydrophilic plants. In: *Uranium in the Aquatic Environment (*pp. 479–488). Springer, Berlin, Heidelberg.

Ceschin, R., Kurland, B. F., Abberbock, S. R., Ellingson, B. M., Okada, H., Jakacki, R. I., & Panigrahy, A., (2015). Parametric response mapping of apparent diffusion coefficient as an imaging biomarker to distinguish pseudo progression from true tumor progression in peptide-based vaccine therapy for pediatric diffuse intrinsic pontine glioma. *American Journal of Neuroradiology, 36*(11), 2170–2176.

Ceschin, S., Salerno, G., Bisceglie, S., & Kumbaric, A., (2010). Temporal floristic variations as indicator of environmental changes in the Tiber River in Rome. *Aquatic Ecology, 44*, 93–100.

Chambers, Patricia, A., et al., (2008). "Eutrophication of agricultural streams: defining nutrient concentrations to protect ecological condition." *Water Science and Technology, 58*(11), 2203–2210.

ChamRezania, S., Ponraj, M., Talaiekhozani, A., Mohamad, S. E., Din, M. F. M., Taib, S. M., & Sairan, F. M., (2015). Perspectives of phytoremediation using water hyacinth for removal of heavy metals, organic and inorganic pollutants in wastewater. *Journal of Environmental Management, 163*, 125–133.

Choi, C., Bareiss, C., Walenciak, O., & Gross, E. M., (2002). Impact of polyphenols on growth of the aquatic herbivore *Acentria ephemerella*. *Journal of Chemical Ecology, 28*(11), 2245–2256.

Cole, J. J., Prairie, Y. T., Caraco, N. F., McDowell, W. H., Tranvik, L. J., Striegl, R. G., & Melack, J., (2007). Plumbing the global carbon cycle: integrating inland waters into the terrestrial carbon budget. *Ecosystems, 10*(1), 172–185.

Conley, D. J., Paerl, H. W., Howarth, R. W., Boesch, D. F., Seitzinger, S. P., Havens, K. E., & Likens, G. E., (2009). Controlling eutrophication: nitrogen and phosphorus. *Science, 323*(5917), 1014–1015.

Correll, D. L., (1998). The role of phosphorus in the eutrophication of receiving waters: a review. *J. Environ. Qual., 27*, 261–266.

Cuassolo, F., Villanueva, V. D., & Modenutti, B., (2020). Litter decomposition of the invasive Potentilla anserina in an invaded and non-invaded freshwater environment of North Patagonia. *Biological Invasions, 22*(3), 1055–1065.

Dar, G. H., & Khuroo, A. A., (2020). *Biodiversity of the Himalaya: Jammu and Kashmir State* (Vol. 18). Springer Nature.

Dar, G. H., Khuroo, A. A., Reddy, C. S., & Malik, A. H., (2012). Impediment to taxonomy and its impact on biodiversity science: an Indian perspective. *Proc. Nat. Acad. Sci. India Sect. B., 82*, 235–240.

Dhir, B., Sharmila, P., & Saradhi, P. P., (2009). Potential of aquatic macrophytes for removing contaminants from the environment. *Critical Reviews in Environmental Science and Technology, 39*(9), 754–781.

Dibble, E. D., & Thomaz, S. M., (2009). Use of fractal dimension to assess habitat complexity and its influence on dominant invertebrates inhabiting tropical and temperate macrophytes. *Journal of Freshwater Ecology, 24*, 93–102.

Dodkins, I., Aguiar, F., Rivaes, R., Albuquerque, A., RodríguezGonzález, P., & Ferreira, M. T., (2012). Measuring ecological change of aquatic macrophytes in Mediterranean rivers. *Limnologica, 42*, 95–107.

Dou, M., Zhao, P., Wang, Y., & Li, G., (2017). Health risk assessment of cadmium pollution emergency for urban populations in Foshan City, China. *Environmental Science and Pollution Research, 24*(9), 8071–8086.

Downing, A. L., & Leibold, M. A., (2010). Species richness facilitates ecosystem resilience in aquatic food webs. *Freshwater Biology, 55*(10), 2123–2137.

Downing, J. A., Prairie, Y. T., Cole, J. J., Duarte, C. M., Tranvik, L. J., Striegl, R. G., & Middelburg, J. J., (2006). The global abundance and size distribution of lakes, ponds, and impoundments. *Limnology and Oceanography, 51*(5), 2388–2397.

Duke, J. A., (2019). *Handbook of Edible Weeds*. CRC Press.

Ekperusi, A. O., Sikoki, F. D., & Nwachukwu, E. O., (2019). Application of common duckweed (Lemna minor) in phytoremediation of chemicals in the environment: State and future perspective. *Chemosphere, 223*, 285–309.

Etesami, H., (2018). Bacterial mediated alleviation of heavy metal stress and decreased accumulation of metals in plant tissues: mechanisms and future prospects. *Ecotoxicol. Environ. Saf., 147*, 175–191.

Filho, P. J. S., Nunes, L. V., Da Rosa, N. N., Betemps, G. R., & Pereira, R. S., (2015). Comparison among native floating aquatic macrophytes for bioconcentration of heavy metals. *Ecotoxicol. Environ. Contam., 10*(1), *1*–6.

Ganie, A. H., Reshi, Z. A., & Wafai, B. A., (2016). Reproductive ecology of *Potamogeton pectinatus* L.(= *Stuckenia pectinata* (L.) Börner) in relation to its spread and abundance in freshwater ecosystems of the Kashmir Valley, India. *Tropical Ecology, 57*(4), 787–803.

Ganie, A. H., Reshi, Z. A., Wafai, B. A., & Puijalon, S., (2015). Phenotypic plasticity: cause of the successful spread of the genus *Potamogeton* in the Kashmir Himalaya. *Aquatic Botany, 120*, 283–289.

Gimenes, L. L. S., Freschi, G. P. G., Júnior, B. I., & Da Cunha, S. M. B., (2020). Growth of the aquatic macrophyte *Ricciocarpos natans* (L.) corda in different temperatures and in distinct concentrations of aluminum and manganese. *Aquatic Toxicology,* 105484.

Goswami, L., Kumar, R. V., Pakshirajan, K., & Pugazhenthi, G., (2019). A novel integrated biodegradation-microfiltration system for sustainable wastewater treatment and energy recovery. *J. Hazard Mater., 365*, 707–715.

Goswami, L., Manikandan, N. A., Pakshirajan, K., & Pugazhenthi, G., (2017). Simultaneous heavy metal removal and anthracene biodegradation by the oleaginous bacteria *Rhodococcus opacus*. *Biotech, 7*(1), 37.

Grant, C. J., Lutz, A. K., Kulig, A. D., & Stanton, M. R., (2016). Fracked ecology: Response of aquatic trophic structure and mercury biomagnification dynamics in the Marcellus shale formation. *Ecotoxicology, 25*(10), 1739–1750.

Hamidian, A. H., Atashgahi, M., & Khorasani, N., (2014). Phytoremediation of heavy metals (Cd, Pb and V) in gas refinery wastewater using common reed (*Phragmites australis*). *Int. J. Aquat. Biol., 2*(1), 29–35.

Han, F., et al., (2019). "High electronic conductivity as the origin of lithium dendrite formation within solid electrolytes." *Nature Energy, 4*(3), 187–196.

Harguinteguy, C. A., Cirelli, A. F., & Pignata, M. L., (2014). Heavy metal accumulation in leaves of aquatic plant *Stuckenia filiformis* and its relationship with sediment and water in the Suquía river (Argentina). *Microchemical Journal, 114*, 111–118.

Hayati, A. A., & Proctor, M. C., (1991). Limiting nutrients in acid mire vegetation: peat and plant analysis and experiments on plant responses to added nutrients. *J. Ecol., 79*, 75–95.

He, W., et al., (2018). "Unified iterative learning control for flexible structures with input constraints." *Automatica, 96*, 326–336.

Hofstra, D., Schoelynck, J., Ferrell, J., Coetzee, J., De Winton, M., Bickel, T. O., & Matheson, F., (2020). On the move: New insights on the ecology and management of native and alien macrophytes. *Aquatic Botany, 162*, 103190.

Hussner, A., Stiers, I., Verhofstad, M. J. J. M., Bakker, E. S., Grutters, B. M. C., Haury, J., & Anderson, L. W. J., (2017). Management and control methods of invasive alien freshwater aquatic plants: a review. *Aquatic Botany, 136*, 112–137.

Janse, J. H., Kuiper, J. J., Weijters, M. J., Westerbeek, E. P., Jeuken, M. H. J. L., Bakkenes, M., & Verhoeven, J. T. A., (2015). GLOBIO-aquatic, a global model of human impact on the biodiversity of inland aquatic ecosystems. *Environmental Science & Policy, 48*, 99–114.

Jordan, S. J., Stoffer, J., & Nestlerode, J. A., (2011). Wetlands as sinks for reactive nitrogen at continental and global scales: a meta-analysis. *Ecosystems, 14*, 144–155.

Kak, A. M., (1990). Aquatic and wetland vegetation of Kashmir Himalaya. *J. Econ. Tax. Bot., 14*(1), 1–14.

Kamarudzaman, A. N., Aziz, R. A., & Jalil, M. F., (2011). Removal of heavy metals from landfill leachate using horizontal and vertical subsurface flow constructed wetland planted with *Limnocharis flava. International Journal of Civil & Environmental Engineering, 11*(5), 85–91.

Kamiyango, M. W., Sajidu, S. M. I., & Masamba, W. R. L., (2011). Removal of phosphate ions from aqueous solutions using bauxite obtained from Mulanje, Malawi. *African Journal of Biotechnology, 10*(56), 11972–11982.

Kantawanichkul, S., Kladprasert, S., & Brix, H., (2009). Treatment of high-strength wastewater in tropical vertical flow constructed wetlands planted with *Typha angustifolia* and *Cyperus involucratus. Ecological Engineering, 35*(2), 238–247.

Kardol, P., Cregger, M. A., Campany, C. E., & Classen, A. T., (2010). Soil ecosystem functioning under climate change: plant species and community effects. *Ecology, 91*(3), 767–781.

Kaul, V., & Zutshi, D. P., (1967). A study of aquatic and marshland vegetation of Srinagar. *Proceedings of the Natural Institute of Science, India, B, 33*(3, 4), 111–127.

Khan, M. A., & Shah, M. A., (2010). Studies on biomass changes and nutrient lock-up efficiency in a Kashmir Himalayan wetland ecosystem, India. *Journal of Ecology and the Natural Environment, 2*(8), 147–153.

Khan, M. A., Shah, M. A., Mir, S. S., & Bashir, S., (2004). The environmental status of a Kashmir Himalayan wetland game reserve: Aquatic plant communities and eco-restoration measures. *Lakes & Reservoirs: Research & Management, 9*(2), 125–132.

Khurana, M. P. S., & Pritpal, S., (2012). Waste water use in crop production: a review. *Resour. Environ., 2*, 116–131.

Khuroo, A. A., Rashid, I., Reshi, Z., Dar, G. H., & Wafai, B. A., (2007). The alien Flora of Kashmir Himalaya. *Biological Invasions, 9*(3), 269–292.

Khuroo, A. A., Weber, E., Malik, A. H., Reshi, Z. A., & Dar, G. H., (2011). Altitudinal distribution patterns of the native and alien woody flora in Kashmir Himalaya, India. *Environmental Research, 111*(7), 967–977.

Klapper, H., (1991). *Control of Eutrophication in Inland Waters*. Prentice Hall, Chichester.

Kushwaha, A., Rani, R., Kumar, S., & Gautam, A., (2015). Heavy metal detoxification and tolerance mechanisms in plants: implications for phytoremediation. *Environ. Rev., 23*, 1–13.

Kushwaha, D. A., (2020). *Phytoremediation: Role of Aquatic Plants to Treatment of Effluent.* Available from: https://www.amazon.in/PHYTOREMEDIATION-AQUATIC-PLANTS-TREATMENT-EFFLUENT-ebook/dp/B087J8ZCXT ().

Lacoul, P., & Bill, F., (2006). "Environmental influences on aquatic plants in freshwater ecosystems." *Environmental Reviews, 14*(2), 89–136.

Lal, M., Nozawa, T., Emori, S., Harasawa, H., Takahashi, K., Kimoto, M., et al., (2001). Future climate change: implications for Indian summer monsoon and its variability. *Curr. Sci.,* 1196–1207.

Lavnert, E., (1981). *Edible and Medicinal Plants.*Himalayan Publishers, ISBN: 0-600-37216-2.

Lee, D. A., Chen, A., & Schroeder, J. I., (2003). Ars1, an arabidopsis mutant exhibiting increased tolerance to arsenate and increased phosphate uptake. *Plant J., 35*, 637–646.

Li, B., Gu, B., Yang, Z., & Zhang, T., (2018). The role of submerged macrophytes in phytoremediation of arsenic from contaminated water: A case study on *Vallisneria natans* (Lour.) hara. *Ecotoxicology and Environmental Safety, 165*, 224–231.

Lin, Y., Jing, S., Wang, T., & Lee, D., (2002). Effects of macrophytes and external carbon sources on nitrate removal from groundwater in constructed wetlands. *Environ. Pollut., 119,* 413.

Liu, H., Zhou, W., Li, X., Chu, Q., Tang, N., Shu, B., & Xing, W., (2020). How many submerged macrophyte species are needed to improve water clarity and quality in Yangtze floodplain lakes? *Science of the Total Environment,* 138267.

Lu, C. H., (2010). Wavelet fuzzy neural networks for identification and predictive control of dynamic systems. *IEEE Transactions on Industrial Electronics, 58*(7), 3046–3058.

Lu, Q., He, Z. L., Graetz, D. A., Stoffella, P. J., & Yang, X., (2011). Uptake and distribution of metals by water lettuce (*Pistia stratiotes* L.). *Environmental Science and Pollution Research, 18*(6), 978–986.

Martin, A. P., Turnbull, R. E., Rissmann, C. W., & Rieger, P., (2017). Heavy metal and metalloid concentrations in soils under pasture of southern New Zealand. *Geoderma Reg., 11*, 18–27.

Medina-Sánchez, J. M., Carrillo, P., Delgado-Molina, J. A., Bullejos, F. J., & Villar-Argaiz, M., (2010). Patterns of resource limitation of bacteria along a trophic gradient in Mediterranean inland waters. *FEMS Microbiology Ecology, 74*(3), 554–565.

Melo, M. S. C., Rocha, L. J. F. N., Brunharo, C. A. C. G., Nicolai, M., Tornisiello, V. L., Nissen, S. J., & Christoffoleti, P. J., (2019). Sour grass resistance mechanism to the herbicide glyphosate. *Planta Daninha, 37.*

Miretzky, P., Andrea, S., & Alicia, F. C., (2004). "Aquatic macrophytes potential for the simultaneous removal of heavy metals (Buenos Aires, Argentina)." *Chemosphere, 57*(8), 997–1005.

Mitsch, W. J., Zhang, L., Anderson, C. J., Altor, A. E., & Hernández, M. E., (2005). Creating riverine wetlands: ecological succession, nutrient retention, and pulsing effects. *Ecological Engineering, 25*(5), 510–527.

Mohamed, Z. A., (2017). Macrophytes-cyanobacteria allelopathic interactions and their implications for water resources management—A review. *Limnologica, 63*, 122–132.

Mosoarca, G., Vancea, C., Popa, S., & Boran, S., (2018). Adsorption, bioaccumulation and kinetics parameters of the phytoremediation of cobalt from wastewater using Elodea canadensis. *Bulletin of Environmental Contamination and Toxicology, 100*(5), 733–739.

Mukesh, B., Dwivedi, D. H., & Rubee, L., (2011). Distribution of heavy metals in edible aquatic plant: water chestnut (*Trapa natans* var. bispinosa Roxb.). *International Journal of Plant Sciences (Muzaffarnagar), 6*(2), 270–273.

Mushtaq, F., & Nee, L. M. G., (2016). *Remote Estimation of Water Quality Parameters of Himalayan Lake (Kashmir) Using Landsat 8 OLI Imagery*. Geocar Internat.

Mushtaq, F., & Pandey, A. C., (2014). Assessment of land use/land cover dynamics vis-a-vis hydrometeorological variability in Wular Lake environs Kashmir Valley, India using multitemporal satellite data. *Arab J. Geosci., 7*(11), 4707–4715.

Nahlik, A. M., Blocksom, K. A., Herlihy, A. T., Kentula, M. E., Magee, T. K., & Paulsen, S. G., (2019). Use of national-scale data to examine human-mediated additions of heavy metals to wetland soils of the US. *Environ. Monit. Assess, 191*, 336.

Naveed, N., & Alka, Y., (2017). *Carbon Sequestration Potential of Wetlands Across Punjab and Possible Climate Mitigation Strategy (*Doctoral dissertation) Lovely Professional University, India. Retrieved from *http://dspace.lpu.in.*

Paccagnella, Y. C., Bianchini, I., & Da Cunha-Santino, M. B., (2020). Decomposition dynamics of two aquatic macrophytes: response of litter interaction with temperature and dissolved oxygen availability. *Brazilian Journal of Botany,* 1–1335.

Pandit, S. V., Berenfeld, O., Anumonwo, J. M., Zaritski, R. M., Kneller, J., Nattel, S., &Jalife, J., (2005). Ionic determinants of functional reentry in a 2-D model of human atrial cells during simulated chronic atrial fibrillation. *Biophysical Journal, 88*(6), 3806–3821.

Pandit, S. V., Clark, R. B., Giles, W. R., & Demir, S. S., (2001). A mathematical model of action potential heterogeneity in adult rat left ventricular myocytes. *Biophysical Journal, 81*(6), 3029–3051.

Panigrahy, S., Singh, T. S., Patel, J. G., Romshoo, S. A., Qadri, T., Rashid, I., & Muslim, M., (2010). *National Wetland Atlas: Jammu and Kashmir*. Ministry of Environment and Forests Government of India.

Parcerisas, L., Marull, J., Pino, J., Tello, E., Coll, F., & Basnou, C., (2012). Land use changes, landscape ecology and their socioeconomic driving forces in the Spanish Mediterranean coast (El Maresme County, 1850–2005). *Environ. Sci. Pol., 23, 120*–132.

Passos, A. B. R., Souza, M. F., Saraiva, D. T., Da Silva, A. A., Queiroz, M. E. L., Carvalho, F. P., & Silva, D. V., (2019). Effects of liming and *Urochloa brizantha* management on leaching potential of picloram. *Water, Air, & Soil Pollution, 230*(1), 12.

Petersen, N. R., & Jensen, K., (1997). Nitrification and denitrification in the rhizosphere of the aquatic macrophyte *Lobelia dortmanna* L. *Limnology and Oceanography, 42*(3), 529–537.

Petrović, D., Jančić, D., Furdek, M., Mikac, N., & Krivokapić, S., (2016). Aquatic plant *Trapa natans* L. as bioindicator of trace metal contamination in a freshwater lake (Skadar Lake, Montenegro). *Acta Botanica Croatica, 75*(2), 236–243.

Phillips, G., Willby, N., & Moss, B., (2016). Submerged macrophyte decline in shallow lakes: what have we learnt in the last forty years? *Aquatic Botany, 135*, 37–45.

Pimentel, D., Zuniga, R., & Morrison, D., (2005). Update on the environmental and economic costs associated with alien-invasive species in the United States. *Ecological Economics, 52*, 273–288.

Pozzobom, U. M., Heino, J., Da Silva, B. M. T., & Landeiro, V. L., (2020). Untangling the determinants of macrophyte beta diversity in tropical floodplain lakes: insights from ecological uniqueness and species contributions. *Aquatic Sciences, 82*, 56.

Pretty, J. N., Mason, C. F., Nedwell, D. B., Hine, R. E., Leaf, S., & Dils, R., (2003). Environmental costs of freshwater eutrophication in England and Wales. *Environ. Sci. Technol., 37*, 201–208.

Puche, E., Jordán, F., Rodrigo, M. A., & Rojo, C., (2020). Non-trophic key players in aquatic ecosystems: a mesocosm experiment. *Oikos.*

Pysek, P., & Richardson, D. M., (2010). Invasive species, environmental change and management, and health. *Annu. Rev. Env. Resour., 35*, 25–55.

Qomariyah, S., Ramelan, A. H., & Setyono, P., (2017). Use of macrophyte plants, sand & gravel materials in constructed wetlands for greywater treatment. In: *IOP Conference Series: Materials Science and Engineering* (Vol. 176, No. 1, p. 012018). IOP Publishing.

Qu, M., Li, H., Li, N., Liu, G., Zhao, J., Hua, Y., & Zhu, D., (2017). Distribution of atrazine and its phytoremediation by submerged macrophytes in lake sediments. *Chemosphere, 168*, 1515–1522.

Racchetti, E., Longhi, D., Ribaudo, C., Soana, E., & Bartoli, M., (2017). Nitrogen uptake and coupled nitrification–denitrification in riverine sediments with benthic microalgae and rooted macrophytes. *Aquatic Sciences, 79*(3), 487–505.

Rahman, M. A., & Hasegawa, H., (2011). Aquatic arsenic: phytoremediation using floating macrophytes. *Chemosphere, 83(*5), 633–646.

Ramachandra, T. V., Sudarshan, P. B., Mahesh, M. K., & Vinay, S., (2018). Spatial patterns of heavy metal accumulation in sediments and macrophytes of Bellandur wetland, Bangalore. *J. Environ. Manag., 206*, 1204–1210.

Rejmankova, E., (2011). The role of macrophytes in wetland ecosystems. *Journal of Ecology and Environment, 34*(4), 333–345.

Rezania, S., Taib, S. M., Din, M. F. M., Dahalan, F. A., & Kamyab, H., (2016). Comprehensive review on phytotechnology: heavy metals removal by diverse aquatic plants species from wastewater. *Journal of Hazardous Materials, 318*, 587–599.

Ribeiro, V. H. V., Alencar, B. T. B., Dos, S. N. M. C., Da Costa, V. A. M., Dos, S. J. B., Francino, D. M. T., & Silva, D. V., (2019). Sensitivity of the macrophytes *Pistia stratiotes* and *Eichhornia crassipes* to hexazinone and dissipation of this pesticide in aquatic ecosystems. *Ecotoxicology and Environmental Safety, 168*, 177–183.

Rojo, C., Puche, E., & Rodrigo, M. A., (2019). The antagonistic effect of UV radiation on warming or nitrate enrichment depends on ecotypes of freshwater macroalgae (Charophytes). *Journal of Phycology, 55*(3), 714–729.

Romshoo, S. A., & Rashid, I., (2014). Assessing the impacts of changing land cover and climate on Hokersar wetland in Indian Himalayas. *Arabian Journal of Geosciences, 7*(1), 143–160.

Romshoo, S. A., et al., (2015). "Implications of shrinking cryosphere under changing climate on the stream flows in the Lidder catchment in the Upper Indus Basin, India." *Arctic. Antarctic. and Alpine Research, 47*(4), 627–644.

Romshoo, S. A., Mohammd, R., & Irfan, R., (2018). "Spatio-temporal variation of land surface temperature and temperature lapse rate over mountainous Kashmir Himalaya." *Journal of Mountain Science, 15*(3), 563–576.

Sahu, Y. K., Deb, M. K., Patel, K. S., Martín-Ramos, P., Towett, E. K., & Tarkowska-Kukuryk, M., (2020). Bioaccumulation of nutrients and toxic elements with macrophytes. *Journal of Hazardous, Toxic, and Radioactive Waste, 24*(1), 05019007.

Samecka-Cymerman, A., & Kempers, A. J., (2002). Aquatic macrophytes as biomonitors of pollution by textile industry. *Bulletin of Environmental Contamination and Toxicology, 69*(1), 82–96.

Samecka-Cymerman, A., Kolon, K., & Kempers, A. J., (2005). A comparison of native and transplanted *Fontinalis antipyretica* hedw. as biomonitors of water polluted with heavy metals. *Science of the Total Environment, 341*(1–3), 97–107.

Sand-Jensen, K., Prahl, C., & Stokholm, H., (1982). Oxygen release from roots of submerged aquatic macrophytes. *Oikos,* 349–354.

Saraswat, S., & Rai, D. J., (2018). Aquatic macrophytes mediated remediation of toxic metals from moderately contaminated industrial effluent. *International Journal of Phytoremediation, 20*(9), 876–884.
Schneider, B., Cunha, E. R., Marchese, M., & Thomaz, S. M., (2015). Explanatory variables associated with diversity and composition of aquatic macrophytes in a large subtropical river floodplain. *Aquatic Botany, 121*, 67–75.
Schneider, B., Cunha, E. R., Marchese, M., & Thomaz, S. M., (2018). Associations between macrophyte life forms and environmental and morphometric factors in a large sub-tropical floodplain. *Frontiers in Plant Science, 9*, 195.
Scholz, M., (2010). *Wetland Systems: Storm Water Management Control.* Springer Science & Business Media.
Sharma, S., Singh, B., & Manchanda, V. K., (2015). Phytoremediation: role of terrestrial plants and aquatic macrophytes in the remediation of radionuclides and heavy metal contaminated soil and water. *Environmental Science and Pollution Research, 22*(2), 946–962.
Shin, J., Keum-Chul, Y., Myung-Hun, Y., & Jae-Kuk, S., (2007). Decomposition and nutrient dynamics of aquatic macrophytes in Lake Paldang. *J. Ecol. Field Biol., 30*(3), 231–236.
Silva, T. S., De Freitas, S. M., Da Silva, T. T. M., Dos, S. M. S., Porto, M. A. F., Souza, C. M. M., & Silva, D. V., (2019). Use of neural networks to estimate the sorption and desorption coefficients of herbicides: A case study of diuron, hexazinone, and sulfometuron-methyl in Brazil. *Chemosphere, 236*, 124333.
Singh, P. K., Kushwaha, A., Hans, N., Gautam, A., & Rani, R., (2019). Evaluation of the cytotoxicity and interaction of lead with lead resistant bacterium *Acinetobacter junii* Pb1. *Braz. J. Microbiol.*
Sood, A., Uniyal, P. L., Prasanna, R. S., & Ahluwalia, A., (2012). Phytoremediation potential of aquatic macrophyte, Azolla. *R. Swed. Acad. Sci., 41*, 122–137.
Soons, M. B., Brochet, A. L., Kleyheeg, E., & Green, A. J., (2016). Seed dispersal by dabbling ducks: an overlooked dispersal pathway for a broad spectrum of plant species. *J. Ecol., 104*, 443–455.
Stewart, J. M., Kruger, G. J., Ammon, H. L., Dickinson, C. T., & Hall, S. R., (1972). *The XRAY System.* Tech. Rep. TR-192. Computer Science Center, Univ. of Maryland, College Park, Maryland.
Suárez, M. L., Mellado, A., Sánchez-Montoya, M. M., & Vidal-Abarca, M. R., (2005). Propuesta de un índice de macrófitos (IM) para evaluar la calidad ecológica de los ríos de la cuenca del Segura. *Limnetica, 24*, 305–318. https://www.limnetica.com/es/propuesta-de-un-%C3%ADndice-de-macr%C3%B3fitos-im-para-evaluar-la-calidad-ecol%C3%B3gica-de-los-r%C3%ADos-de-la-cuenca.
Swan, C. M., & Kominoski, J. S., (2012). *Biodiversity and Ecosystem Function of Decomposition.* eLS.
Szoszkiewicz, K., et al., (2017). "Is the macrophyte diversification along the trophic gradient distinct enough for river monitoring?" *Environmental Monitoring and Assessment, 189*(1), 4.
Szoszkiewicz, K., Ferreira, T., Korte, T., Baattrup-Pedersen, A., Davy-Bowker, J., & O'Hare, M., (2006). European river plant communities: the importance of organic

pollution and the usefulness of existing macrophyte metrics. *Hydrobiologia, 566*, 211–234.

Thiébaut, G., (2008). Phosphorus and aquatic plants. In: *The Ecophysiology of Plant-Phosphorus Interactions* (pp. 31–49). Springer, Dordrecht.

Ting, W. H. T., Tan, I. A. W., Salleh, S. F., & Wahab, N. A., (2018). Application of water hyacinth (*Eichhornia crassipes*) for phytoremediation of ammoniacal nitrogen: A review. *Journal of Water Process Engineering, 22*, 239–249.

Tombolini, I., Caneva, G., Cancellieri, L., Abati, S., & Ceschin, S., (2014). Damming effects on upstream riparian and aquatic vegetation: the case study of Nazzano (Tiber River, central Italy). *Knowledge and Management of Aquatic Ecosystems*, (412), 3.

Tranvik, L. J., Downing, J. A., Cotner, J. B., Loiselle, S. A., Striegl, R. G., Ballatore, T. J., & Kortelainen, P. L., (2009). Lakes and reservoirs as regulators of carbon cycling and climate. *Limnology and Oceanography, 54*(6 part 2), 2298–2314.

Vanhoudt, N., Vandenhove, H., Leys, N., & Janssen, P., (2018). Potential of higher plants, algae, and cyanobacteria for remediation of radioactively contaminated waters. *Chemosphere, 207*, 239–254.

Velthuis, M., Kosten, S., Aben, R., Kazanjian, G., Hilt, S., Peeters, E. T., & Bakker, E. S., (2018). Warming enhances sedimentation and decomposition of organic carbon in shallow macrophyte-dominated systems with zero net effect on carbon burial. *Global Change Biology, 24*(11), 5231–5242.

Verma, R., & Suthar, S., (2018). Performance assessment of horizontal and vertical surface flow constructed wetland system in wastewater treatment using multivariate principal component analysis. *Ecological Engineering, 116*, 121–126.

Vila-Costa, M., Pulido, C., Chappuis, E., Calvino, A., Casamayor, E. O., & Gacia, E., (2016). Macrophyte landscape modulates lake ecosystem-level nitrogen losses through tightly coupled plant-microbe interactions. *Limnology and Oceanography, 61*(1), 78–88.

Vymazal, J., (2008). Constructed wetlands for wastewater treatment: a review. In: *Proceedings of Taal2007: The 12th World Lake Conference* (Vol. 965, p. 980).

Vymazal, J., (2011). Plants used in constructed wetlands with horizontal subsurface flow: a review. *Hydrobiologia, 674*(1), 133–156.

Wang, J., Zhou, W., Chen, H., Zhan, J., He, C., & Wang, Q., (2019). Ammonium nitrogen tolerant chlorella strain screening and its damaging effects on photosynthesis. *Frontiers in Microbiology, 9*, 3250.

Williamson, C. E., Saros, J. E., Vincent, W. F., & Smol, J. P., (2009). Lakes and reservoirs as sentinels, integrators, and regulators of climate change. *Limnology and Oceanography, 54*(6 part 2), 2273–2282.

Xin, J., Ma, S., Li, Y., Zhao, C., & Tian, R., (2020). Pontederia cordata, an ornamental aquatic macrophyte with great potential in phytoremediation of heavy-metal-contaminated wetlands. *Ecotoxicology and Environmental Safety, 203*, 111024.

Yu, L., Li, H. G., & Liu, F. C., (2017). Pollution in the urban soils of Lianyungang, China, evaluated using a pollution index, mobility of heavy metals, and enzymatic activities. *Environ. Monit. Assess, 189*(1), 34.

Zhang, J., Lü, F., Shao, L., & He, P., (2014). The use of biochar-amended composting to improve the humification and degradation of sewage sludge. *Bioresource Technology, 168*, 252–258.

Zhang, P., Grutters, B., Van, L. C. H., Xu, J., Petruzzella, A., Van, D. B. R. F., & Bakker, E. S., (2019). Effects of rising temperature on the growth, stoichiometry, and palatability of aquatic plants. *Frontiers in Plant Science, 9*, 1947.

Zhang, X., Uroic, M. K., Xie, W. Y., Zhu, Y. G., Chen, B. D., McGrath, S. P., & Zhao, F. J., (2012). Phytochelatins play a key role in arsenic accumulation and tolerance in the aquatic macrophyte *Wolffia globosa*. *Environmental Pollution, 165*, 18–24.

Zhou, X., He, Z., Jones, K. D., Li, L., & Stoffella, P. J., (2017). Dominating aquatic macrophytes for the removal of nutrients from waterways of the Indian River Lagoon basin, South Florida, USA. *Ecological Engineering, 101*, 107–119.

CHAPTER 4

Genus Schizothorax in the Northwestern Himalayas

UZMA NASEER,[1] IBRAQ KHURSHID,[1] and MUSHEERUL HASSAN[2]

[1]*Department of Zoology, Central University of Kashmir, Jammu and Kashmir – 191201, India*

[2]*Clybay Private Limited, Bangalore – 560114, Karnataka, India, E-mail: musheer123ni@gmail.com*

ABSTRACT

Schizothorax is morphologically, meristically, and economically a very viable and valuable fish genus of Kashmir valley. Wide diversity bolstered with a high market demand makes it a very dynamic entity in the research realms. *Schizothorax* is intrinsically an herbivore growing to an appreciable 500 odd millimeters in length. It is an abounding source of protein, low saturated fatty acids with a significant omega-3 fatty acid content besides some minerals. However, the genus' reported attrition is attributed to the gross trespassing of the fish ecosystem by humans through a range of means. The vital cogs accelerating this reported recession in fish numbers is primarily the use of pesticides, the introduction of exotic fish species, construction of dams, and eutrophication of water bodies. Eutrophication, a multipronged factor, has rendered a powerful jolt to aquatic ecosystems impacting the fish at large. There is an urge to redress the decline from the cardinals by comprehensive introspection while chalking out SOPs

Biodiversity of Freshwater Ecosystems: Threats, Protection, and Management. Suhaib A. Bandh, PhD, Basharat Mushtaq, PhD, Mohammad Farooq Mir, PhD, & Shahid A. Wani (Eds.)

associated with chemicals, fish conservation methods, and breeding techniques to replenish the dwindling genus from the valley.

4.1 INTRODUCTION

The valley of Kashmir (Figure 4.1) situated between 33°01'–34°08' N latitude and 74° 47'–74° 49' E longitude with an average altitude of 1,550 meters above sea level amid Himalayas (Figure 4.1) often referred to as a 'paradise' around the earth (Bhat et al., 2010). One of the leading attributes contributing to Kashmir's worldwide reputation is the rich biodiversity that embellishes its captivating landscapes. The valley also possesses many water bodies such as rivers, lakes, and fast running streams embroider the valley. These water bodies inhabit many indigenous and exotic fish species, with *Schizothorax* (Genus) comprising cyprinid fish species as the best-known indigenous ones (Rather et al., 2015). The water bodies like Dal Lake, Wular Lake, Mansbal Lake, and river Jhelum are reported to inhabit these species. Additionally, other small water bodies in the valley like Arapat (Anantnag), Lidder stream, Vishu, Rambiara stream, Lam stream, Aripal stream (Pulwama). Kishan Ganga, Nallah Kehmil, Nalla Lolab, Shalliganga stream (Budgam), Tailbal Nalla (Ganderbal), Phoru (Kauser, 2008) also affirm the presence of *Schizothorax* species. Fish is known as an essential single source of high-quality protein consumed globally (Barik, 2017). In India, Fish constitutes around 12.8% of the total consumed animal protein (Barik, 2017). Fish has large nutritional values, including a low level of saturated fats and a significant amount of essential (omega-three) fatty acids (Kris-Etherton et al., 2003). Fish are also rich in other nutrient constituents like *n*-3 fatty acids, selenium, Vitamins-D, and Iodine (Skare et al., 2014). These nutrients play a crucial role in healthier metabolic profiles (Potenza et al., 2009). The fisheries sector occupies an important place in the socio-economic development of the valley. The current fish production from "Jammu and Kashmir" is 20,700 metric tons, and its volume of output over the last decade has varied between 19,000 to 20,000 metric tons (Sandhya, 2019). It contributes a total revenue of roughly 5.5 crore rupees in Jammu and Kashmir (Economic Survey, 2017). Out of the total fish production in Jammu and Kashmir, about 70% comes from the Dal and the Wular lakes (Qureshi et al., 2016).

Nearly 16 species of fish are found in Kashmir waters, out of which 10 belong to cyprinids (Heckel, 1838). The fish accounted for 15% of the total catch in 2005, which was reduced to 10% in 2015 (Bhat, 2016). Not only this, but the biomass of the species also decreased (Bhat, 2016). This reduction in population and deterioration in fish quality is attributed to various anthropogenic activities like inordinate and unchecked pollution-causing herbicides, pesticides, and fertilizers dumped in to the water bodies (Yousuf, 2011). A decline has been reported in the population of algaad (*Schizothorax niger*), besides other species like churn (*Schizothorax esocinus*), chosh (*Schizothorax labiatus*), khont (*Schizothorax plageostomus*), and satter gaad (*Schizothorax curvifrons*), which are also under some severe threats (Yousuf, 2011). Pesticides are extensively being sprayed in paddy fields during the crops' flowering stages to prevent multiple disease outbreaks (Rossi et al., 2020). Hence, pesticides are the most persistent pollutants in water bodies near paddy fields (Elfman et al., 2011). In this sense, various chemicals are accumulated through the water and food chain by the fish (Arisekar et al., 2019; Clasen et al., 2018), which brings about the deterioration of their health as well as a serve and advanced ecological risk (Cazenave et al., 2019; Perez et al., 2018).

Accelerated eutrophication due to human causes has led to the propagation of many bacteria like *Klebsiella pneumonia, Escherichia coli, Klebsiella aerogenes, Citrobacter freundii, Pseudomonas aeruginosa,* and *Vibrio cholerae* (Saleem et al., 2011), which are considered as reliable indicators of contamination (Clark and Pagel, 1977). The world-famous "Dal lake" is suffering from the same problem (contamination), causing the disappearance of biota (fish) (Saleem et al., 2013). Mansbal Lake, the deepest freshwater valley lake, fed by groundwater covers about 2.81 Km (Lawrence, 1895; Raina 1971). It is famous for economically important fish like *Schizothorax niger* and *Schizothorax esocinus*. Owing to stone quarrying and fortuitous urbanization in its surroundings, this freshwater lake is plagued by factors like nutrient and silt load from the catchment (Rashid et al., 2013). The distress brought by anthropogenic activities on the biophysical and socio-economic span of Manasbal catchment directly affects the trophic status and biota of Manasbal Lake (Rashid et al., 2013). Along with an increase in pollution, these polluted water bodies harbor many pathogens like helminths, protozoans, bacteria, viruses, etc. This has led to a terrible impact on the fish population, primarily the indigenous *Schizothorax* species (Qayoom and Shah, 2018). From the perspective of microbiology, the pathogens like bacteria are a potential risk not only to the fish but also to the humans who handle and consume them.

Apart from this, the introduction of the non-native common Carp also has a detrimental effect on Schizothoracine fish population in Kashmir valley (Yousuf et al., 1992). The predominance of this non-native fish over the native one was that the former's food and feeding habits were identical to the later, i.e., feeding on Detritus and Benthos and that the common Carp has higher fecundity and breeds in confined waters (Sunder et al., 1984). Moreover, this freshwater fish has also shown a decline in numbers due to dams, including habitat fragmentation and blocking migration routes (Wu et al., 2019). Fragmentation of the freshwater habitats has some significant implications on freshwater fish as dams block the migration routes crucial for spawning or feeding and curb dispersal (Carvajal-Quintero et al., 2019; Fuller et al., 2015; van et al., 2019).

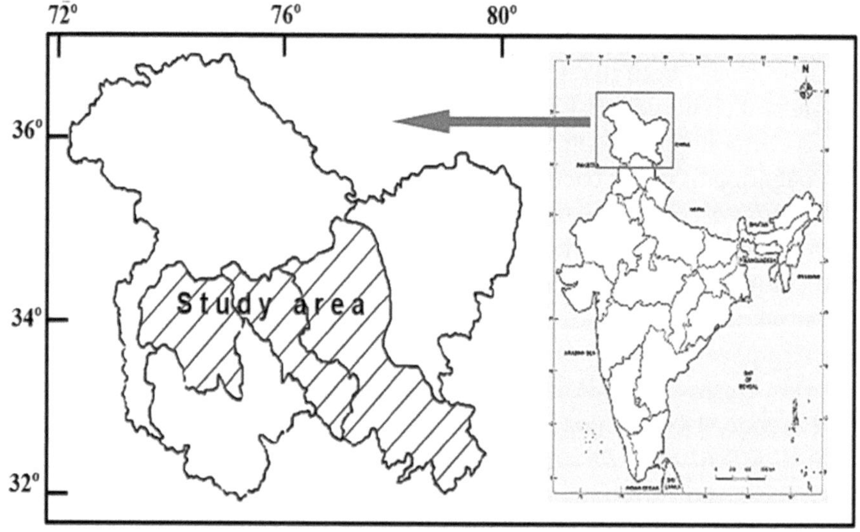

FIGURE 4.1 Map of study area (Kashmir valley).

4.2 SYSTEMATIC ACCOUNT OF SCHIZOTHORAX

- **Phylum:** *Chordata*
- **Class:** *Actinopterygii*
- **Order:** *Cypriniformes*

- **Family:** *Cyprinidae*
- **Genus:** *Schizothorax* (Heckel, 1838).

With around 65 species of *Schizothorax* (Figure 4.2) known in the world (Froese et al., 2019), the reported species from the valley of Kashmir are *Schizothorax esocinus, Schizothorax curvifrons, Schizothorax richardsonii, Schizothorax planifrons, Schizothorax micropogan, Schizothorax plageostomus, Schizothorax longipinnis,* and *Schizothorax niger* (Zutshi et al., 1970).

FIGURE 4.2 *Schizothorax* species.

4.3 MORPHOMETRIC CHARACTERISTICS

Owing to their morphometric resemblance, anonymity is seen between snow trouts (Bashir et al., 2016). Bashir et al. (2016) carried out a study based on molecular and morphometric characteristics on five different species of genus *Schizothorax* in the valley. They revealed a significant grouping of species to individual clusters. Genetic divergence was observed among the selected species by Mitochondrial DNA cytochrome oxidase I (COI) gene analysis. Two distinct clads of *Schizothorax* were seen; the first one included *Schizothorax niger, Schizothorax curvifrons* and *Schizothorax plagiostomus*, and second included only two species

(*Schizothorax esocinus* and *Schizothorax labiatus*). In another study (Bhat et al., 2010), the biology of fish of river Lidder investigated between 2003 and 2005, showed almost seven species out of which three commercially significant species (*Schizothorax plagiostomus, Schizothorax esocinus,* and *Schizothorax labiatus)* dominated the catch both by number as well as by weight. Around 133 samples of *Schizothorax plagiostomus* with the total length ranging from 96–520 mm and weight ranging from 7–948 g, 70 samples of *Schizothorax esocinus* having total length range from 57 mm to 420 mm and total weight from 1 g to 644 g and 40 samples of *Schizothorax labiatus* with the total length ranging from 92 mm to 255 mm and total weight from 7 g to 179 g were taken for the study. Maximum growth (head, pre-dorsal length, pre-pectoral length, and eye diameter) was seen in *Schizothorax esocinus* followed by *Schizothorax. Schizothorax plagiostomus* was seen to have maximum caudal fin length and pre-anal length. *Schizothorax labiatus* was observed to have a maximum body and peduncle depth.

4.4 FEEDING HABITS

The fish food habits vary every month, and this variation is due to the composition of food organisms occurring in the different seasons of the year (Manon et al., 2011). A study carried out by Sabha et al. (2017) on the feeding habits of *Schizothorax niger* in Nigeen Lake revealed that *Schizothorax niger* feeds on green algae, plant fragments, diatoms, detritus. *Schizothorax plagiostomus* is a bottom-feeding herbivorous fish and feeds on algae, diatoms, and other plant materials. For bottom-feeding, it possesses several adaptive modifications, e.g., the lower jaw of the fish is modified into a stony pad to scrap the algae and diatoms from the stone and rocks. The mouth is ventrally suctorial and can suck the scraped material; also, the pharyngeal teeth are relatively well-formed (Singh, 2016).

Jan and Das (1970) reported *Schizothorax niger* and *Schizothorax esocinus* as herbivorous fishes. The average diet of *Schizothorax curvifrons* comprised of dissolved organic matter (40.33%), phytoplankton (38.78%), mud (17.51%), zooplankton (2.0%), and miscellaneous matter (1.38%) (Sunder et al., 1984, 1985). Kauser et al. (2010), while investigating the feeding habits of *Schizothorax curvifrons* and

Schizothorax esocinus designated the former as a phytophagous fish and the later as an omnivorous fish. *Schizothorax* species are generally surface feeders but occasionally feed at the lower levels due to a shortage of food or interference in the upper water strata (Jan et al., 1970). In nearly all cases, the percentage of vegetable matter is estimated to be higher than the animal matter. Hence the fish can be declared as herbivorous. Very specialized alimentary systems are present in herbivorous fishes, which indicate such fishes' food habits from their appearance only (Hilderbrand, 1974). The true stomach is absent; instead, they possess a long thin-walled intestine often known as pseudogaster (Fouche et al., 2011).

4.5 BREEDING BIOLOGY

Understanding the reproductive behavior of fish is important for studying their fundamental biology and their management and conservation (Hussain et al., 2018). Nasreen et al. (2017), while conducting experiments on the natural condition breeding biology *of Schizothorax esocinus* and *Schizothorax curvifrons*, found that the breeding season in the former starts from March till July with the peak breeding season as April-May and the later breeds from May till August with the peak breeding season as May-June. It is also demonstrated that the fecundity of *Schizothorax esocinus* is remarkably greater than *Schizothorax curvifrons.* YR (1970) undertook a detailed study on the breeding biology of *Schizothorax* niger (Dal lake) and observed that the breeding season of *Schizothorax niger* starts from mid of April to the middle of June. Based on the oocytes' histological studies, the studies on the diameter of oocytes and gonadosomatic index (GSI) indicate that breeding season depends upon the optimum external factors like light, food, and temperature. Jyoti and MK (1972) confirmed that *Schizothorax niger* breeds once a year, and the peak spawning is observed in March-May. Jan and Ahmed (2019) investigated the reproductive biology of *Schizothorax plagiostomus* and revealed that it breeds only once a year in a very synchronized manner. The investigations also revealed that the ripening and maturation of ovary are influenced by the climatic conditions, the health status of the fish, harsh winters, infections, etc.

4.6 NUTRITIONAL VALUE AND ECONOMIC STATUS

Fishes are intrinsically loaded with balanced nutrients and happen to be the food at almost every table across the lines of geography, religion, age, and income, laced with overstocked proteins, lipids, necessary vitamins, minerals (Gormley, 2006; Lund, 2013) besides some other vital components particularly LC-PUFaS, they are seen to alleviate troubles related to heart, pancreas (diabetes) and cancer (Tocher, 2015). Samples of snow trout were seen as an exemplary nutritional bed with *Schizothorax esocinus* and *Schizothorax niger* showing a richer carriage of polyunsaturated fatty acids, *Schizothorax plagiostomus* rich in calcium and phosphorus and *Schizothorax progastus* rich in iron and copper (Joshi et al., 2017). A nexus between nutritional content and certain optimum factors such as the nature of feed served, the water body's temperature, and water quality was well-conceived. The frequent and robust gage of these factors would bumper fish production with long terms in quelling the menace of hunger and malnutrition.

Besides a rich nutritional quotient, the fish industry is a pivotal contributor to our country's agricultural economy. In Jammu and Kashmir alone, throughout approximately 27,800 km of water bodies, a rough figure of roughly 2 lac quintals of fish come under mobilized fish farming (market fishing). A study supported by empirical evidence revealed an overall surge in fish in J&K over the last couple of years (Rather et al., 2020). The research reported a population bulge in only trout species, while mirror carps and country fish were featured on a negative compound growth curve (Rather et al., 2020). The decline is attributed to various significant factors, particularly to the mushrooming of anthropogenic activities in and around water bodies coupled with other social hurdles (Baba et al., 2019).

4.7 ELEMENTS ACCOUNTABLE FOR STRESS AND DECLINE OF SCHIZOTHORAX

The Government's laxity, coupled with human negligence, has brought *Schizothorax* – otherwise considered the fishing industry's backbone in Kashmir valley, to extinction. The breeding grounds have been annihilated, appreciably affecting the very nascent phases of *Schizothorax* life cycle. Some of the significant elements leading to the decline of fish are mentioned in subsections.

4.7.1 INTRODUCTION OF EXOTIC FISH SPECIES

Increasing the fish production and to meet the demand by introducing Carp in the valley lakes is seen to cost the gradual replacement of *Schizothorax* (Qureshi and Krishnan, 2015). In Dal Lake Kashmir, the common Carp was introduced in 1956, and since then, this fish has shown noticeable adaptations in different water bodies of the valley (Foteder and Qadri, 1974). The establishment of exotic common Carp resulted in a steep decline in the Schizothoracine fishes' population (Yousuf and Qadri, 1992). According to Foteder and Qadri (1974), three varieties of *Cyprinus carpio* (mirror carp, leather carp, and scale carp) are responsible for the decline of *Schizothorax*. The predominance can be attributed to the fact that unlike *Schizothorax*, which undergoes breeding migration in streams, the Carp breeds in confined waters, which leads to their higher fecundity (Sunder et al., 1984). Another reason is the feeding competition, as both share the same food and feeding habits (Naik et al., 2015). The introduction of these species has been done without considering its effects on endemic fish. Besides this, to fulfill sport fishing demands, two species of trout (*Salmo trutta* and *Oncorhynchus mykiss*) were made known to Kashmir waters. Out of the two, brown trout has well flourished with many self-subsisting populations in Kashmir's streams. However, though cultured in fish farms, the rainbow trout could not be set up in the stream ecosystems. It is also reported that the brown trout preys upon the earlier stages of the *Schizothorax,* which is also one of the reasons for the sharp decline of this endemic fish. (Kauser et al., 2017).

4.7.2 EUTROPHICATION

Inadvertent introduction of *Cyprinus carpio* in Kashmir's inland waters is not the only reason for the sharp decline of *Schizothorax*. Issues like pollution, poor maintenance of lakes, sewage runoff from urban areas, and agricultural practices in rural areas have intensified the nutrient inputs many folds, resulting in accelerated eutrophication and pollution (Zan et al., 2011). The endemic Schizothoracine fish that loves to live in cold and clean water finds it hard to endure eutrophication issues and usually suffer heavy losses. The increased eutrophication can be appraised by the growth of red bloom, which has the inherent potential to produce toxicity in aquatic environments and endanger native fish populations (Kahn, 2000). Along with an increase in pollution, these polluted water bodies harbor many pathogens like helminths, protozoans, bacteria, viruses, etc., which has led to a terrible impact on the fish population, primarily the indigenous *Schizothorax* species. Helminths like *Pomphorhyncus kashmerensis* and *Neoechinorhynchus* are found in bulk in the fish found in water-bodies with high pollution levels (Qayoom and Shah, 2018). Dar et al. (2014) reported the infestation of a protozoan parasite *Chilodonella* sp. from Kashmir's fishes. Chalkoo et al. (2007) reported the incidences of furunculosis in *Schizothorax niger, Schizothorax esocinus, Schizothorax curvifrons Schizothorax. labiatus* in Wular Lake. From the perspective of microbiology, the pathogens like bacteria and viruses are a potential risk not only to the fish but also to the humans who handle and consume them.

4.7.3 USE OF PESTICIDES

Under the public health protection program, pesticides find prime recognition when safeguarding the plant and human health. Typical examples of pesticides include rodenticides, herbicides, insecticides, fungicides, and plant growth regulators (WHO, 1990). A broad span survey under Kashmir's Agricultural department's aegis has yielded a 20% surge in the consumption of pesticides per annum by Kashmiri farmers. The excessive use of organophosphorus and organochlorine pesticides in agriculture has a drastic impact on *Schizothorax*. The pesticides enter the fish body by ingestion, dermal absorption, and respiration (Gold et al., 1955). Unlike *Cyprinus carpio,* which has a thick, scaly layer over its body and does not show traces of pesticide, the *Schizothorax* shows an increased susceptibility for chronic dermal pesticide absorption owing to its delicately thin layer of body scales (Muddasir et al., 2012). Exposure of *Schizothorax niger* to sub-lethal concentrations of "butachlor," an herbicide used in rice fields, leads to a metabolic crisis in fish. Such changes prove detrimental to the fish, and consequently, the fish easily falls prey to secondary infections (Habiba, 2012).

4.7.4 CONSTRUCTION OF DAMS

Schizothorax is potamodromous because its entire life cycle occurs in freshwaters (Northcote, 1998). It undertakes migration from upstream to downstream to avoid low water temperatures of winter and vice versa for breeding in summer (Petr, 1999). Any obstruction on its migratory routes will harm its population. The construction of dams on rivers has negatively impacted the fish migrations (Baxter, 1997). There is an interruption in aquatic migratory species' movement between feeding and spawning sites (Kline et al., 1990). In the case of potamodromous fish like *Schizothorax*, the competition for spawning sites and food intensifies as the dam's isolate, disconnect, and reduce the number and size of their habitats (Cambray et al., 1997). While passing through the hydraulic turbines, fish face various forms of stress resulting in excessive mortality. The parts in the motion of a turbine (guide vanes, vanes, or blades on the wheel) cause sudden changes in pressure cavitation, shocks, unpredicted acceleration, or deceleration (Bell and Delacy, 1972). The timing of fish migration is greatly affected by impoundments. Such fish migration delays have dire effects because the fish is exposed to intensive predation, nitrogen supersaturation, and parasites (Ebel, 1977). The construction of fish ladders to permit fish to pass dams is the best concept (Clay, 1961). However, doubt has been whether such structures are adequate in all cases (Geen, 1975). Poaching in the fish ladder during fish migration is a serious concern faced at Lower Jhelum barrages and Uri Hydropower Projects (Nyman, 1995). Habitat alteration (Light et al., 2007), damage due to spillways (Bell and Delacy, 1972), increased predation (Ebel et al., 1973) and thermal pollution (Qureshi, 2008), etc., are also some of the possible dam related causes that are responsible for *Schizothorax* decline.

4.7.5 CLIMATE CHANGE

Climate change is one of the greatest threats to global biodiversity (Thomas et al., 2004; Pachauri et al., 2014). According to Morrongiello et al. (2011), freshwater ecosystems are particularly in peril due to climate change because aquatic habitats mainly depend upon the climate and hydrologic system. Climate change is not only an issue at the global level but also a major trouble maker at the local level, particularly in third world countries like India (Gerlitz, 2015). In Kashmir valley, climatic changes can be seen via rising temperature, loss of glaciers and water bodies, erratic rainfall, and biological diversity erosion (Ishtiyak et al., 2016). Reports from 1980 to 2014 generated by the Indian Metrological Department (IMD) revealed that the Kashmir valley's temperature has increased and rainfall significantly decreased (Shafiq et al., 2019). As *Schizothorax* is a cold-water fish and prefers to live in running waters, the hike in temperature and lack of precipitation could attenuate its performance at various biological levels (Kamalam et al., 2019). The temperature has an obvious effect on the growth rate and thermal tolerance of *Schizothorax,* which exhibits reduced growth rate and narrow thermal tolerance when exposed to a higher temperature (He et al., 2014).

4.7.6 INTRADEPARTMENTAL ENIGMA

In valley Kashmir, the issue of fisheries is complex. Like fishermen, direct stakeholders favor *Schizothorax* fishery because its market value is much higher than other fishes. Meanwhile, there is also an urgent need to hike fish production so the market demand can be contented, regardless of fish species (Qureshi et al., 2013). This conflict gets more intricate with grace disagreement seen in the Department of Fisheries and Department of tourism's objective framing. The question as to which side must be prioritized over the other is still an open debate. However, the evidence is more skewed towards revenue generation and paying minimal heed to the conservative developmental process about lake fisheries, which in the long haul would yield better dividends to the involved individuals both in terms of capital and time employed in the process.

4.8 CARDINAL LINE UP TO STOP EROSION OF SCHIZOTHORAX

Since the decline in fish is related to the factors mentioned above, we need to take proper measures to overcome these catastrophic factors. By employing different stakeholders (government/private), different remedial measures can be suggested as given in subsections.

4.8.1 DEVELOPMENT OF SEWAGE TREATMENT PLANS TO STOP EUTROPHICATION

Sir Walter Lawrence (Settlement Commissioner of Kashmir in British India) quotes in his book "The Valley of Kashmir" (1985) that Kashmiri craftsmen are saying that "Shawls of Kashmir owe much of their excellence to being washed in the soft waters of the Dal Lake." Praising the beauty of the lake himself, Lawrence writes: "Perhaps in the whole world there is no corner as pleasant as the Dal Lake." Unfortunately, uncontrolled inter-city development over time has not only diminished the beauty of the lake. Still, it has also changed it into a vast sump in which untreated sewage from the city (old Srinagar) is deposited, resulting in the lake's eutrophication. The same is the case with other water bodies like Wular, Khushalsar, Mansbal Lake, Anchar Lake, etc. The improvement of the lake water quality can be one of the remedies to restore the *Schizothorax* population. The diversion of effluents/sewage coming from the adjoining areas and nutrient removal from the wastewater before it is discharged into a particular water body could act as an impactful tool in this direction (Vass, 1980).

4.8.2 ESTABLISHMENT OF SEPARATE AQUACULTURE FOR SCHIZOTHORAX AND CYPRINUS CARPIO

The Lakes and Waterways Development Authority (LAWDA), Kashmir, should chalk out a strategic plan for *Schizothorax* and *Cyprinus carpio's* steady development that inhabit the same water space. Moreover, the Carp should be segregated from the *Schizothorax* by promoting carp culture as distinct freshwater aquaculture (Qureshi et al., 2013). Apart from this, the Government should make combined attempts to lessen the Carp's predominance and uplift the institutionalization of *Schizothorax* via the construction of an appropriate end to end chain arrangements. The Carp, along with trout, must be treated as different entities. A subsequently determined approach must be adopted to address the technical aspects like constructing hatcheries and raceways to foster culturing of native fish.

4.8.3 INDUCED BREEDING

Administering gonadotrophin into the fish to forcibly mature/breed is called induced breeding (Agarwal, 2007). Induced breeding is difficult in Indian cold-water fishes and is limited to a few species of masheer (Sehgal and Kumar, 1977; Agarwal, 2001; Ogale, 2000). The spawning is achieved in captivity by induced breeding through hypophysation in various teleosts. Though pituitary extracts are very effective in inducing maturation and spawning in many fishes, these preparations are found ineffective (Goswami and Sundararaj, 1971). A similar situation has been noticed in schizothoracids (Raina et al., 1985). Failure of spawning after administration of gonadotrophin in many species is due to the dopamine inhibitor activity (Richard et al., 1986; Nandeesha et al., 1991). Gharaei et al. (2011) investigated the breeding performance of *Schizothorax zarudnyi* under partially controlled conditions after inducing breeding using synthetic hormone (Ovaprim) and suggested that ovaprim is suitable for inducing breeding in snow trout.

The first report of successful induced breeding in *Schizothorax niger* from Kashmir waters came in 2014. Ovatide, an injectable solution containing a synthetic peptide analogous to gonadotrophic releasing hormone (GnRH) with a dopamine antagonist, was found to promote reproduction in *Schizothorax niger* with a fertilization percentage of 85% by inducing ovulation and spermiation (Najar et al., 2014).

4.8.4 EX-SITU AND IN-SITU CONSERVATION

According to the International Union for the Conservation of Nature and Natural Resources *Schizothorax* is a vulnerable species (IUCN, 2001). Therefore, there is an urgent need for genetic conservation and management. Singh and Sharma (1998) studied the different aspects (ecological status, biodiversity) of fishes in the Alaknanda River concluded that *ex-situ* conservation could be in the form of living collection. A live gene bank is one of the pillars of *ex-situ* conservation. Here species are bred in captivity with proper care to avoid interbreeding, depression, and unintended selection (Minckley and Deacon, 1991). Meanwhile, such collections are expensive and can be attacked by pests easily; hence proper management is essential (Singh, 1998).

In-situ conservation can be accomplished through the maintenance of *Schizothorax* in its natural ecosystem, where fishing is banned and protected with the support of community participation (Agarwal and Singh, 2009). The idea of the creation of fish sanctuaries can be implemented for *the in-situ* conservation of *Schizothorax* in Kashmir Himalayas (Singh and Sharma, 1998). These fish sanctuaries should be governed by the National Bureau of Fish Genetic Resources (Agarwal and Singh, 2009).

4.8.5 *TAKING HEED AT THE DAMS*

Fish ladders, also known as "fish-way/fish-pass/fish-steps," being used at various places on the dams, enable fishes to migrate easily. However, poaching by humans and predatory birds at these ladders is a problem of serious concern. To avoid this problem, a course metal grid can be welded across the fish ladder's entire length (Nyman, 1995). Pool type fish passes (Travade et al., 1998), Denil fish passes (Larinier, 1992), Bypass channels (Parasiewicz et al., 1998), Fish locks (Travade and Larinier, 1992) can be used to ensure safe migration of the fish from its feeding to breeding grounds and vice versa. These passes' effectiveness can be analyzed by visual inspections and checks (Travade et al., 1998).

4.8.6 *LEGAL MEASURES*

Besides the above measures in place, there is a serious need for legal measures that act as a proper bulwark to audit and analyze the diminution of both the number and quality of freshwater fish. One such stride by the union territory (UT) is implementing the J&K Fisheries Act (2018) mentioned below in brevity:

- Use of proper fishing gear to avoid the mutilation of fish;
- Higher inclination towards wastewater management to mitigate the damage done to fish colonies due to hazardous chemicals;
- Checks and sanctions against over-exploitation of fish;
- Prohibition of the introduction of any fish that may be harmful to the existing fish species without obtaining prior permission of the Government;
- A complete prohibition on fishing and fish marketing during peak breeding seasons.

4.9 CONCLUSION

Schizothorax is a versatile freshwater fish type with nutritional, economic attribution/values playing a vital role in our day-to-day lives. As per the International Union for Conservation of Nature and Natural Resources (IUCN), the conservation status of *Schizothorax* is vulnerable. Currently, the genus is facing enormous anthropogenic stress resulting in its decline. The anthropogenic activities of prime concern include indiscriminate use of pesticides, sewage runoff resulting in eutrophication, introduction of exotic fish species like "common carp," and developmental projects like construction of dams for hydroelectricity on rivers are also pushing back the said genus. Keeping all these limiting factors in mind Government, NGOs, and other stakeholders need to take appropriate measures to protect this valuable genus. Ordinary people should be provided with the proper knowledge for pesticides via print and electronic media. Educational institutions like Universities and colleges should come forward in this regard. Other aspects include improving the water quality, which can be done by diversifying effluents/sewage coming from the adjoining areas and nutrient removal from the polluted water before its release into a particular water body. Induced breeding should be adopted by using synthetic hormones like "Ovaprim." *Ex-situ* (conservation in the form of the living collection) and *In-situ* conservation (maintenance in a natural ecosystem where fishing is totally banned) should be initiated. Besides all these measures, there is a vital need for a legitimate system that can examine the ecological status of *Schizothorax*.

ACKNOWLEDGMENT

The authors of this chapter appreciate the Department of the Zoology Central University of Kashmir for providing the necessary literature and support.

KEYWORDS

- **conservation**
- **eutrophication**
- **gonadosomatic index**
- **Kashmir**
- **pollution**
- ***Schizothorax***

REFERENCES

Agarwal, N. K., & Singh, H., (2009). Snow trout fishery in Uttaranchal: causes of depletion & strategy for propagation. *Environment Conservation Journal., 10*(1, 2), 135–140.

Agarwal, N. K., (2001). Reproductive strategies of snow trout (*Schizothorax* sp.) with reference to the environment of the river Bhagirathi & Bhilangana of Tehri Garhwal. *A final Project Report to Indian Council of Agriculture Research, 78.*

Agarwal, N. K., Thapliyal, B. L., Raghuvanshi, S. K., (2007). Induced breeding and artificial fertilization of snowtrout, *Schizothorax richardsonii* through the application of ovaprim. *J. Inland Fish. Soc. India., 39*(1), 12–19.

Arisekar, U., Shakila, R. J., Jeyasekaran, G., Shalini, R., Kumar, P., Malani, A. H., & Rani, V., (2019). Accumulation of organochlorine and pyrethroid pesticide residues in fish, water, and sediments in the Thamirabarani river system of southern peninsular India. *Environmental Nanotechnology Monitoring & Management, 11*, 100194.

Baba, S. H., Husain, N., Zargar, B. A., Bhat, I. F., & Malik, I., (2019). Growth of fisheries sector in Jammu & Kashmir. *Journal of Pharmacognosy and Phytochemistry, 8*(4), 909.

Barik, N. K., (2017). Freshwater fish for nutrition security in India: Evidence from FAO data. *Aquaculture Reports, 7*, 1–6.

Bashir, A., Bisht, B. S., Mir, J. I., Patiyal, R. S., & Kumar, R., (2016). Morphometric variation and molecular characterization of snow trout species from Kashmir valley, India. *Mitochondrial DNA Part, 27*(6), 4492–4497.

Bashir, A., Sharma, N. K., Bisht, B. S., Sing, R., Mir, J. I., Akhtar, M. S., (2016). Length–weight relationship of five commercially important freshwater fish species in the Kashmir Valley, India. *Journal of Applied Ichthyology, 32*(4), 740–741.

Baxter, R. M., (1977). Environmental effects of dams and impoundments. *Annual Review of Ecology and Systematics,* 225–283.

Bell, M. C., & DeLacy, A. C., (1972). *A Compendium on the Survival of Fish Passing Through Spillways and Conduits*. Report Fish. Eng. Res. Prog., US Army Corps of Eng., North Pcific Div., Portland, Oregon, USA.

Bhat, F. A., & Farooq, I., (2016). *Estimation of Length-Weight Relationship and Condition Factor of Crossocheilus Diplochilus (Heckel, 1838)*. A freshwater benthopelagic fish from Wular Lake in Kashmir Himalaya.

Bhat, F. A., Yousuf, A. R., Balkhi, M. H., Mahdi, M. D., Shah, F. A., (2010). Length-weight relationship and morphometric characteristics of *Schizothorax* spp. in the River Lidder of Kashmir. *Indian Journal of Fisheries, 57*(2), 73–76.

Bhat, F. A., Yousuf, A. R., Balkhi, M. H., Mahdi, M. D., Shah, F. A., (2010). Length-weight relationship and morphometric characteristics of *Schizothorax* spp. in the River Lidder of Kashmir. *Indian Journal of Fisheries, 57*(2), 73–76.

Cambray, J. A., King, J. M., & Bruwer, C., (1997). Spawning behavior and early development of the *Clanwillian yellowfish* (*Barbus capensis; Cyprinidae*), linked to experimental dam releases in the Olifants River, South Africa. *Regulated Rivers: Research & Management: An International Journal Devoted to River Research and Management, 13*(6), 579–602.

Carvajal-Quintero, J., Villalobos, F., Oberdorff, T., Grenouillet, G., Brosse, S., Hugueny, B., Jèzèquel, C., & Tedesco, P. A., (2019). Drainage network position and historical connectivity explain global patterns in freshwater fishes' range size. *Proceedings of the National Academy of Sciences, 116*(27), 13434–13439.

Cazenave, J., De Los, A. B. M., Pesce, S. F., Wunderlin, D. A., (2006). Differential detoxification and antioxidant response in diverse organs of *Corydoras paleatus* experimentally exposed to microcystin-RR. *Aquatic Toxicology, 76*(1), 1–2.

Chalkoo, S. R., Najar, A. M., Qureshi, T. A., & Shafi, A., (2007). Furunculosis in snow trout (*Schizothoracinae*) in Kashmir: First report. *Journal of the Indian Fisheries Association, 34*, 59–73.

Clark, J. A., & Pagel, J. E., (1977). pollution indicator bacteria associated with municipal raw and drinking water supplies. *Canadian Journal of Microbiology, 23*(4), 465–470.

Clasen, B., Loro, V., Murussi, C. R., Tiecher, T. L., Moraes, B., & Zanella, R., (2018). Bioaccumulation and oxidative stress caused by pesticides in *Cyprinus carpio* reared in a rice-fish system. *Science of the Total Environment, 626*, 737–743.

Clay, C.H. (1961). Design of fishways and other fish facilities, Department of Fisheries of Canada, Ottawa.

Dar, S. A., Chishti, M. Z., Kaur, H., & Dar, G. H., (2014). Parasites of some fishes (*Labeo rohita* and *Schizothorax Niger*) of Jammu and Kashmir in India. *International Journal of Fisheries and Aquaculture, 6*(9), 104–107.

Ebel, W. J., (1977). *Major Passage Problems and Solutions: Major Passage Problems*, pp. 33–39: In E. Schweibert (ed). Columbia River Salmon and Steelhead. American Fisheries Society. Special Publication No. 10. Washington D.C.

Elfman, L., Tooke, N., Patring, J. D., (2011). Detection of pesticides used in rice cultivation in streams on the island of Leyte in the Philippines. *Agricultural Water Management, 101*(1).

Fotedar, D. N., Qadri, M. Y., (1974). Fish and fisheries of Kashmir and the impact of Carp, *Cyprinus carpio*, on the endemic fishes. *Journal of Science, 2*, 79–89.

Fouche, P. S., & Gaigher, I. G., (2001). Niche differentiation in the rheophilic fishes of the Mutshindudi River. A Socio-biological study of the aquatic resources and their utilization in an underdeveloped rural region, the Mutshindudi River Catchment. *WRC Project Report, 714*(*3*), 1.

Froese, R. & Pauly, D. (2019). *Fish Base.* World Wide Web electronic publication. http://www.fishbase.org.

Fuller, M. R., Doyle, M. W., Strayer, & D, L., (2015). Causes and consequences of habitat fragmentation in river networks. *Annals of the New York Academy of Sciences, 1355*(1), 31–51.

Geen, G. H., (1975). Ecological consequences of the proposed Moran Dam on the Fraser River. *J. Fish. Res. Board Can., 32*(1), 126–135.

Gerlitz, J. Y., Banerjee, S., Brooks, N., Hunzai, K., & Macchi, M., (2015). An approach to measure vulnerability and adaptation to climate change in the Hindu Kush Himalayas. *Handbook of Climate Change Adaptation*, 151–176.

Gharaei, A., Rahdari, A., & Ghaffari, M., (2011). Induced spawning of *Schizothorax zarudnyi* (Cyprinidae) by using synthetic hormones (*Ovaprim* and HCG). *World Journal of Fish and Marine Sciences, 3*(6), 518–522.

Gold-Bouchat, G., Silva-Herrea, T., & Zapata-Pèrez, O., (1995). Organochlorine pesticides residue concentration in biota and sediments from Riopalizada, Mexico. *Bulletin of environmental Contamination and Toxicology, 54*(4), 554–561.

Gormley, R., (2006). Fish as a functional food. *Food Science and Technology, 20*(3), 27. London.

Goswami, S. V., & Sundararaj, B. I., (1971). *In vitro* maturation and ovulation of oocytes of the catfish, *Heteropneustes fossilis* (Bloch): Effects of mammalian hypophyseal hormones, catfish pituitary homogenate, steroid precursors and metabolites, and gonadal and adrenocortical steroids. *Journal of Experimental Zoology, 178*(4), 467–477.

Government of India (2017). *Economic Survey 2016–17.* p. 177.

Habiba, B., & Yousuf, A. R. (2013). *Butachlor Induced Alterations in Haematological Parameters of Schizothorax Niger Heckel.* (Doctoral dissertation) University of Kashmir.

He, Y., Wu, X., Zhu, Y., Li, H., Li, X., & Yang, D., (2014). Effect of rearing temperature on growth and thermal tolerance of *Schizothorax* (Racoma) kozlovi larvae and juveniles. *Journal of Thermal Biology, 46,* 24–30.

Heckel, J. J. (1838). *Fish from Kashmir Collected and edited by Carl Freiherrn v. Hill, Described by Joh. Jacob Heckel.* Pressed by the PP Mechitaristen, Vienna, 112 pp., 13 pls.

Hildebrand, M., & Goslow, G., (1974). *Analysis of Vertebrate Structure*. New York.

Hussain, S., Bhat, F. A., Maqsood, H. M., Balkhi, M. H., Majid, I., Najar, A. M., (2018). Present status of breeding biology of *Schizothorax Niger* in Dal Lake Kashmir. *J. Ento. Zool. Stud., 6*, 930–935.

International Union for Conservation of Nature, IUCN. (2012). IUCN Red List Categories and Criteria: Version 3.1. Second edition. Gland, Switzerland and Cambridge, UK: IUCN. iv + 32pp.

Ishtiyak, P., Reddy, M., Panse, S., Wani, I., Peer, Q. J., (2016). Impact of climate change and anthropogenic interventions on natural vis-à-vis human resources in Kashmir, India-An overview. *Journal of Applied and Natural Science,* (1), 489–493.

J&K Fisheries Act, (2018). https://www.indiacode.nic.in/handle/123456789/16467?locale =en (accessed on 13 December 2021).

Jan, M., & Ahmed, I., (2019). Reproductive biology and histological studies of ovarian development of *Schizothorax plagiostomus* in river Lidder from Kashmir Himalaya. *Journal of Applied Ichtyology, 35*(2), 512–519.

Jan, N. A., Das, S. M., (1970). Qualitative and quantitative studies on the food of eight fishes of Kashmir valley. *Ichthyologica., 10*(1, 2), 20–26.

Jayedi, A., Shab-Bidar, S., Eimeri, S., & Djafarian, K., (2018). Fish consumption and risk of all-cause and cardiovascular mortality: a dose–response meta-analysis of prospective observational studies. *Public Health Nutrition, 21*(7), 1297–1306.

Joshi, V., Akhtar, M. S., Sharma, P., Singh, K. S., Baruah, D., Ciji, A., Pande, V., & Sarma, D., (2017). Himalayan fish manifest higher potential of quality nutrients for human health. *Journal of Aquatic Food Product Technology, 26*(7), 843–855.

Jyoti, M. K., & MK, J., (1972). *Studies on the Fecundity of Schizothorax Niger.* Heckel from Dal Lake. Kashmir.

Kahn, M. A., (2000). Anthropogenic eutrophication and red tide outbreak in lacustrine systems of the Kashmir Himalaya. *Acta Hydrochimica et Hydrobiologica, 28*(2), 95–101.

Kamalam, B. S., Mahija, J., Baral, P., Pandey, A., Akhtar, M. S., Ciji, A., & Rajesh, M., (2019). Temperature and oxygen related ecophysiological traits of snow trout (*Schizothorax richardsonii*) are sensitive changes in a Himalayan stream environment. *Journal of Thermal Biology, 83*, 22–29.

Kausar, N., (2008). *Studies on Digestibility of Selected and Naturally Occurring Feed Stuffs their Effect on Growth and Spawning of Schizothorax Esocinus and Schizothorax Curvifrons.* http://hdl.handle.net/10603/94415 *(accessed on 13 December 2021).*

Kauser, N., Shah, G. M., & Jan, U., (2010). Seasonal fluctuations in the gut contents of *Schizothorax esocinus* and *Schizothorax curvifrons*. *IJSN., 3*, 928–930.

Kline, Jr. T C., Goering, J. J., Mathisen, O. A., Poe, P. H., Parker, P. L., (1990). Recycling of elements transported upstream by runs of Pacific salmon: I, δ15N and δ13C evidence in Sashin Creek, Southeastern Alaska. *Canadian Journal of Fisheries and Aquatic Sciences, 47*(1), 136–144.

Kris-Etherton, P. M., Harris, W. S., Appel, L. J., (2003). Fish consumption, fish oil, omega-3 fatty acids, and cardiovascular disease. *Arteriosclerosis, Thrombosis, and Vascular Biology, 23*(2), e20–e30.

Larinier, M., (1992). Passes to successive basins, pre-dams and artificial rivers. *French Bulletin of Fishing and Fish Farming,* (326, 327), 45–72.

Lawrence, W. R., (1895). *The Valley of Kashmir*. H. Frowde.

Light, T., Marchetti, M. P., (2007). Distinguishing between invasions and habitat changes as drivers of diversity loss among California's freshwater fishes. *Conservation Biology, 21*(2), 434–446.

Lund, E. K., (2013). Health benefits of seafood; is it just the fatty acids? *Food Chemistry., 140*(3), 413–420.

Manon, M. R., & Hossain, M. D., (2011). Food and feeding habit of *Cyprinus carpio* var. specularis. *Journal of Science Foundation, 9*(1, 2), 163.

Minckley, W. L., Marsh, P. C., Brooks, J. E., Johnson, J. E., & Jensen, B. L. (1961). Management toward recovery of the razorback sucker. Chapter 17. In: Minckley, W. L., & Deacon, J. E., (eds.), *Battle Against Extinction: Native Fish Management in the American West.* https://doi.org/10.2307/j.ctt1rfzxt0.28.

Morrongiello, J. R., Beatty, S. J., Bennett, J. C., Crook, D. A., Ikedife, D. N., Kennard, M. J., Kerezsy, A., et al., (2011). Climate change and its implications for Australia's freshwater fish. *Marine and Freshwater Research, 62*(9), 1082–1098.

Muddasir, B., Dhar, J. K., Shafiqa, A., & Qureshi, S. I. A., (2012). Contamination of fresh water fish "*Schizothorax Niger*" with chloropyrifos from "Dal Lake" basins, India. *International Journal of Pharma sciences and Research, 3*(12), 535–542.

Naik, G., Rashid, M., Balkhi, M. H., & Bhat, F. A., (2015). Food and feeding habits of *Cyprinus carpio* var. communis: A reason that decline *Schizothoracine* fish production from Dal Lake of Kashmir Valley. *Fisheries and Aquaculture Journal, 6*(4).

Najar, A. M., Bhat, F. A., Balkhi, M. H., Samoon, M. H., Dar, S. A., & Wani, (2014). Induced breeding and larval rearing of snow trout (Schizothorax niger Heckel) in Kashmir Himalaya with the application of Ovatide. *Fishery Technology, 51*(1).

Nandeesha, M. C., Varghese, T. J., (1991). *Further Observations on Breeding of Carps with Ovaprim.* Asian Fisheries Society, Indian Branch.

Nasreen, K. D., Reyaz, A. M. D., & Ulfat, J. G. M., (2017). Effect of soybean incorporated diets on the performance of *Schizothorax curvifrons*. *International Journal of Fauna and Biological Studies, 4*(4), 01–03.

Northcote, T. G., (1998). Migratory behavior of fish and its significance to movement through riverine fish passage favilities. *Fish Migration and Fish Bypasses,* 3.

Nyman, L., (1995). *Environmental Impact Assessment: Uri Hydroelectric Power Project. On River Jhelum in Kashmir, India*. Swedmar.

Ogale, S. N., (2000). Mahseer hatchery: planning and management. *Fishing Chimes., 19*(10–11), 69–77.

Pachauri, R. K., Allen, M. R., Barros, V. R., Broome, J., Cramer, W., Christ, R., Church, J. A., et al., (2014). *Climate Change 2014: Synthesis Report*. Contribution of working groups I, II and III to the fifth assessment report of the intergovernmental panel on climate change. IPCC.

Parasiewicz, P., Eberstaller, J., Weiss, S., & Schmutz, S., (1998). Conceptual guidelines for nature-like bypass channels. *Fish Migration and Fish Bypasses,* 348.

Pérez, M. R., Rossi, A., Elorriaga, Y., Bacchetta, C., Carriquiriborde, P., & Cazenave, J., (2018). *In situ* evaluation of the toxicological impact of a wastewater effuent on the fish *Prochilodus lineatus*: biochemical and histological assessment. *Ecological Indicators, 84*, 345–353.

Petr, T., (1999). *Fish and Fisheries at Higher Altitudes: Asia*. Food & Agriculture Org.

Potenza, M. V., Mechanick, J. I., (2009). The metabolic syndrome: definition, global impact, and pathophysiology. *Nutrition in Clinical Practice, 24*(5), 560–577.

Qayoom, I., Balkhi, M. H., Shah, F. A., Bhat, B. A., (2018). Toxicological evaluation and effects of oganophosphate compounds on hematological profile of juvenile common carps (*Cyprinus carpio* var. Communis). *Indian Journal of Animal Research, 52*(10), 1469–1475.

Qureshi, M., Shakir, A. S., & Lesleighter, E., (2016). channel forming discharge in rivers: a case study of Jhelum river in Pakistan. *Pakistan Journal of Engineering and Applied Sciences*.

Qureshi, N. W., Krishnan, M., Sundaramoorthy, C., Vasisht, A. K., Baba, S. H., Kumar, N. R., & Sharma, R., (2013). Truncated growth and compromised sustainability: The case of lake fisheries in Kashmir. *Agricultural Economics Research Review, 26*(347-2016-17109), 57–66.

Qureshi, N. W., & Krishnan, M., (2015). *Lake Fisheries in Kashmir: A Case More Undone Than Done*. Economic and Political Weekly.

Qureshi, T. A., (2008). Effect of thermal pollution on the hydrological parameters of River Jhelum (J&K). *Journal of the Indian Fisheries Association, 35*, 159–166.

Raina, H. S., Vass, K. K., Sunder, S., Moza, U., Langer, R. K., (1985). Prospects and problems of snow trout culture in Kashmir. *Zoologica Orientalis., 2*(1, 2), 24–30.

Rashid, I., Farooq, M., Muslim, M., Romshoo, S. A., (2013). Assessing the impact of anthropogenic activities on Manasbal Lake in Kashmir Himalayas. *International Journal of Environmental Sciences, 3*(6), 2036–2047.

Rather, M. A., Baba, S. A., (2015). Traditional use of medicinal plants in Kashmir: A review. *Res. Rev. J. Biol., 3,* 26–32.

Rather, T., (2020). Fisheries sector in Jammu and Kashmir -growth and perspectives. *IJARW, 1*(7), 2582–1008.

Richard, E. P., Jhon, P. C., Carol, S. N., Robert, J. O., Sokolowska, M., Stephen, H. S., & Roland, B., (1986). Interactions of catecholamines and GnRH in regulation of gonadotropin secretion in teleost fish. In: *Proceedings of the 1985 Laurentian Hormone Conference* (pp. 513–548). Academic press.

Rossi, A. S., Fantón, N., Michlig, M. P., Repetti, M. R., & Cazenave, J., (2020). Fish inhabiting rice fields: Bioaccumulation, oxidative stress and neurotoxic effects after pesticides application. *Ecological Indicators, 113*, 106186.

Sabha, K. K., Najar, A. M., Bhat, F. A., Shah, T. H., Balkhi, M. H., & Faisal, R., (2017). Food and feeding habits of snow trout (*Schizothorax Niger*) inhabiting Nigeen Lake, Kashmir. *Journal of Experimental Zoology, India, 20*(1), 635–637.

Saleem, S., Kamili, A. N., Kakru, D. K., Bandh, S. A., (2013). Water quality assessment of Dal Lake, Kashmir using the coliforms as indicator bacteria. *J. New Biol. Rep.,* 2(1), 30–35.

Saleem, S., Kamili, A. N., Kakru, D. K., Bandh, S. A., Ganai, B. A., (2011). Isolation, identification and seasonal distribution of bacteria in Dal Lake, Kashmir. *International Journal of Environmental Sciences, 2*(1), 185–193.

Sandhya, K., (2019). *Fish Production Volume Across Jammu and Kashmir in India*. Statista.

Sehgal, K. L., & Kumar, K., (1977). *Final Project Report on Induced Breeding and Rearing of Mahseer (Tor putitora) Seed in Running Water Ponds*. Central Inland Fisheries Research Institute, Barrackpore.

Shafiq, M. U., Rasool, R., Ahmed, P., Dimri, A. P., (2019). Temperature and precipitation trends in Kashmir Valley, north western Himalayas. *Theoretical and Applied Climatology, 135*(1, 2), 293–304.

Singh, D., Sharma, R. C., (1998). Biodiversity, ecological status and conservation priority of the fish of the River Alaknanda, a parent stream of the River Ganges (India). *Aquatic Conservation: Marine and Freshwater Ecosystems, 8*(6), 761–772.

Singh, Y., Dewan, S., Bhatt, J. P., (2016). Ascertaining spawning behavior of a Himalayan trout, *Schizothorax plagiostomus* using combinatorial approach. *Int. J. Res. Aquacul., 6,* 39–46.

Skare, J. U., Brantsæter, A. L., Frøyland, L., Hemre, G. I., Knutsen, H. K., Lillegaard, I. T. L., Andreassen, Å. K., et al., (2014). Benefit-risk assessment of fish and fish products in the Norwegian diet–an update. *Opinion of the Scientific Steering Committee of the Norwegian Scientific Committee for Food Safety*. VKM Report.

Sunder, S., Kumar, K., Raina, H. S., (1984). Food and feeding habits and length weight relationship of *Cyprinus carpio* specularis Linnaeus of Dal Lake, Kashmir. Indian. *Journal of Fisheries, 31*(1), 90–99.

Thomas, C. D., Cameron, A., Green, R., Bakkenes, M., Beaumont, L. J., Collingham, Y. C., Erasmus, B. F., et al., (2004). Extinction risk from climate change. *Nature, 427*(6970), 145–148.

Tocher, D. R., (2015). Omega-3 long-chain polyunsaturated fatty acids and aquaculture in perspective. *Aquaculture, 449*, 94–107.

Travade, F., & Larinier, M., (1992). Locks and fish lifts. *French Bulletin of Fishing and Fish Farming,* (326, 327), 95–110.

Travade, F. Larinier, M. Boyer Bernard, S., & Dartiguelongue, J. (1996). Performance of four fish pass installations recently built on two rivers in South-west France. *Fish Migration and Fish Bypass Channels, Vienne, AUT, 24–26 September 1996*, Austria. pp. 146–170.

Van, P. P. J., Buijse, A. D., Kraak, M. H., Verdonschot, P. F., (2019). Species and river specific effects of river fragmentation on European anadromous fish species. *River Research and Applications, 35*(1), 68–77.

Vass, K. K., (1980). On the trophic status and conversation of Kashmir lakes. *Hydrobiologia, 68*(1), 9–15.

World Health Organization, (1990). *Public Health Impact of Pesticides Used in Agriculture*. World Health Organization.

Wu, H., Chen, J., Xu, J., Zeng, G., Sang, L., Liu, Q., Yin, Z., et al., (2019). Effects of dam construction on biodiversity: A review. *Journal of Cleaner Production, 221*, 480–489.

Yousuf, A. R., (1992). Current trends in fish and fishery biology and aquatic ecology. In: *National Seminar on Fish and Their Environment 1989.* Srinagar, Jammu and Kashmir, India. PG Dept. of Zoology, University of Kashmir.

Yousuf, R., Mir, S. H., Shah, A. W., Ganaie, G. A., (2011). A comparative study of parasites infecting some selected fishes from the water bodies of Kashmir valley. *Ind. J. Fund Appl. Life Sci., 1,* 150–153.

YR, M., (1970). Studies on the seasonal changes in the ovary of *Schizothorax Niger* Heckel from Dal Lake in Kashmir. *Japanese Journal of Ichthyology, 17*(3), 110–116.

Zan, F., Huo, S., Xi, B., Su, J., Li, X., Zhang, J., & Yeager, K. M., (2011). A 100 year sedimentary record of heavy metal pollution in a shallow eutrophic lake, Lake Chaohu, China. *Journal of Environmental Monitoring, 13*(10), 2788–2797.

Zutshi, D. P., Vass, K. K., (1970). High altitude lakes of Kashmir. *Ichthiologica,* 10, 12–15.

CHAPTER 5

Influence of Catchment on Freshwater Biodiversity

LONE RAFIYA MAJEED,[1] SUMAIRA RASHID,[2] HEENA NISAR PAHLAVI,[2] and BISMA NISAR[2]

[1]*Vivekananda Global University, Jaipur, Rajasthan – 303012, India, E-mail: rafiyaevs@gmail.com*

[2]*Department of Environmental Science, University of Kashmir, Jammu and Kashmir – 190006, India*

ABSTRACT

Water pollution is recognized as a global problem in the twenty-first century. Pure water is becoming increasingly scarce as a result of water contamination. The existence of life is the most distinctive aspect of the aquatic environment, and the most unique element of life is its biodiversity. Water contamination occurs for a variety of reasons, all of which have a negative impact on biodiversity. Any human action has a significant impact on the planet's biological variety. Catchment has a major role to play in defining the pollution of freshwater ecosystems. Acidification, climate change, overexploitation, water pollution, habitat degradation, flow modification, and exotic species invasion are all challenges to aquatic biodiversity. Biodiversity conservation is regarded as one of the most important factors in the preservation of ecosystem services. As a result, it might be claimed that today's greatest problem is the preservation of freshwater biodiversity.

Biodiversity of Freshwater Ecosystems: Threats, Protection, and Management. Suhaib A. Bandh, PhD, Basharat Mushtaq, PhD, Mohammad Farooq Mir, PhD, & Shahid A. Wani (Eds.)

5.1 INTRODUCTION

Freshwater systems are among the most miscellaneous and prolific ecosystems worldwide (Flitcroft et al., 2019). Freshwater fishes are the chief source of protein for people worldwide (Tickner et al., 2020). Rivers deposit silt on flood plains and deltas and result in the best fertile agricultural lands. Freshwater biologist E F. Stoermer who coined the term Anthropocene in 1980s (Luke, 2020), has written about the human-induced planetary destruction, among which freshwater biodiversity is facing decline to the extent that we are facing sixth mass extinction (Ceballos et al., 2017). Global Wetlands Outlook has highlighted the freshwater biodiversity loss resulting from changes in the catchment (Gardner et al., 2018). Factors that are directly responsible for the loss of freshwater biodiversity include changes in the physical regime of the catchment (Anderson et al., 2019), the harvest of species, introduction of invasive species (Reid et al., 2019), extraction of material like gravel, stones, and sand, pollution from different sources like pesticides and, microplastics, etc. (Chen et al., 2020). Other factors include Hydropower generation and climate change (Reid et al., 2019). It is also affected by indirect factors like fundamental systems for energy generation, food production (livestock, plantation), urbanization of catchment with unsustainable infrastructure development, and water supply (Zarfl et al., 2015). All the direct and indirect drivers of freshwater biodiversity have resulted in an average 1% annual increase in global water demand and if continues, in the same way, would amount to about 20%–30% above current water utilization (Anna et al., 2019). The acuteness of the freshwater biodiversity crisis is widely acknowledged (Cooke et al., 2021). Loss of biodiversity in freshwater ecosystems is far worse than in marine or terrestrial ecosystems (Tickner et al., 2020). Fishes in the freshwater ecosystem are believed to be worst hit by extinction than other freshwater species (Tavares-Dias et al., 2017). More than 60% of freshwater habitats are believed to be moderately or highly threatened (Grill et al., 2019). Populations of freshwater megafauna, defined as animals that reach a body mass of 30 kilograms, declined by 88% from 1970 to 2012, with the highest declines in the Indomalaya and Palearctic realms (−99% and −97%, respectively) (He et al., 2019). Out of the 29,500 species of freshwater assessed so far by IUCN, the class of species assessed and the percentage of threat with extinction is given in Table 5.1. Targeting freshwater habitats with conservative efforts often fail (Darwall et al., 2018) partly because the protection of freshwater environments often require large-scale multi-sectoral efforts (WRT et al., 2011).

TABLE 5.1 Risk Percentage of Freshwater Species

Species Assessed	Percentage of Threat
Turtle	62%
Gastropods	47%
Mammals	42%
Amphibians	33%
Decapod crustaceans	30%
Fishes	28%
Birds	20%

Source: IUCN (2019); www.iucnredlist.org.

5.2 IMPACT OF LAND USE LAND COVER CHANGE (LULCC) IN CATCHMENT ON BIODIVERSITY

Fast track growth in anthropogenic activities has increased biodiversity loss (Ceballos et al., 2015; Steffen et al., 2015). LULCC at the global level has altered both availability and competition for freshwater (Luck et al., 2015). It is now a well-established global issue that impacts earth surface processes and the minute ecological factors that result in the loss of freshwater biodiversity (Foley et al., 2005). Climate change aggravated with LULCC in the catchment area possesses a great potential to impose excessive pressure on both availability and accessibility of water globally, resulting in the disruption of the water cycle (Shawul, 2019). When the plantation of riparian areas is converted to agronomic land, it creates pressure on the headwater stream ecosystems. For instance, algae in freshwater ecosystems are highly affected by amplified heat and light in the absence of riparian vegetation, resulting in altered community composition (Allan, 2004). Amplified temperature due to the removal of vegetation from the riparian zone could result in reduced performance of ectotherms that alters the community composition. An increase in the number of grazers and decreased shredders change the production base from allochthonous to autochthonous (Delong et al., 1998). LULCC in the catchment area may result in the change in the composition of dominant flora resulting in the change in runoff that has an incidental effect on the freshwater biodiversity.

Changes in catchment LULC is also portrayed in physical, chemical, and biological aspects of freshwater resources (Allan, 2004), enlarged piles of nutrients to freshwater ecosystems, with effects on water quality, habitat characteristics, and biodiversity (Richards et al., 1996, Sponseller et al., 2001).

5.2.1 *ACIDIFICATION*

Local biotic and abiotic factors play a significant role in freshwater chemistry (Heino et al., 2018). The influence of catchment freshwater chemistry is relatively less studied due to its multi-parameter involvement (Carpenter et al., 2009). The reaction of H_2O with CO_2 (Phillips et al., 2015) responsible for acid production is affected by temperature. The metabolism of aquatic animals is adversely affected in such acidified environments—the majority of non-chordate animal skeletons, such as corals, crustaceans, molluscs, and echinoderms.

There would be a reduction in biodiversity with increased acidification of freshwater habitats, with an increased extinction of acid-sensitive plants. A decrease in pH to 6 will dramatically impact the freshwater populations of both snails and crustaceans. These species constitute 45% of the trout's food supply within the Norwegian lakes, resulting in a 10–30% decline in trout due to freshwater acidification (Muniz, 1990).

5.2.2 *FLOW REGULATION, DAMS, AND WATER ABSTRACTION*

Accommodation of the social needs has compelled humans to modify the flow regime of freshwater sources that have resulted in a total change in hydro geomorphological processes of ecosystem functioning (Anderson and Jackson, 2019). The biodiversity of a particular habitat directly reflects the local conditions it has. Any ecological degradation proves more harmful for freshwater diversity than terrestrial ecosystems (Sala et al., 2000). Both natural and human-induced have a deeps impact on freshwater species. The watercourse size also affects biodiversity (Vander et al., 2017). Anthropogenic load changes the primary support habitat conditions of the freshwater ecosystems, leading to the perishing of sensitive species (Vörösmarty et al., 2010). Dam building on the watercourses alters the standard hydrological regime in which different species occupy different areas. It is believed that plankton species greatly vary from dam reservoir to downstream, resulting in an entire change in aquatic ecosystems' food chain (Alhassan et al., 2015). Sediment and pollution accumulation in dam reservoirs lead to a sharp decline in plankton species compared to downstream plankton species. Species richness in backwaters depends on the specific habitat conditions (Mbaka, 2015). Zoobenthos are negatively affected by this kind of habitat disturbance (Horsak et al., 2009). Different forms of cultivation and areas impervious to runoff, including urban areas, are involved in disrupting drainage basins. Urban development exerts various burdens on river ecosystems and declines their biodiversity (Rusanov et al., 2017).

5.2.3 OVEREXPLOITATION OF BIOLOGICAL RESOURCES

Overexploitation is the second most significant threat to freshwater biodiversity. Excavation of a 9,200-year-old Early Mesolithic site in Sweden, where there were signs of fermentation and preservation of more than 60% of food, provides evidence of large-scale early processing of freshwater fish. Cyprinids subjugated the fastener, essentially *Rutilus rutilus*, although *Perca fluviatilis*, *Esox lucius* were also exploited, as were European eel (*Anguilla anguilla*), European smelt *Osmerus eperlanu*, and salmonids (Boethius, 2016). Overfishing is a pervasive delinquent in numerous aquatic fisheries (Link and Watson, 2019). The exploitation of fisheries primarily disturbs large, late-maturing species, causing 'fishing' down the food chain and manipulating small, rapidly maturing species (Taylor et al., 2005). Not all merchandisable domestic fisheries are under jeopardy of exploitation (Funge, 2019). Small and promptly reproducing planktivorous fish are known to withstand a very high fishing burden (Sarvala et al., 1999). Other treasured goods include crocodile skins and body parts of turtles used in traditional Chinese medicine; cumulative scarcity of target species drives up their price and inspires more exploitation.

5.2.4 ALIEN (INTRODUCED OR NON-NATIVE) INVASIVE SPECIES

One of the major factors responsible for freshwater biodiversity deterioration is invasive species (Reid et al., 2019). Species distribution across the worlds' diverse geographical areas is mixed up by anthropogenic actions (Su et al., 2016) through intentional or unintentional introductions (Minchin et al., 2013) due to new trade routes, food, recreation, tourism, etc., across the world. It is seen as invasive where a non-native species causes the damage of any kind (Sultana et al., 2015). These adverse effects are seen in three manifestations: economic disruption, public health risks, and ecosystem resources impact (Chown et al., 2015). Some animals can cause more than one type of damage, and over time, the kind of damage can change. A tiny population of alien species can result in large-scale invasive species obtaining favorable local habitat conditions. At times invasive species are so small that they go unnoticed. Also, the damage an alien species causes depends on the invader's nature and the receiving population's principal characteristics (Strayer, 2010). In some instances, when the invader is carnivorous, native species perish when they lack anti-predatory adaptations. In the presence of invasive species competition for food, space may occur if invaders are vicarious eaters (Sardain et al., 2019). In freshwater ecosystems, alien species can be pathogenic and may bring diseases with them. Hybridization stances a menace if there is a close evolutionary relationship between exotic and native species (Robertson et al., 2019).

5.2.5 GLOBAL CLIMATE CHANGE

Climate change is an extensively documented risk to freshwater inhabitants (Lynch et al., 2016). Climate change backed by anthropogenic factors represents a profound and devious threat to freshwater biodiversity. Rising temperatures result in fast-melting that changes the flow rate and shifts the rainfall pattern, increasing the frequency of extreme events that affect freshwater biodiversity (Milly et al., 2008). The 'fingerprint' of climate change is now clear (Scheffers et al., 2016), as 31 freshwater ecological processes were measured, out of which 23 showed the negative effect of climate change.

5.3 THREAT INTERACTIONS

Threats of freshwater biodiversity that originate from catchment aren't independent but highly impact each other; for example, increased temperatures increase toxicity and acidification in freshwater (Leung et al., 2020). Dam building and water abstraction for energy generation reduce rivers' pollutant dilution capacity (Wen et al., 2017). Climate change and pollutant inflow from catchment changes the freshwater habitat and favors invasive species that tend to be more tolerant than existing ones (Strayer and Dudgeon, 2010). The unending global rampant of dam building (Lumsdon and Berlekamp, 2015) will increase 40% of reservoir storage by 2030 with a direct impact on freshwater biodiversity. Dam building has eliminated various fish species like the one in northwest Europe (salmon fish) has been completely eradicated (Limburg and Waldman, 2009). Species like potamodromous fishes that show breeding migrations within habitats are in perils and species that migrate from freshwater to marine habitats like eels, river prawns, and sturgeon (Winemiller et al., 2016).

5.4 CONCLUSIONS

Freshwater on a global scale is recognized both as a biodiversity hotspot. Primary hydrological factors that aid the species richness initially become a reason for biodiversity decline when subjected to anthropogenic alterations; this array has no alternative as for now or soon. Global climate change has the capacity to amplifying the hydrological and chemical factors related to biodiversity loss. The modern water engineering schemes are becoming very challenging to freshwater biodiversity. Amelioration of threats to biodiversity must include a combined approach with large-scale measures like reducing emissions to limit climate change. Concentration on single threats will not be fruitful for future times, so a blended approach must meet the goal. However, any failure in alleviating the proximate threats to biodiversity will compromise the whole freshwater system and its biota that will result in biodiversity loss.

KEYWORDS

- **biodiversity**
- **climate change**
- **ecological factors**
- **ecosystems**
- **freshwater**
- **land use land cover change**

REFERENCES

[IUCN] International Union for Conservation of Nature, (2019). *The IUCN Red List of Threatened Species.* Version 2019–1. IUCN. www.iucnredlist.org (accessed on 13 December 2021).

Alhassan, E., Ofori-Danson, P., & Nunoo, F., (2015). *Physicochemical characteristics of the Bui Dam Area of the Black Volta River*. Ghana.

Allan, J. D., (2004). Landscapes and riverscapes: The influence of land use on stream ecosystems. *Annu. Rev. Ecol. Evol. Syst., 35*, 257–284.

Allan, J. D., Abell, R., Hogan, Z., Revenga, C., Taylor, B. W., Welcomme, R. L., & Winemiller, K., (2005). Overfishing of inland waters. *BioScience, 55*(12), 1041–1051.

Anderson, E. P., Jackson, S., Tharme, R. E., Douglas, M., Flotemersch, J. E., Zwarteveen, M., Lokgariwar, C., et al., (2019). Understanding rivers and their social relations: A critical step to advance environmental water management. *Wiley Interdisciplinary Reviews: Water, 6*(6), e1381.

Anna, J, Schuijt, K., & Zwaal, N., (n.d.). *A Deep Dive into Freshwater.* www.hubspotusercontent20.net/hubfs/4783129/LPR/PDFs/Living_Planet_Report_Freshwater_Deepdive.pdf

Bickford, D., Howard, S. D., Ng, D. J., & Sheridan, J. A., (2010). Impacts of climate change on the amphibians and reptiles of Southeast Asia. *Biodiversity and Conservation, 19*(4), 1043–1062.

Boethius, A., (2016). Something rotten in Scandinavia: The world's earliest evidence of fermentation. *Journal of Archaeological Science, 66*, 169–180.

Carpenter, S. R., Fisher, S. G., Grimm, N. B., & Kitchell, J. F., (1992). Global change and freshwater ecosystems. *Annual Review of Ecology and Systematics, 23*(1), 119–139.

Ceballos, G., Ehrlich, P. R., & Dirzo, R., (2017a). Biological annihilation via the ongoing sixth mass extinction signaled by vertebrate population losses and declines. *Proceedings of the National Academy of Sciences, 114*(30), E6089–E6096.

Ceballos, G., Ehrlich, P. R., Barnosky, A. D., García, A., Pringle, R. M., & Palmer, T. M., (2015). Accelerated modern human–induced species losses: Entering the sixth mass extinction. *Science Advances, 1*(5), e1400253.

Chen, X., Chen, X., Zhao, Y., Zhou, H., Xiong, X., & Wu, C., (2020). Effects of microplastic biofilms on nutrient cycling in simulated freshwater systems. *Science of the Total Environment, 719*, 137276.

Chown, S. L., Hodgins, K. A., Griffin, P. C., Oakeshott, J. G., Byrne, M., & Hoffmann, A. A., (2015). Biological invasions, climate change and genomics. *Evolutionary Applications, 8*(1), 23–46.

Cooke, S. J., Twardek, W. M., Lynch, A. J., Cowx, I. G., Olden, J. D., Funge-Smith, S., Lorenzen, K., et al., (2021). A global perspective on the influence of the COVID-19 pandemic on freshwater fish biodiversity. *Biological Conservation, 253*, 108932.

Darwall, W. R., Holland, R. A., Smith, K. G., Allen, D., Brooks, E. G., Katarya, V., Pollock, C. M., et al., (2011). Implications of bias in conservation research and investment for freshwater species. *Conservation Letters, 4*(6), 474–482.

Darwall, W., Bremerich, V., De Wever, A., Dell, A. I., Freyhof, J., Gessner, M. O., Grossart, H., et al., (2018). The alliance for freshwater life: A global call to unite efforts for freshwater biodiversity science and conservation. *Aquatic Conservation: Marine and Freshwater Ecosystems, 28*(4), 1015–1022.

Delong, M. D., & Brusven, M. A., (1998). Macroinvertebrate community structure along the longitudinal gradient of an agriculturally impacted stream. *Environmental Management, 22*(3), 445–457.

Flitcroft, R., Cooperman, M. S., Harrison, I. J., Juffe-Bignoli, D., & Boon, P. J., (2019). Theory and practice to conserve freshwater biodiversity in the Anthropocene. *Aquatic Conservation: Marine and Freshwater Ecosystems, 29*(7), 1013–1021.

Foley, J. A., DeFries, R., Asner, G. P., Barford, C., Bonan, G., Carpenter, S. R., Chapin, F. S., et al., (2005). Global consequences of land use. *Science, 309*(5734), 570–574.

Funge-Smith, S., & Bennett, A., (2019). A fresh look at inland fisheries and their role in food security and livelihoods. *Fish and Fisheries, 20*(6), 1176–1195.

Gardner, R. C., & Finlayson, C., (2018). *Global Wetland Outlook: State of the World's Wetlands and their Services to People*. Ramsar Convention Secretariat.

Gassert, F., Luck, M., & ISciences, L. L. C. (2015). Aqueduct water stress projections: decadal projections of water supply and demand using CMIP5 GCMs. https://www.wri.org/research/aqueduct-water-stress-projections-decadal-projections-water-supply-and-demand-using-cmip5.

Grill, G., Lehner, B., Thieme, M., Geenen, B., Tickner, D., Antonelli, F., Babu, S., Borrelli, P., Cheng, L., & Crochetiere, H., (2019). Mapping the world's free-flowing rivers. *Nature, 569*(7755), 215–221.

Grizzetti, B., Pistocchi, A., Liquete, C., Udias, A., Bouraoui, F., & Van De Bund, W., (2017). Human pressures and ecological status of European rivers. *Scientific Reports, 7*(1), 1–11.

Hasler, C. T., Jeffrey, J. D., Schneider, E. V., Hannan, K. D., Tix, J. A., & Suski, C. D., (2018). Biological consequences of weak acidification caused by elevated carbon dioxide in freshwater ecosystems. *Hydrobiologia, 806*(1), 1–12.

Heino, J., (2010). Are indicator groups and cross-taxon congruence useful for predicting biodiversity in aquatic ecosystems? *Ecological Indicators, 10*(2), 112–117.

Heino, J., Virkkala, R., & Toivonen, H., (2009). Climate change and freshwater biodiversity: Detected patterns, future trends and adaptations in northern regions. *Biological Reviews, 84*(1), 39–54.

Horsak, M., Bojková, J., Zahrádková, S., Omesová, M., & Helešic, J., (2009). Impact of reservoirs and channelization on lowland river macroinvertebrates: A case study from Central Europe. *Limnologica, 39*(2), 140–151.

IUCN International Union for Conservation of Nature, (2019). *The IUCN Red List of Threatened Species, Version 2019–1*. www.iucnredlist. org.

Kernan, M., Battarbee, R. W., & Moss, B. R., (2011). *Climate Change Impacts on Freshwater Ecosystems*. John Wiley & Sons.

Kuzmanovic, M., Dolédec, S., De Castro-Catala, N., Ginebreda, A., Sabater, S., Muñoz, I., & Barceló, D., (2017). Environmental stressors as a driver of the trait composition of benthic macroinvertebrate assemblages in polluted Iberian rivers. *Environmental Research, 156*, 485–493.

Lakew, A., & Moog, O., (2015). A multimetric index based on benthic macroinvertebrates for assessing the ecological status of streams and rivers in central and southeast highlands of Ethiopia. *Hydrobiologia, 751*(1), 229–242.

Lee, J. K., Kang, S. I., Kim, Y. J., Kim, M. J., Heu, M. S., Choi, B. D., & Kim, J. S., (2016). Comparison of collagen characteristics of sea-and freshwater-rainbow trout skin. *Food Science and Biotechnology, 25*(1), 131–136.

Leung, K. M., Yeung, K. W., You, J., Choi, K., Zhang, X., Smith, R., Zhou, G., et al., (2020). Toward sustainable environmental quality: Priority research questions for Asia. *Environmental Toxicology and Chemistry, 39*(8), 1485–1505.

Limburg, K. E., & Waldman, J. R., (2009). Dramatic declines in North Atlantic diadromous fishes. *BioScience, 59*(11), 955–965.

Link, J. S., & Watson, R. A., (2019). Global ecosystem overfishing: Clear delineation within real limits to production. *Science Advances, 5*(6), eaav0474.

Lowenstam, H. A., & Weiner, S., (1989). Biomineralization processes. In: *On Biomineralization*. Oxford University Press.

Luck, G. W., (2007). A review of the relationships between human population density and biodiversity. *Biological Reviews, 82*(4), 607–645.

Luke, T. W., (2020). Staying Alert during the Great Acceleration: Anthropocene Ahead. *New Political Science, 42*(4), 595–601.

Lynch, A. J., Myers, B. J., Chu, C., Eby, L. A., Falke, J. A., Kovach, R. P., Krabbenhoft, T. J., et al., (2016). Climate change effects on North American inland fish populations and assemblages. *Fisheries, 41*(7), 346–361.

Mbaka, J. G., & Wanjiru, M. M., (2015). A global review of the downstream effects of small impoundments on stream habitat conditions and macroinvertebrates. *Environmental Reviews, 23*(3), 257–262.

Milly, P., Betancourt, J., Falkenmark, M., Hirsch, R. M., Kundzewicz, Z. W., Lettenmaier, D. P., & Stouffer, R. J., (2008). Stationarity is dead: Whither water management? *Earth, 4*, 20.

Minchin, D., Cook, E. J., & Clark, P. F., (2013). Alien species in British brackish and marine waters. *Aquatic Invasions, 8*(1).

Muniz, I. P., (1990). Freshwater acidification: Its effects on species and communities of freshwater microbes, plants and animals. *Proceedings of the Royal Society of Edinburgh, Section B: Biological Sciences, 97*, 227–254.

Opiso, E. M., Quimpang, V. T., LeaÃ±o, E. P., Galan, G. L., Acma, F. M., Coritico, F. P., Labadan, A., et al., (2014). Assessment of biodiversity and water quality in association with land use in the Alanib River, Mt. Kitanglad Range Park, Philippines. *Asian Journal of Biodiversity, 5*(1).

Phillips, J. C., McKinley, G. A., Bennington, V., Bootsma, H. A., Pilcher, D. J., Sterner, R. W., & Urban, N. R., (2015). The potential for CO_2-induced acidification in freshwater: A Great Lakes case study. *Oceanography, 28*(2), 136–145.

Reid, A. J., Carlson, A. K., Creed, I. F., Eliason, E. J., Gell, P. A., Johnson, P. T., Kidd, K. A., et al., (2019). Emerging threats and persistent conservation challenges for freshwater biodiversity. *Biological Reviews, 94*(3), 849–873.

Richards, C., Johnson, L. B., & Host, G. E., (1996). Landscape-scale influences on stream habitats and biota. *Canadian Journal of Fisheries and Aquatic Sciences, 53*(S1), 295–311.

Robertson, P., Roy, S., Mill, A. C., Shirley, M., Adriaens, T., Ward, A., Tatayah, V., & Booy, O., (2019). *Invasive Species Removals and Scale–Contrasting Island and Mainland Experience,* 687–691.

Rusanov, A., & Khromov, V., (2016). Longitudinal distribution of periphyton algae in the Moskva River under eutrophication. *Water Resources, 43*(3), 513–521.

Sala, O. E., Chapin, F. S., Armesto, J. J., Berlow, E., Bloomfield, J., Dirzo, R., Huber-Sanwald, E., et al., (2000). Global biodiversity scenarios for the year 2100. *Science, 287*(5459), 1770–1774.

Sardain, A., Sardain, E., & Leung, B., (2019). Global forecasts of shipping traffic and biological invasions to 2050. *Nature Sustainability, 2*(4), 274–282.

Scheffers, B. R., De Meester, L., Bridge, T. C., Hoffmann, A. A., Pandolfi, J. M., Corlett, R. T., Butchart, S. H., et al., (2016). The broad footprint of climate change from genes to biomes to people. *Science, 354*(6313).

Shawul, A. A., Chakma, S., & Melesse, A. M., (2019). The response of water balance components to land cover change based on hydrologic modeling and partial least squares regression (PLSR) analysis in the Upper Awash Basin. *Journal of Hydrology: Regional Studies, 26,* 100640.

Sponseller, R., Benfield, E., & Valett, H., (2001). Relationships between land use, spatial scale and stream macroinvertebrate communities. *Freshwater Biology, 46*(10), 1409–1424.

Steffen, W., Broadgate, W., Deutsch, L., Gaffney, O., & Ludwig, C., (2015). The trajectory of the Anthropocene: The great acceleration. *The Anthropocene Review, 2*(1), 81–98.

Strayer, D. L., & Dudgeon, D., (2010). Freshwater biodiversity conservation: Recent progress and future challenges. *Journal of the North American Benthological Society, 29*(1), 344–358.

Strayer, D. L., (2010). Alien species in fresh waters: Ecological effects, interactions with other stressors, and prospects for the future. *Freshwater Biology, 55,* 152–174.

Su, H., Hyde, K. D., Maharachchikumbura, S. S., Ariyawansa, H. A., Luo, Z., Promputtha, I., Tian, Q., et al., (2016). The families *Distoseptisporaceae* fam. Nov., *Kirschsteiniotheliaceae*, *Sporormiaceae* and *Torulaceae*, with new species from freshwater in Yunnan Province, China. *Fungal Diversity, 80*(1), 375–409.

Sultana, M., & Hashim, Z., (2015). Invasive alien fish species in freshwater of the continents. *Journal of Environmental Science and Natural Resources, 8*(2), 63–74.

Tavares-Dias, M., & Martins, M. L., (2017). An overall estimation of losses caused by diseases in the Brazilian fish farms. *Journal of Parasitic Diseases, 41*(4), 913–918.

Tickner, D., Opperman, J. J., Abell, R., Acreman, M., Arthington, A. H., Bunn, S. E., Cooke, S. J., et al., (2020a). Bending the curve of global freshwater biodiversity loss: An emergency recovery plan. *BioScience, 70*(4), 330–342.

Tickner, D., Opperman, J. J., Abell, R., Acreman, M., Arthington, A. H., Bunn, S. E., Cooke, S. J., et al., (2020b). Bending the curve of global freshwater biodiversity loss: An emergency recovery plan. *BioScience, 70*(4), 330–342.

Vander, V. R., McElmurray, P., Bell, S., Eliason, K. M., & Brown, B. L., (2017). Does stream size really explain biodiversity patterns in lotic systems? A call for mechanistic explanations. *Diversity, 9*(3), 26.

Vörösmarty, C. J., McIntyre, P. B., Gessner, M. O., Dudgeon, D., Prusevich, A., Green, P., Glidden, S., et al., (2010). Global threats to human water security and river biodiversity. *Nature, 467*(7315), 555–561.

Wang, B., Liu, D., Liu, S., Zhang, Y., Lu, D., & Wang, L., (2012). Impacts of urbanization on stream habitats and macroinvertebrate communities in the tributaries of Qiangtang River, China. *Hydrobiologia, 680*(1), 39–51.

Wen, Y., Schoups, G., & Van De Giesen, N., (2017). Organic pollution of rivers: Combined threats of urbanization, livestock farming and global climate change. *Scientific Reports, 7*(1), 1–9.

Winemiller, K. O., McIntyre, P. B., Castello, L., Fluet-Chouinard, E., Giarrizzo, T., Nam, S., Baird, I., et al., (2016). Balancing hydropower and biodiversity in the Amazon, Congo, and Mekong. *Science, 351*(6269), 128–129.

Zarfl, C., Lumsdon, A. E., Berlekamp, J., Tydecks, L., & Tockner, K., (2015). A global boom in hydropower dam construction. *Aquatic Sciences, 77*(1), 161–170.

CHAPTER 6

Pesticide Pollution in Freshwater Environs: Impacts on Aquatic and Terrestrial Life

ZIA UR RAHMAN FAROOQI,[1] MUHAMMAD SAAD BIN ZAFAR,[1] MOAZ KHURSHEED,[1] MUHAMMAD USAMA RAHEEM,[2] NOUMAN GULZAR,[1] and PREDRAG ILIC[3]

[1]*Institute of Soil and Environmental Sciences, University of Agriculture, Faisalabad – 38040, Pakistan, Mobile: +923336809609, E-mail: ziaa2600@gmail.com (Z. U. R. Farooqi)*

[2]*Institute of Soil and Environmental Sciences, University of Agriculture, Faisalabad – 38040, Pakistan*

[3]*PSRI Institute for Protection and Ecology of the Republic of Srpska, Vidovdanska 43, Banja Luka – 78000, Republic of Srpska, Bosnia and Herzegovina*

ABSTRACT

Chemical-based pesticides are commonly used to improve agricultural yield and protect crops from various insects and pests. Misuse of these chemicals by farmers has been known in rural and urban settings due to their general uses and easy availability. Pesticide poisoning is now a significant public health issue worldwide due to its widespread use. Unfortunately, this problem continues to go unaddressed, owing to a lack of awareness of its consequences. There is little legislation outside

Biodiversity of Freshwater Ecosystems: Threats, Protection, and Management. Suhaib A. Bandh, PhD, Basharat Mushtaq, PhD, Mohammad Farooq Mir, PhD, & Shahid A. Wani (Eds.)

the standardized approval process for uses, though most countries have established implementation guidelines. Pesticide misuse pollutes the land, climate, and waterways. Contamination in water sources has negative consequences for marine species, people, and animals since aquatic life is mainly eaten. It causes various diseases, varying from mild to severe and as a result, proper monitoring and remediation procedures are needed. This chapter discusses many of the problems around the chemical pollution of freshwater sources, remediation strategies and alternative solutions, and potential study needs.

6.1 INTRODUCTION

Over the past century, pesticides have been an integral part of the global agriculture system, allowing for significant increases in crop yields and food output. Nonetheless, the ever-increasing human population emphasizes the need to increase the food supply (Carvalho, 2017). Conflicts that halt food processing and displace millions of migrants and the impact of climate change on agriculture exacerbate food shortages in several regions and necessitate renewed food production efforts. Simultaneously, we discovered in recent decades that agrochemical emissions have penetrated through the atmosphere, contaminating terrestrial environments, and poisoning human foods. Furthermore, chemical residue degradation in aquatic environments worldwide – as shown here with case studies in tropical coastal habitats – has consistently harmed marine food supplies, fisheries, and aquaculture (McKnight et al., 2015).

Paths are available for trial and evaluation as an alternative to the extensive usage of crop safety chemicals. Future pathways for increased food output, on the other hand, must be chosen by wise and scientific decision-making processes (Cycoń et al., 2017). Therefore, scientific study is a crucial part of this mechanism for developing food security and improving food safety and environmental conservation. To keep up with the demands of population, agricultural productivity has risen dramatically since the turn of the century. Population figures soared from 1.5 billion in 1,900 to about 7.3 billion in contemporary times, corresponding to a multi-fold growth in the global population throughout humanity's existence. Since 2003, the world's population has grown by 1 billion, and at present rates, the world's population would be about 9.4–10 billion by 2050 (Van Bavel, 2013).

The Stockholm Convention, which was accepted in 2002, prohibited persistent and bioaccumulative chemical compounds, including DDT, HCH, toxaphene, aldrin, and dieldrin, and substituted them with environmentally safe and less bioaccumulative chemicals. This has been the theme for decades. It has been fueled by concerns about the toxicity of chemical contaminants in food to people and the presence of chemicals in the atmosphere and their toxicity to nonhuman biota. However, because of their widespread use in the past, poisonous OCs are now found in soils, sediments, and the biosphere. Toxaphene in cotton fields, for example, is no longer utilized in Nicaragua. Still, the residue in agricultural soils remains a source of pollutants transferred to the aquatic ecosystems via surface drainage, posing a danger to shrimp farming in coastal lagoons (Tsai, 2010).

Indeed, the prohibition on resistant organic compounds in cultivation reduced OC pesticides in certain areas, but it did not eliminate questions regarding their toxic effects. As a result of previous applications, these OC's can still be used in environmental compartments. Soil degradation, land drainage, and river discharges bring and cycle significant permanent OCs in these areas. These compounds (e.g., DDT, chlordane) were still present in the coastal ecosystem several years after their prohibition, where they declined steadily, as shown by declining biota concentrations over time. A similar gradual decrease in residue concentrations was observed in Mediterranean Sea coastal ecosystems in Europe (Gaur et al., 2018). To overcome the pollution and associated risks, different pollution remediation techniques and pesticides alternatives are being studied and used for their efficacy checks enlisted in the chapter.

6.2 USE OF PESTICIDES: BENEFITS AND GREEN REVOLUTION

Pesticides are chemical substances that have been an essential part of agriculture to protect crops from pest infections and the subsequent reduction in crop yield (Özkara et al., 2016). Cotton is Pakistan's most affected crop, led by rice, the country's second most significant foreign exchange source.

Over the last four or five decades, insecticides, and herbicides are considered to have played a significant role in improving crop yields all over the world. The late 1960s green revolution included fertilizers and pesticides, high fertilizer tolerant dwarf crop varieties with high yield capacity, the production of water supplies, intensive cropping, and best management practices. Fertilizers and insecticides are critical resources and a crucial part of the seed development chain. These inputs are crucial in closing the yield gap between potential yield and yield realized on the field, especially in cotton and rice. Insect pest assault on various crops has been described as one of the main constraints to exploiting crop genetic ability and realizing potential yield and quality of produce over the last three to four decades. The pest issue has developed due to the development of high-yielding, fertilizer-responsive varieties that are vulnerable to a variety of pests.

Pesticide use is one of the most critical aspects of current farming methods for defending crops from various pests. As a result, pesticide usage has risen gradually and significantly over the years (Habib, 1996; Khan et al., 2010).

6.3 TYPES OF PESTICIDES

There are mainly six types of pesticides:

- Insecticides;
- Herbicides;
- Rodenticides;
- Bactericides;
- Fungicides; and
- Larvicides.

6.3.1 INSECTICIDES

Those pesticides that are used to destroy dangerous insects which trigger a reduction in crop production. Insecticides kill insects by penetrating their bodies and acting as poison. When selecting an insecticide, it is critical to understand the behaviors of the target insect. Moving bugs, such as adult mosquitoes, react best to aerosol sprays or fogs (airborne droplets) while crawling insects respond best to surface powders, sprays, or granules for dermal and/or oral entry.

6.3.2 *HERBICIDES*

The herbicide is a kind of pesticide that is used to eliminate undesirable plants. Herbicides that destroy specific targets while keeping the desired crop largely unharmed are known as selective herbicides. They work by inhibiting weed growth and are often dependent on plant hormones.

6.3.3 *RODENTICIDES*

Pesticides that destroy mice are known as rodenticides. Rodents include squirrels, woodchucks, chipmunks, porcupines, nutria, and beavers, in addition to rats and rodents. While rodents play an essential role in nature, they can often be challenging to manage. They harm crops, break building codes, spread disease, and, in certain situations, harm the environment.

6.3.4 *BACTERICIDES*

Bacteriocide is a drug that destroys bacteria. Disinfectants, antiseptics, and antibiotics are an example of bactericides. Material surfaces, such as biomaterials like insect wings, may also have bactericidal properties dependent purely on their physical surface composition.

6.3.5 *FUNGICIDES*

Fungicides are chemicals that destroy or block fungi and their spores from growing. They may be used to monitor plant-damaging fungi such as rusts, mildews, and blights. They may even be used to keep mold and mildew at bay in some places. Although the majority of them destroy fungal cell membranes or prevent fungal cells from producing energy, Fungicides function in several ways.

6.3.6 *LARVICIDE*

A larvicide (also known as a larvacide) is an insecticide designed to kill insects during their larval stage. Mosquitoes are their most popular target.

Touch poisons, stomach poisons, growth inhibitors, and (increasingly) biological prevention agents are all examples of larvicides.

6.4 PESTICIDES CONTAMINATION OF FRESHWATER ENVIRONS: BIOMAGNIFICATION AND AQUATIC LIFE

Pesticides can infiltrate water in several ways: they can drift from general aerial spraying or be swept into lakes and waterways from the ground. Pesticides can be introduced into water sources by living species (Table 6.1). Pesticide application in freshwater ecosystems is influenced by the formulations in which they are transported and the chemistry and physics of the climate. Many pesticides are water-soluble; certain inorganic salts, such as copper sulfate and sodium arsenite, and organic pesticides, such as the herbicide diquat and the fungicide formaldehyde, are highly soluble and dissipate evenly throughout the pond or stream. On the other hand, inorganic salts may precipitate out as they come in contact with high-pH waters; for example, copper-containing chemicals can precipitate out as copper carbonate (Cope, 1966).

TABLE 6.1 Effects of Pesticides on Plant Germination, Physiology, and Yield

Pesticides	Plant Attributes	References
	Germination	
Chlorpyrifos, Cypermethrin, fenvalerate.	Delay in seed germination at higher concentrations, i.e., 75–100 mg/kg.	Dubey and Fulekar (2011)
Paraquat, glyphosate, glufosinate ammonium	No effect	Wibawa et al. (2009)
Menazon, disulfoton, GS-1425	Reduced seed germination.	Dalvi et al. (1972)
Bayleton, topsin-M	Reduced seed germination of *Vigna radiate*.	Siddiqui and Ahmed (1996)
Pendimethalin	Reduced seed germination.	Rajashekhar et al. (2012)
Morphological Traits		
Chlorpyrifos	Increase in plant height, number of branches, leaves per plant, total leaf area, plant biomass.	Parween et al. (2011b)

TABLE 6.1 *(Continued)*

Pesticides	Plant Attributes	References
	Germination	
Chlorotoluron	Reduced root, leaves, and biomass production.	Song et al. (2007)
Metobromuron, metribuzinand chlomazon	Decreased number of leaves, shoot length, fresh, dry shoot, and tuber weight.	Dobozi and Lehoczky (2002)
Lorsban, Decis, Pyrifos, Karate, Ripcord	Reduced nodulation and grain yield.	Khan et al. (2009); Mahmood and Shah (2003)
Acetochlor and bensulfuron methyl	Decreased root and shoot biomass.	Huang and Xiong (2009)
Atrazine, iso-proturon, metribuzin	Decrease in yield, root, and shoot length, plant dry weight, number, and dry weight of nodules.	Khan et al. (2006)
Fusilade	Decreased root and shoot growth, and high chlorosis and curling.	Aksoy and Dane (2007)
Paclobutrazol	Decreased plant dry matter.	Gopi et al. (2007)
Dimethoate	Decrease in root and shoot length, leaf area, fresh, and dry mass of shoot, root, and leaves.	Mishra et al. (2008, 2009)
Biochemical Parameters		
Chlorpyrifos	Decreased nitrate, soluble sugar, and protein contents.	Parween et al. (2011a)
Chlorotoluron	Decreased chlorophyll contents.	Song et al. (2007)
Acetochlor and bensulfuron methyl	Decreased nitrate, sugar, and protein contents, and NR-activity.	Huang and Xiong (2009)
Methyl viologen	Reduced chlorophyll and protein contents.	Chagas et al. (2008)
Paraquat, Alachlor, and metolachlor	Reduced chlorophyl *a* and *b*, and carotenoids contents.	Stajner et al. (2003, 2004)

TABLE 6.1 *(Continued)*

Pesticides	Plant Attributes	References
	Germination	
Clomazone	Reduced chlorophyl *a* and *b*, carotenoid content.	Kana et al. (2004)
Captan	Reduced chlorophyll contents.	Tort and Turkyilmaz (2003)
Paraquat	Loss of chlorophyl *a* and *b* and carotenoid contents.	Ekmekci and Terzioglo (2005)
Atrazine, isoproturon, metribuzin	Decrease in seed protein and total chlorophyll contents.	Khan et al. (2006)
Dimethoate	Decrease chl *a* and *b*, carotenoids, photosynthetic oxygen yield, photo-fixation of carbon (14-CO_2).	Mishra et al. (2008)
Metribuzin, butachlor, and chlorimuron ethyl	Affects the nitrate reductase, nitrite reductase (NiR), glutamine synthetase and glutamate synthase.	Nemat et al. (2008)
Rimsulfuron, imazethapyr, alachlor, atrazine	Inhibited the C4 photosynthetic.	Nemat et al. (2007)
Flumioxazin	Negative impact on photosynthesis and accumulation of soluble sugar and starch.	Saladin et al. (2003)
Plant Yield		
Chlorpyrifos	Reduced number of pods per plant, number of seeds per pod, and 100 grains weight.	Parween et al. (2011b)
Chitosan	Reduced number of fruits/plant, fruit size.	Mondal et al. (2012)
Lorsban, Decis, Karate, Ripcord	Lowered grain yield.	Khan et al. (2009); Mahmood and Shah (2003)
Paclobutrazol	Fresh and dry weight of leaves and tuber decreased.	Gopi et al. (2007)

TABLE 6.1 *(Continued)*

Pesticides	Plant Attributes	References
	Germination	
Dimethoate	Significant decrease in leaf area, shoot, and root length, fresh, and dry mass of shoot, root, and leaf.	Mishra et al. (2008)

6.5 SOURCES OF CONTAMINATION

A variety of man-made toxins may contaminate water bodies. We distinguish it into two types when it comes to its origin: point and diffuse. Industrial sites, cities, farm installations, manure collection, and landfills are all examples of significant point sources. Diffuse (non-point) causes, such as fertilizer leaching into the surface and groundwater due to rainfall, soil infiltration, and groundwater runoff from agricultural fields, are more challenging to identify and monitor. These types of sources induce significant fluctuations in the contaminant load of water over time (EPA, 2007) (Figure 6.1).

6.6 IMPACTS ON AQUATIC LIFE

6.6.1 FISH

Industrial pollution has long been recognized as a significant pollutant of the environments. Pesticides harm organisms that are subjected to them. Pesticides over allowable levels have been found in water all over the world. Industrial effluents have severely poisoned the majority of the lake. Petrochemical complexes, pesticide plants, oil refineries, pulp paper mills, sugar mills, steel mills, and tanneries, among other businesses, emit them. Both industrial waste chemicals are poisonous to animals, and several cases of death or sub-lethal pathology in the liver, kidney, gonads, and nervous tissues of fish have been recorded. The volume of pesticides in water has increased as a result of effluent inflows. Pesticide compounds have poor solubility, meaning they are only found in water in the dissolved form at low temperatures. Pesticides have also been observed in aquatic species. Pesticides cause changes in behavior by bioaccumulating in the body of fish (Napit, 2013).

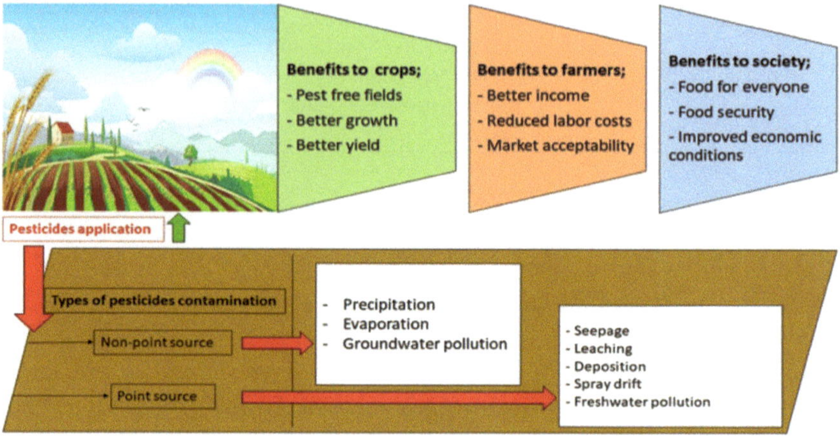

FIGURE 6.1 Advantages and fate of pesticides application in the environment.

Pesticides (insecticides, herbicides, fungicides) and other chemical contaminants such as polynuclear aromatic hydrocarbons (PAH), polychlorinated biphenyls (PCB), and tributyltin are the most common xenobiotics used as immunomodulators (TBT). Immunotoxicology has been a very active field in mammals, but there is still a lack of knowledge on fish immunotoxicology. The evidence on the impact of such toxins in the marine ecosystem on fish humoral and cellular immunity was gathered in this study (Dunier and Siwicki, 1993).

Future research into the importance of these factors about different stressors may help to improve fish populations and the resilience of Mediterranean river habitats in the face of future water scarcity. Research is also needed to prioritize specific chemicals and recommend practical basin-scale conservation steps (Belenguer et al., 2014).

Acute toxicity can have a minor impact on fish stocks than chronic exposure to low levels of pesticides. Pesticide doses that are not large enough to destroy fish are linked to gradual shifts in behavior and physiology that make it difficult for fish to survive and reproduce. Pesticidal stress causes biochemical alterations that cause metabolic disruptions, inhibition of essential enzymes, growth retardation, and a decrease in the organism's fecundity and lifespan (Khan and Law, 2005).

6.6.2 *AQUATIC PLANTS*

Aquatic plants have a lot of promise as on-site bio-sinks and biofilters of aquatic contaminants (Gao et al., 2000). They've been used to sequester heavy metals and nutrients through their root systems and their plant bodies (Bragato et al., 2006). Pollutants have well-known negative impacts on marine life and habitats (He et al., 2005). It is critical to implement advanced solutions to clean polluted water to reduce contamination from marine systems. Physical and/or electrochemical treatments are used in existing procedures for cleaning up wastewater (Jia et al., 2006).

6.7 IMPACTS ON HUMANS

Acute and persistent neurotoxicity, lung injury, chemical burns, and baby methemoglobinemia are all common adverse effects of pesticide exposure in humans. Pesticide exposure has also been linked to numerous cancers, including hematopoietic cancer (Farooqi et al., 2020). Immunologic defects and harmful reproductive and developmental impacts have been identified as a result of pesticides (Sankhla et al., 2018). Pesticide residues in water are always negligible, so the proportion of food eaten can be decreased even in areas polluted by water, according to the National Academy of Sciences (NAS) study on pesticides in the diets of infants and children (Holliger et al., 1993).

6.8 REMEDIATION OF PESTICIDES CONTAMINATED WATER ENVIRONS: TECHNIQUES AND PROSPECTS

Various remedial steps, such as leaching and landfilling, have been taken to reduce these toxic chemical contaminants (Table 6.2). However, scraping harmful contaminants from soil and debris disposal is time-consuming and costly (Kumar et al., 2018). To conserve agriculture and reduce the usage of multiple hazardous substances, encourage agricultural waste/earth and composting technologies as soil amendments, provides a diverse range of bacteria capable of producing biosurfactants (Rajmohan et al., 2020).

6.8.1 PHYSICOCHEMICAL TREATMENTS

The oxidation technology is very relevant for using pesticides, remarkably free radical hydroxylation, and oxidants (Sillanpää et al., 2018). According to the method by which the hydroxyl radical arises, AOPs can be divided into four categories: homogeneous, photochemical, sonochemical, and electrochemical oxidation processes (Rodrigo et al., 2014). Because of its versatility, performance, and ease of use, hydrogen peroxide-based technology has gotten a lot of publicity in AOP (Micó et al., 2010).

TABLE 6.2 Remedial Measure for Pesticides Contaminated Soil and Water Environs

Treatment Technologies	Process	Advantages	Disadvantages	References
Screening	Remove the solid waste present in the wastewater.	Remove suspended and soluble organic and inorganic substances.	Replacement cost and disposal of wastes generated.	Tchobanoglous et al. (1991)
Filtration	Passing water through fine pores, i.e., about 0.1 to 0.5 mm.	Remove suspended and soluble organic and inorganic substances.	Replacement cost and disposal of wastes generated.	Nemerow and Dasgupta (1991)
Centrifugal separation	Rotation in equipment to remove suspended non-colloidal solids.	Remove suspended and soluble organic and inorganic substances.	Low powered suction.	Tchobanoglous et al. (1991)
Sedimentation and gravity separation	Allowing water to be stand and settling of contaminants.	Remove suspended organic and inorganic substances.	Slow process/ cross-contamination	Nemerow and Dasgupta (1991)
Coagulation	non-settlable solids are allowed to settle by the addition of certain chemicals	Remove soluble and suspended inorganic substances.	Requires adjunction of non-recyclable chemicals.	Latifossglu et al. (1997)

TABLE 6.2 *(Continued)*

Treatment Technologies	Process	Advantages	Disadvantages	References
Flotation	Contaminants are removed through adhesion.	Remove suspended organic and inorganic substances.	High initial capital, energy, maintenance, and operational costs.	Sinev et al. (1997)
Aerobic decomposition	Biodegradable contaminants undergo aerobic decomposition by aerobic and facultative bacteria.	Remove soluble and suspended organic matter.	Slow/management and maintenance of microorganisms/ uncontrolled degradation/ necessary to create optimal conditions.	Barragán et al. (2007)
Anaerobic decomposition	Biodegradable contaminants undergo anaerobic decomposition by anaerobic bacteria.	Remove soluble and suspended organic matter.	–	Mohan et al. (2007)
Distillation	Heating up to 100°C (boiling point) at which liquid water is vaporized, leaving behind the pollutants.	Remove soluble organic and inorganic substances.	Energy consumption.	Gupta et al. (2012)
Crystallization	Pollutants removal by raising their concentrations up to a point where they start to crystallize out.	Remove soluble organic and inorganic substances.	Investment and maintenance costs are often too high for small and medium industries.	Van der Ham et al. (1998)
Evaporation	A natural process to reduce the wastewater volume.	Remove suspended and soluble organic and inorganic substance.	High pollution load in the concentrates/ crystallization/ problem with free cyanide containing effluents.	Gupta et al. (2012)

TABLE 6.2 *(Continued)*

Treatment Technologies	Process	Advantages	Disadvantages	References
Solvent extraction	Organic solvents, immiscible with water and having the capacity to dissolve pollutants is added to wastewater.	Remove soluble organic substances.	Uneconomic in contaminant concentrations/use of large volume of potential toxic solvents/cross contamination and emulsification.	Ahn and Ahn (1997)
Oxidation	Oxidized into water and carbon dioxide and alcohols, aldehydes, ketones, and carboxylic acids.	Remove soluble organic and inorganic substances.	Production, transport, and management of oxidants (chemicals).	Gupta et al. (2012)
Precipitation	Dissolved contaminants are converted into solid precipitates by reducing their solubilities.	Remove soluble organic and inorganic substances.	Ineffective in removal of the metal ions at low concentration/chemical consumption.	Gupta et al. (2012)
Ion exchange	Toxic ions present in wastewater are exchanged with the non-toxic ions from a solid material used as ion exchanger.	Remove soluble organic and inorganic substances.	Large volume requires large columns/saturation of exchanger/matrix degrades with time and with radioactive and strong oxidant wastes/pH sensitive.	Dąbrowski et al. (2004)
Micro- and ultra-filtration	Removal of contaminants through 0.04 to 1 mm size filters.	Remove soluble organic and inorganic substances.	High maintenance and operation costs/limited flow rates/low throughput/Elimination of the concentrate.	Bick et al. (2003)
Reverse osmosis	Uses membranes for pollutants removal.	Remove soluble organic and inorganic substances.	Biofouling of reverse osmosis membranes.	Greenlee et al. (2009)

TABLE 6.2 *(Continued)*

Treatment Technologies	Process	Advantages	Disadvantages	References
Adsorption	Increase in the concentration of a particular pollutant at the surface or interface between two phases.	Remove soluble organic and inorganic substances.	Requirement for specific types of adsorbents.	Gupta et al. (2007)
Electrolysis	Pollutants deposition on the surface of the electrodes by electrochemical redox reaction.	Remove soluble organic and inorganic substances.	High initial cost of the equipment and its maintenance and requires post-treatment.	Tongwen (2002)
Electrodialysis	Water soluble ions passed through ion selective semi-permeable membranes.	Remove soluble organic and inorganic substances.	–	Jüttner et al. (2000)
Bioreactors	Use of biological cultures.	Simple, economically attractive, well accepted.	Slow process/ necessary to create optimal conditions/ management and maintenance of microorganisms/ uncontrolled degradation.	Crini and Lichtfouse (2019)
Biological activated sludge	Use of biological cultures (pure or mixed).	Simple, economic, large amount of species can be used.	–	Tałałaj et al. (2019)
Microbial treatments	Use of biological cultures.	Simple, economic, large amount of species can be used.	–	Shahid et al. (2020)

TABLE 6.2 *(Continued)*

Treatment Technologies	Process	Advantages	Disadvantages	References
Enzymatic decomposition	Use of biological substances	Economic, simple with high degradation capacity	Necessary to create optimal conditions/ management and maintenance of microorganisms/ uncontrolled degradation	Tałałaj et al. (2019)
Incineration	Destruction by combustion.	Simple process/ production of energy/ useful for concentrated effluents.	Initial investment costs/transport and storage of the effluents.	Crini and Lichtfouse (2019)
Thermal and catalytic oxidation	Temperature and catalyst (chemical/ biological) controlled oxidation of compounds.	Highly efficient/ useful for concentrated effluents.	Initial investment costs/transport and storage of the effluents/formation of dioxins.	Crini and Lichtfouse (2019)

6.8.2 BIOLOGICAL TREATMENT

Bioremediation is a beneficial, cost-effective, and environmentally sustainable *in-situ* treatment system that enhances the natural biodegradation mechanism by increasing microorganisms' metabolic function, thus decreasing pesticide emissions in agricultural soils (Reddy and Antwi, 2016) (Hussain et al., 2020). Phytoremediation and microbial remediation are two forms of bioremediation. Bean uptake, vegetation decay, volatilization, and rhizobiatin degradation by rhizobiatin are all used in phytoremediation of pesticides from the soil (SYJZJ, 2016; Ismail et al., 2015). The aerobic bioremediation of pesticides is improved by bacterial colonies (Doolotkeldieva et al., 2018). Vegetables (*Chrysopogon zizanoides*) help phytostabilization in moderately dioxin-contaminated areas (Mohiddin et al., 2015).

6.9 ALTERNATIVES TO PESTICIDES

There are a variety of barriers to overcome to improve the effectiveness of biopesticides. Many farming cultures have adopted and introduced other alternate conditions, such as existing behavioral policies, usage, and strategies, biomarkers, and chemical pheromones, and genetically engineered products, to address these challenges (Hossain et al., 2017).

6.9.1 BIOCONTROL

Biological regulation is an important research topic since it was only recently recognized. Biological protection products are materials and goods that utilize natural processes to treat pests and crops in an automated manner. Macroscopic cells, crop protection materials, microorganisms, chemical media (such as pheromone), or [other] natural plants, species, or minerals are among them (Villemaine et al., 2020). As a result, the word 'biocontrol' encompasses adaptive pest management strategies like mating disturbance and the development of beneficial organisms and other types of approaches (Kogan, 1998). Some biocontrol drugs are considered to plant defense stimulators or resistance inducers, whereas others aren't. In general, certified biocontrol chemicals do not always work in the same way as conventional crop defense materials do (Villemaine et al., 2020).

6.9.2 PLANT GROWTH-PROMOTING BACTERIA (PGPB)

Plant growth-promoting bacteria (PGPB) is a healthy and sustainable agricultural development tool used as a biological fertilizer and biological power. Thanks to its low environmental effects, PGPB is commonly used as a plant and soil inoculant to increase plant production and disease control (Singh et al., 2019) in the current times.

Endophytic symbiotic PGPB found in legume roots (*Azorhizobium* sp., *Bradyrhizobium* sp., *Mesorhizobium* sp., *Rhizobium* sp., *Sinorhizobium* sp., *Sinorhizobium* sp., Sinorhizobium sp., *Sinorhizobium* sp. (Gupta et al., 2012) and other different PGPB's such as *Bacillus pumilus*, *Mesorhizobium* sp., *Variovorax paradoxus*, *Burkholderia phytofirmans*, *P. libanensis*, *P. reactants*, and *Azotobacter* species have been confirmed for possible abilities of bioremediation of heavy metals from infected soils (Kong et al., 2017).

6.9.3 ALLELOPATHY

Allelopathy is a normal biological process in which living organisms create and release biochemical compounds (allelochemicals) into the atmosphere, influencing their growth, development, reproduction, and survival in the environment (Hussain and Farooqi, 2021). Allelopathy may be used in field crops to ensure good pest and disease management by interplanting, covering crops, mulch, seed rotation, and using plant water extracts alone or in conjunction with plants or with decreased doses of herbicides. Allelopathic plant water extracts and allelopathic plant powder can be very effective in combating insect pests (field and storage insect pests). Cell structure modifications, cell elongation/division reduction, membrane structure disruption, water, and nutrient absorption destruction, and the photosynthesis mechanism are all effects of allelochemicals on undesirable plants' growth (Nawaz et al., 2020).

6.9.4 GENETIC IMPROVEMENTS IN PLANT VARIETIES

Drought, cold, and salinity tolerance have been demonstrated in genetically modified (GM) crops, expressing several stress-inducible genes (Shinozaki and Yamaguchi-Shinozaki, 2007). Stress-specific genes, such as AP2/ERF, bZIP, MYB, MYC, NAC, zinc-finger Cys2His2, and WRKY, may either be active in signal transduction pathways or belong to classes of transcription factors (TFs) involved in regulating plant stress responses. Since TFs are excellent candidates for modifying complex characteristics like drought tolerance in crop plants that are multigenic, TF-based genetic engineering is an ideal crop improvement technique (Tran et al., 2010).

Rice plants that express the transcription factor HYR (higher yield rice) have been shown to produce higher grain yields under typical drought and high-temperature stress conditions (Ambavaram et al., 2014). HYR is a master regulator that enhances photosynthesis by directly activating photosynthesis genes, TFs, and other downstream genes involved in photosynthetic carbon metabolism (Kamthan et al., 2016).

6.9.5 POLYCULTURE

Polyculture is a way of growing many crops in the same space and at the same time. Plant interaction offers valuable knowledge that can be used to help choose organisms for different polyculture structures. A polyculture framework for legume and non-legume crops is a widely accepted cultivation method (Adamczewska-Sowińska and Sowiński, 2020).

6.10 CONCLUSIONS

Pesticides have become an essential component of the global green revolution soon after its start. Their use results in a significant increase in crop yields, food production and food security problem. However, the exponential use of pesticides has resulted in environmental deterioration and negative impacts on humans and other organisms. Foods that are contaminated with toxic pesticides have severe effects on human health as they are a prerequisite for life. They can contaminate soil, water, pastures, and plants. Their pollution is posing a severe threat, not only to humans but also to other factors of the environment. So, to remediate Ps, different approaches have been used, but every method has its limitations. The need of the time is to stop using such potential threats and look for viable alternatives.

KEYWORDS

- **genetically modified**
- **higher yield rice**
- **national academy of sciences**
- **plant growth-promoting bacteria**
- **polychlorinated biphenyls**

REFERENCES

Adamczewska-Sowińska, K., & Sowiński, J., (2020). Polyculture management: A crucial system for sustainable agriculture development. In: *Soil Health Restoration and Management* (pp. 279–319). Springer.

Ahn, J., & Ahn, J., (1997). Solvent extraction for the treatment of industrial wastes. *Chawn Risaikring, 6*, 48–54.

Akhtar, N., Kayani, S. A., Ahmad, M. M., & Shahab, M., (1996). Insecticide-induced changes in secretory activity of the thyroid gland in rats. *J. Appl. Toxicol., 16*, 397–400.

Aksoy, O., & Dane, F., (2007). The effects of fusillade (Fluazifop-p-butyl) on root and shoot growth of lentil (Lens culinaris Medik.) seedlings. *J. App. Boil. Sci., 1*(3), 9–13.

Ambavaram, M. M., Basu, S., Krishnan, A., Ramegowda, V., Batlang, U., Rahman, L., Baisakh, N., & Pereira, A., (2014). Coordinated regulation of photosynthesis in rice increases yield and tolerance to environmental stress. *Nature Communications, 5*, 1–14.

Amitai, G., Moorad, D., Adani, R., & Doctor, B. P., (1998). Inhibition of acetylcholinesterase and butyrylcholinesterase by chlorpyrifos-oxon. *Biochem. Pharmacol., 56*, 293–299.

Anway, M. D., Cupp, A. S., Uzumcu, M., & Skinner, M. K., (2005). Epigenetic transgenerational actions of endocrine disruptors and male fertility. *Science, 308*, 1466–1469. New York, N.Y.

Barragán, B. E., Costa, C., & Marquez, M. C., (2007). Biodegradation of azo dyes by bacteria inoculated on solid media. *Dyes and Pigments, 75*, 73–81.

Belenguer, V., Martinez-Capel, F., Masiá, A., & Picó, Y., (2014). Patterns of presence and concentration of pesticides in fish and waters of the Júcar river (eastern spain). *Journal of Hazardous Materials, 265*, 271–279.

Betarbet, R., Sherer, T. B., MacKenzie, G., Garcia-Osuna, M., Panov, A. V., & Greenamyre, J. T., (2000). Chronic systemic pesticide exposure reproduces features of Parkinson's disease. *Nat. Neurosci., 3*, 1301–1306.

Bick, A., Oron, G., Gillerman, L., & Manor, Y., (2003). Data envelopment analysis for assessing optimal operation of ultra-filtration systems for effluent polishing. *Water Science and Technology: Water Supply., 3*, 379–384.

Bragato, C., Brix, H., & Malagoli, M., (2006). Accumulation of nutrients and heavy metals in phragmites australis (cav.) trin. Ex steudel and *Bolboschoenus maritimus* (l.) palla in a constructed wetland of the Venice lagoon watershed. *Environmental Pollution, 144*, 967–975.

Branchet, P., Cadot, E., Fenet, H., Sebag, D., Ngatcha, B. N., Borrell-Estupina, V., Ngoupayou, J. R. N., et al., (2018). Polar pesticide contamination of an urban and peri-urban tropical watershed affected by agricultural activities (yaoundé, center region, cameroon). *Environmental Science and Pollution Research, 25*, 17690–17715.

Brouwer, M., Huss, A., Van, D. M. M., Nijssen, P. C. G., Mulleners, W. M., Sas, A. M. G., Van, L. T., et al., (2017). Environmental exposure to pesticides and the risk of Parkinson's disease in the Netherlands. *Environ. Int., 107*, 100–110.

Carvalho, F. P., (2017). Pesticides, environment, and food safety. *Food and Energy Security, 6*, 48–60.

Chagas, R. M., Silveira, J. A. G., Ribeiro, R. V., Vitorello, A. V., & Carrer, H., (2008). Photochemical damage and comparative performance of superoxide dismutase and ascorbate peroxidase in sugarcane leaves exposed to paraquat-induced oxidative stress. *Pestic. Biochem. Physiol., 90*, 181–188.

Choudhary, S., Yamini, N. R., Yadav, S. K., Kamboj, M. L., & Amit, S., (2018). A review: Pesticide residue: Cause of many animal health problems. *Journal of Entomology and Zoology Studies, 6*, 330–333.

Cocco, P., (2002). On the rumors about the silent spring. Review of the scientific evidence linking occupational and environmental pesticide exposure to endocrine disruption health effects. *Cad. Saude Publica, 18*, 379–402.

Cope, O. B., (1966). Contamination of the freshwater ecosystem by pesticides. *Journal of Applied Ecology*, 33–44.

Crini, G., & Lichtfouse, E., (2019). Advantages and disadvantages of techniques used for wastewater treatment. *Environmental Chemistry Letters, 17*, 145–155.

Cycoń, M., Mrozik, A., & Piotrowska-Seget, Z., (2017). Bioaugmentation as a strategy for the remediation of pesticide-polluted soil: A review. *Chemosphere, 172*, 52–71.

Dąbrowski, A., Hubicki, Z., Podkościelny, P., & Robens, E., (2004). Selective removal of the heavy metal ions from waters and industrial wastewaters by ion-exchange method. *Chemosphere, 56*, 91–106.

Dalvi, R. R., Singh, B., & Salunkhe, D. K., (1972). Influence of selected pesticides on germination and associated metabolic changes in wheat and mungbean seeds. *J. Agric. Food Chem., 20*(5), 1000–1003.

Dobozi, M., & Lehoczky, E., (2002). Influence of soil herbicides on the growth of potato. *Acta. Biol. Szegediensis, 46*(3, 4), 197–198.

Doolotkeldieva, T., Konurbaeva, M., & Bobusheva, S., (2018). Microbial communities in pesticide-contaminated soils in Kyrgyzstan and bioremediation possibilities. *Environmental Science and Pollution Research, 25*, 31848–31862.

Dubey, K. K., & Fulekar, M. H., (2011). Effect of pesticides on the seed germination of *Cenchrus setigerus* and *Pennisetum pedicellatum* as monocropping and co-cropping system: implications for rhizospheric bioremediation. *Rom. Biotechnol. Lett., 16*(1), 5909–5918.

Dunier, M., & Siwicki, A. K., (1993). Effects of pesticides and other organic pollutants in the aquatic environment on immunity of fish: A review. *Fish & Shellfish Immunology, 3*, 423–438.

Ekmekci, Y., & Terzioglu, S., (2005). Effects of oxidative stress induced by paraquat on wild and cultivated wheats. *Pestic. Biochem. Physiol., 83*, 69–81.

Elbetieha, A., Da'as, S. I., Khamas, W., & Darmani, H., (2001). Evaluation of the toxic potentials of cypermethrin pesticide on some reproductive and fertility parameters in the male rats. *Arch. Environ. Contam. Toxicol., 41*, 522–528.

EPA, U., (2007). *Drinking Water Contaminants*. US EPA, Washington, DC.

Fareed, M., Pathak, M. K., Bihari, V., Kamal, R., Srivastava, A. K., & Kesavachandran, C. N., (2013). Adverse respiratory health and hematological alterations among agricultural workers occupationally exposed to organophosphate pesticides: a cross-sectional study in North India. *PloS One, 8*, e69755.

Farooqi, Z. U. R., Kareem, A., Ayub, M. A., Hussain, M. M., Zeeshan, N., & Shahzad, M. T., (2020). Use of pesticides in agriculture: impacts on soil, plant, and human health. In: Mehmood, et al., (eds.), *Pesticide Contamination in Freshwater and Soil Environs: Impacts, Threats, and Sustainable Remediation.* CRC-Apple Academic Press, USA. ISBN 9781771889537.

Fitzmaurice, A. G., Rhodes, S. L., Lulla, A., Murphy, N. P., Lam, H. A., O'Donnell, K. C., Barnhill, L., et al., (2013). Aldehyde dehydrogenase inhibition as a pathogenic mechanism in Parkinson disease. *Proc. Natl. Acad. Sci. U. S. A., 110*, 636–641.

Gao, J., Garrison, A. W., Hoehamer, C., Mazur, C., & Wolfe, N. L., (2000). Uptake and phytotransformation of organophosphorus pesticides by axenically cultivated aquatic plants. *Journal of Agricultural and Food Chemistry, 48*, 6114–6120.
Gaur, N., Narasimhulu, K., & PydiSetty, Y., (2018). Recent advances in the bio-remediation of persistent organic pollutants and its effect on environment. *Journal of Cleaner Production, 198*, 1602–1631.
Giri, S., Giri, A., Sharma, G. D., & Prasad, S. B., (2002). Mutagenic effects of carbosulfan, a carbamate pesticide. *Mutat. Res., 519*, 75–82.
Gopi, R., Jaleel, C. A., Sairam, R., Lakshmanan, Gomathinayagam, M., & Panneerselvam, (2007). Differential effects of hexaconazole and paclobutrazol on biomass, electrolyte leakage, lipid peroxidation and antioxidant potential of *Daucas carota* L. *Colloids Surf B: Biointerfaces, 60*, 180–186.
Greenlee, L. F., Lawler, D. F., Freeman, B. D., Marrot, B., & Moulin, P., (2009). Reverse osmosis desalination: Water sources, technology, and today's challenges. *Water Research, 43*, 2317–2348.
Gupta, V. K., Ali, I., Saleh, T. A., Nayak, A., & Agarwal, S., (2012). Chemical treatment technologies for wastewater recycling—An overview. *RSC Advances, 2*, 6380–6388.
Gupta, V. K., Jain, R., & Varshney, S., (2007). Removal of reactofix golden yellow 3 RFN from aqueous solution using wheat husk—an agricultural waste. *Journal of Hazardous Materials, 142*, 443–448.
Habib, N., (1996). *Invisible Farmers in Pakistan: A Study on the Role of Women in Agriculture and the Impact of Pesticides on Them*. Khoj, Research and Publication Centre.
Hayden, K. M., Norton, M. C., Darcey, D., Ostbye, T., Zandi, P. P., Breitner, J. C., & Welsh- Bohmer, K. A., (2010). Occupational exposure to pesticides increases the risk of incident AD: The cache county study. *Neurology, 74*, 1524–1530.
He, Z. L., Yang, X. E., & Stoffella, P. J., (2005). Trace elements in agroecosystems and impacts on the environment. *Journal of Trace elements in Medicine and Biology, 19*, 125–140.
Holliger, P., Prospero, T., & Winter, G., (1993). " Diabodies": Small bivalent and bispecific antibody fragments. *Proceedings of the National Academy of Sciences, 90*, 6444–6448.
Hossain, L., Rahman, R., & Khan, M. S., (2017). Alternatives of pesticides. In: *Pesticide Residue in Foods* (pp. 147–165). Springer.
Hu, L., Luo, D., Zhou, T., Tao, Y., Feng, J., & Mei, S., (2017). The association between non- Hodgkin lymphoma and organophosphate pesticides exposure: a meta-analysis. *Environ. Pollut. (Barking, Essex, 1987), 231*, 319–328.
Huang, H., & Xiong, Z. T., (2009). Toxic effects of cadmium, acetochlor and bensulfuron-methyl on nitrogen metabolism and plant growth in rice seedlings. *Pestic. Biochem. Physiol., 94*, 64–67.
Hussain, M. M., & Farooqi, Z. U. R., (2021). Allelopathic bacteria as an alternate weedicide: Progress and future standpoints. In: Hakeem, K. R., Dar, G. H., Mehmood, M. A., & Bhat, R. A., (eds.), *Microbiota and Biofertilizers: A Sustainable Continuum for Plant and Soil Health* (pp. 211–230). Cham: Springer International Publishing.
Hussain, M. M., Farooqi, Z. U. R., Latif, J., Mubarak, M. U., & Younas, F., (2020). Bioremediation: A green solution to avoid pollution of the environment. In: Parry, et al.,

(eds.), *Soil Bioremediation: An Approach Towards Sustainable Technology* (pp. 15–40). John Willey and Sons, USA. ISBN 978-1119547952.

Ismail, B., Mazlinda, M., & Tayeb, M., (2015). The persistence of deltamethrin in Malaysian agricultural soils. *Sains Malaysiana., 44*, 83–89.

Jia, Z., Li, Y., Lu, S., Peng, H., Ge, J., & Chen, S., (2006). Treatment of organophosphate-contaminated wastewater by acidic hydrolysis and precipitation. *Journal of Hazardous Materials, 129*, 234–238.

Jüttner, K., Galla, U., & Schmieder, H., (2000). Electrochemical approaches to environmental problems in the process industry. *Electrochimica Acta, 45*, 2575–2594.

Kabir, A., Zendehdel, R., & Tayefeh-Rahimian, R., (2018). Dioxin exposure in the manufacture of pesticide production as a risk factor for death from prostate Cancer: a meta-analysis. *Iran. J. Public Health, 47*, 148–155.

Kamthan, A., Chaudhuri, A., Kamthan, M., & Datta, A., (2016). Genetically modified (gm) crops: Milestones and new advances in crop improvement. *Theoretical and Applied Genetics, 129*, 1639–1655.

Kana, R., Spundova, M., Ilik, P., Lazar, D., Klem, K., Tomek, P., Naus, J., & Prasil, O., (2004). Effect of herbicide clomazone on photosynthetic processes in primary barley (*Hordeum vulgare* L.) leaves. *Pesti. Biochem. Physiol., 78*, 161–170.

Khan, H., Zeb, A., Ali, Z., & Shah, S. M., (2009). Impact of five insecticides on chickpea (*Cicer arietinum* L.) nodulation, yield and nitrogen fixing rhizospheric bacteria. *Soil & Environ., 28*, 56–59.

Khan, M. J., Zia, M. S., & Qasim, M., (2010). Use of pesticides and their role in environmental pollution. *World Acad. Sci. Eng. Technol., 72*, 122–128.

Khan, M. S., Chaudhary, P., Wani, P. A., & Zaidi, A., (2006). Biotoxic effects of the herbicides on growth, seed yield and grain protein of green gram. *J. Appl. Sci. Environ. Manage, 10*(3), 141–146.

Khan, M. Z., & Law, F. C., (2005). Adverse effects of pesticides and related chemicals on enzyme and hormone systems of fish, amphibians and reptiles: A review. *Proceedings of the Pakistan Academy of Sciences, 42*, 315–323.

Kim, H. J., Park, Y. I., & Dong, M. S., (2005). Effects of 2,4-D and DCP on the DHT-induced androgenic action in human prostate cancer cells. *Toxicol. Sci., 88*, 52–59.

Kogan, M., (1998). Integrated pest management: Historical perspectives and contemporary developments. *Annual Review of Entomology, 43*, 243–270.

Kong, Z., Deng, Z., Glick, B. R., Wei, G., & Chou, M., (2017). A nodule endophytic plant growth-promoting pseudomonas and its effects on growth, nodulation and metal uptake in *Medicago lupulina* under copper stress. *Annals of Microbiology, 67*, 49–58.

Kumar, P. S., Carolin, C. F., & Varjani, S. J., (2018). Pesticides bioremediation. In: *Bioremediation: Applications for Environmental Protection and Management* (pp. 197–222). Springer.

Kuo, H. H., Shyu, S. S., & Wang, T. C., (2008). Genotoxicity of low dose N-nitroso propoxur to human gastric cells. *Food Chem. Toxicol., 46*, 1619–1626.

Latifossglu, A., Surucu, G., & Evirgen, M., (1997). Water pollut. Iv: Model. *Paper Read at Meas., Predict., 4th Int. Conf.*

Lee, W. J., Sandler, D. P., Blair, A., Samanic, C., Cross, A. J., & Alavanja, M. C., (2007). Pesticide use and colorectal cancer risk in the agricultural health study. *Int. J. Cancer Suppl., 121*, 339–346.

Leghait, J., Gayrard, V., Picard-Hagen, N., Camp, M., Perdu, E., Toutain, P. L., & Viguie, C., (2009). Fipronil-induced disruption of thyroid function in rats is mediated by increased total and free thyroxine clearances concomitantly to increased activity of hepatic enzymes. *Toxicology, 255*, 38–44.

Lerro, C. C., Koutros, S., Andreotti, G., Friesen, M. C., Alavanja, M. C., Blair, A., Hoppin, J. A., et al., (2015). Organophosphate insecticide use and cancer incidence among spouses of pesticide applicators in the agricultural health study. *Occup. Environ. Med., 72*, 736–744.

Lim, Y. W., Kim, H. H., Lee, C. S., Yang, S. H., Kim, S. D., Jeon, J. M., Lee, J. Y., Sohn, J. R., Chang, Y. S., & Shin, D. C., (2011). Child-specific health risk assessment of 25 indoor air pollutants in child-care facilities and indoor playgrounds. *Epidemiology, 22*, S277.

Mahmood, K. S., & Shah, Z. A., (2003). Screening of the best insecticide in reducing the chickpea pod borer damage infected by gram pod borer (H. armigera) in Faisalabad. *Pak. J. Biol. Sci., 6*, 1156–1158.

Mamane, A., Raherison, C., Tessier, J. F., Baldi, I., & Bouvier, G., (2015). Environmental exposure to pesticides and respiratory health. *Eur. Respir. Rev., 24*, 462–473.

Marsit, C. J., Eddy, K., & Kelsey, K. T., (2006). MicroRNA responses to cellular stress. *Cancer Res., 66*, 10843–10848.

McKnight, U. S., Rasmussen, J. J., Kronvang, B., Binning, P. J., & Bjerg, P. L., (2015). Sources, occurrence and predicted aquatic impact of legacy and contemporary pesticides in streams. *Environmental Pollution, 200*, 64–76.

Micó, M. M., Bacardit, J., & Sans, C., (2010). Experimental design applied to photo-Fenton treatment of highly methomyl-concentrated water. *Water Science and Technology, 62*, 2066–2074.

Mishra, P. K., Bhargava, A., Raghuram, G. V., Jatawa, S. K., Akhtar, N., Khan, S., Tiwari, A., & Maudar, K. K., (2009). Induction of genomic instability in cultured human colon epithelial cells following exposure to isocyanates. *Cell Biol. Int., 33*, 675–683.

Mishra, V., Srivastava, G., & Prasad, S. M., (2009). Antioxidant response of bitter gourd (*Momordica charantia* L.) seedlings to interactive effect of dimethoate and UV-B irradiation. *Scien. Horti., 120*, 373–378.

Mishra, V., Srivastava, G., Prasad, S. M., & Abraham, G., (2008). Growth, photosynthetic pigments and photosynthetic activity during seedling stage of cowpea (Vigna unguiculata) in response to UV-B and dimethoate. *Pestic. Biochem. Physiol., 92*, 30–37.

Mohan, S. V., Babu, V. L., & Sarma, P., (2007). Anaerobic biohydrogen production from dairy wastewater treatment in sequencing batch reactor (ansbr): Effect of organic loading rate. *Enzyme and Microbial Technology, 41*, 506–515.

Mohiddin, G. J., Srinivasulu, M., Maddela, N., Manjunatha, B., Rangaswamy, V., Kaiser, A. R. K., Asqui, J. C. M., & Rueda, O. D., (2015). Influence of the insecticides acetamiprid and carbofuran on arylamidase and myrosinase activities in the tropical black and red clay soils. *Environmental Monitoring and Assessment, 187*, 1–9.

Mondal, M. M. A., Malek, M. A., Puteh, A. B., Ismail, M. R., Ashrafuzzaman, M., & Naher, L., (2012). Effect of foliar application of chitosan on growth and yield in okra. *Aust. J. Crop Sci., 6*, 918–921.

Napit, M. K., (2013). The effect of pesticides on fish fauna of Bhopal lower lake (mp). *African Journal of Environmental Science and Technology, 7*, 725–727.

Nawaz, A., Sarfraz, M., Sarwar, M., & Farooq, M., (2020). Ecological management of agricultural pests through allelopathy. *Co-evolution of Secondary Metabolites*, 543–574.

Nemat, A. M. M., Badawi, A. M., Hassan, N. M., El-Bastawisy, Z. M., & Badran, E. G., (2008). Effect of metribuzin, butachlor and chlorimuron-ethyl on amino acid and protein formation in wheat and maize seedlings. *Pestic. Biochem. Physiol., 90*(1), 8–18.

Nemat, A. M. M., Hassan, N. M., & El-Bastawisy, Z. M., (2007). Differential influence of herbicide treatments on activity and kinetic parameters of C4 photosynthetic enzymes from Zea mays. *Pestic Biochem. Physio., 89*, 198–205.

Nemerow, N. L., & Dasgupta, A., (1991). *Industrial and Hazardous Waste Treatment.*

Njau, B., (1988). Pesticide poisoning of livestock in northern Tanzania-cases investigated in 1977 and 1978. *Bulletin of animal Health and Production in Africa, 36*, 2.

Oerke, E., (2006). Crop losses to pests. *The Journal of Agricultural Science., 144*, 31.

Özkara, A., Akyıl, D., & Konuk, M., (2016). Pesticides, environmental pollution, and health. In: *Environmental Health Risk-Hazardous Factors to Living Species*: IntechOpen.

Parron, T., Requena, M., Hernandez, A. F., & Alarcon, R., (2011). Association between environmental exposure to pesticides and neurodegenerative diseases. *Toxicol. Appl. Pharmacol., 256*, 379–385.

Parween, T., Jan, S., Mahmooduzzafar, & Fatma, T., (2011a). Alteration in nitrogen metabolism and plant growth during different developmental stages of green gram (*Vigna radiata* L.) in response to chlorpyrifos. *Acta Physiol. Plant, 33*, 2321–2328.

Parween, T., Jan, S., Mahmooduzzafar, & Fatma, T., (2011b). Assessing the impact of Chlorpyrifos on growth, photosynthetic pigments and yield in *Vigna radiata* L. at different phenological stages. *Afr. J. Agric. Res., 6*, 4432–4440.

Petersen, R. C., (2011). Clinical practice. Mild cognitive impairment. *N. Engl. J. Med., 364,* 2227–2234.

Rajashekhar, N., Prakasha, & Murthy, T. C. S., (2012). Seed germination and physiological behavior of maize (cv. Nac-6002) seedlings under abiotic stress (Pendimethalin) condition. *Asian J. Crop Sci., 4*(2), 80–85.

Rajmohan, K., Chandrasekaran, R., & Varjani, S., (2020). A review on occurrence of pesticides in environment and current technologies for their remediation and management. *Indian Journal of Microbiology, 60*, 125–138.

Reddy, G. V., & Antwi, F. B., (2016). Toxicity of natural insecticides on the larvae of wheat head armyworm, *Dargida diffusa* (*Lepidoptera*: *Noctuidae*). *Environmental Toxicology and Pharmacology, 42*, 156–162.

Rios, A. C., Salvadori, D. M., Oliveira, S. V., & Ribeiro, L. R., (1995). The action of the herbicide paraquat on somatic and germ cells of mice. *Mutat. Res., 328*, 113–118.

Rodrigo, M., Oturan, N., & Oturan, M. A., (2014). Electrochemically assisted remediation of pesticides in soils and water: A review. *Chemical Reviews, 114*, 8720–8745.

Ross, J. A., Blackman, C. F., Thai, S. F., Li, Z., Kohan, M., Jones, C. P., & Chen, T., (2010). A potential microRNA signature for tumorigenic conazoles in mouse liver. *Mol. Carcinog., 49*, 320–323.

Rusiecki, J. A., Baccarelli, A., Bollati, V., Tarantini, L., Moore, L. E., & Bonefeld-Jorgensen, E. C., (2008). Global DNA hypomethylation is associated with high serum-persistent organic pollutants in Greenlandic Inuit. *Environ. Health Perspect., 116,* 1547–1552.

Saladin, G., Magne, C., & Clement, C., (2003). Physiological stress responses of Vitis vinifera L. to the fungicides fludioxonil and pyrimethanil. *Pesti. Biochem. Physiol., 77*, 125–137.
Sankhla, M. S., Kumari, M., Sharma, K., Kushwah, R. S., & Kumar, R., (2018). Water contamination through pesticide & their toxic effect on human health. *International Journal for Research in Applied Science and Engineering Technology, 6*, 967–970.
Shahid, A., Malik, S., Zhu, H., Xu, J., Nawaz, M. Z., Nawaz, S., Alam, M. A., & Mehmood, M. A., (2020). Cultivating microalgae in wastewater for biomass production, pollutant removal, and atmospheric carbon mitigation: a review. *Science of the Total Environment, 704*, 135303.
Shinozaki, K., & Yamaguchi-Shinozaki, K., (2007). Gene networks involved in drought stress response and tolerance. *Journal of Experimental Botany, 58*, 221–227.
Shutoh, Y., Takeda, M., Ohtsuka, R., Haishima, A., Yamaguchi, S., Fujie, H., Komatsu, Y., Maita, K., & Harada, T., (2009). Low dose effects of dichlorodiphenyltrichloroethane (DDT) on gene transcription and DNA methylation in the hypothalamus of young male rats: implication of hormesis-like effects. *J. Toxicol. Sci., 34*, 469–482.
Siddiqui, Z. S., & Ahmad, S., (1996). Effect of systemic fungicide on germination, seedling growth and phenolic contents of *Vigna radiata* L. *Pak. J. Bot., 28*, 191–193.
Sillanpää, M., Ncibi, M. C., & Matilainen, A., (2018). Advanced oxidation processes for the removal of natural organic matter from drinking water sources: A comprehensive review. *Journal of Environmental Management, 208*, 56–76.
Sinev, I., Sinev, O., & Linevich, S., (1997). Apparatus of flotation treatment of natural waters and wastewater. *Izobreteniya., 26*, 369–370.
Singh, M., Singh, D., Gupta, A., Pandey, K. D., Singh, P., & Kumar, A., (2019). Plant growth promoting rhizobacteria: Application in biofertilizers and biocontrol of phytopathogens. In: *PGPR Amelioration in Sustainable Agriculture* (pp. 41–66). Elsevier.
Song, C., Kanthasamy, A., Anantharam, V., Sun, F., & Kanthasamy, A. G., (2010). Environmental neurotoxic pesticide increases histone acetylation to promote apoptosis in dopaminergic neuronal cells: relevance to epigenetic mechanisms of neurodegeneration. *Mol. Pharmacol., 77*, 621–632.
Song, C., Kanthasamy, A., Jin, H., Anantharam, V., & Kanthasamy, A. G., (2011). Paraquat induces epigenetic changes by promoting histone acetylation in cell culture models of dopaminergic degeneration. *Neurotoxicology, 32*, 586–595.
Song, N. H., Yin, X. L., Chen, G. F., & Yang, H., (2007). Biological responses of wheat (*Triticum aestivum* L) plants to the herbicide chlorotoluron in soils. *Chemosphere, 68*(9), 1779–1787.
Stajner, D., Popovic, M., & Stajner, M., (2003/04). Herbicide induced oxidative stress in lettuce, beans, pea seeds and leaves. *Biol. Plant., 47*(4), 575–579.
Stevens, M. M., Fox, K. M., Coombes, N. E., & Lewin, L. A., (1999). Effect of fipronil seed treatments on the germination and early growth of rice. *Pest Sci., 55*, 517–523.
Stouder, C., & Paoloni-Giacobino, A., (2011). Specific transgenerational imprinting effects of the endocrine disruptor methoxychlor on male gametes. *Reproduction, 141*, 207–216. Cambridge, England.
Sugiyama, S., Shimada, N., Miyoshi, H., & Yamauchi, K., (2005). Detection of thyroid system disrupting chemicals using *in vitro* and *in vivo* screening assays in *Xenopus laevis*. *Toxicol. Sci., 88*, 367–374.

Sunyer, J., Torrent, M., Garcia-Esteban, R., Ribas-Fito, N., Carrizo, D., Romieu, I., Anto, J. M., & Grimalt, J. O., (2006). Early exposure to dichlorodiphenyldichloroethylene, breastfeeding and asthma at age six. *Clin. Exp. Allergy, 36*, 1236–1241.

Sunyer, J., Torrent, M., Munoz-Ortiz, L., Ribas-Fito, N., Carrizo, D., Grimalt, J., Anto, J. M., & Cullinan, P., (2005). Prenatal dichlorodiphenyldichloroethylene (DDE) and asthma in children. *Environ. Health Perspect., 113*, 1787–1790.

SYJZJ, L., (2016). Research on the phytoremediation for eliminating the pesticide pollution. *Chin. J. Pestic. Marker News*, 25.

Tałałaj, I. A., Biedka, P., & Bartkowska, I., (2019). Treatment of landfill leachates with biological pretreatments and reverse osmosis. *Environmental Chemistry Letters, 17*, 1177–1193.

Tchobanoglous, G., Burton, F. L., & Stensel, H., (1991). Wastewater engineering. *Management,* 7, 4.

Tongwen, X., (2002). Electrodialysis processes with bipolar membranes (edbm) in environmental protection—A review. *Resources, Conservation and Recycling, 37*, 1–22.

Tort, N., & Turkyilmaz, B., (2003). Physiological effects of captan fungicide on pepper (*Capsicum annuum* L.). *Pak J. Biol. Sci., 6*(24), 2026–2029.

Tran, L. S. P., Nishiyama, R., Yamaguchi-Shinozaki, K., & Shinozaki, K., (2010). Potential utilization of NAC transcription factors to enhance abiotic stress tolerance in plants by biotechnological approach. *GM Crops, 1*, 32–39.

Tsai, W. T., (2010). Current status and regulatory aspects of pesticides considered to be persistent organic pollutants (pops) in Taiwan. *Int. J. Environ. Res. Public Health, 7*, 3615–3627.

Van, B. J., (2013). The world population explosion: Causes, backgrounds and -projections for the future. *Facts Views Vis. Obgyn., 5*, 281–291.

Van, D. H. F., Witkamp, G. J., De Graauw, J., & Van, R. G., (1998). Eutectic freeze crystallization: Application to process streams and waste water purification. *Chemical Engineering and Processing: Process Intensification, 37,* 207–213.

Villemaine, R., Compagnone, C., & Falconnet, C., (2020). The social construction of alternatives to pesticide use: A study of biocontrol in Burgundian viticulture. *Sociologia Ruralis*.

Wahlang, B., Petriello, M. C., Perkins, J. T., Shen, S., & Hennig, B., (2016). Polychlorinated biphenyl exposure alters the expression profile of microRNAs associated with vascular diseases. *Toxicol. in Vitro, 35,* 180–187.

Wang, X., Zhou, S., Ding, X., Zhu, G., & Guo, J., (2010). Effect of triazophos, fipronil and their mixture on miRNA expression in adult zebrafish. *J. Environ. Sci. Health B, 45*, 648–657.

Wibawa, W., Mohamad, R. B., Puteh, A. B., Omar, D., Juraimi, A. S., & Abdullah, S. A., (2009). Residual phytotoxicity effects of paraquat, glyphosate and glufosinate-ammonium herbicides in soils from field treated plots. *Int. J. Agric. Biol., 11*, 214–216.

Zaganas, I., Kapetanaki, S., Mastorodemos, V., Kanavouras, K., Colosio, C., Wilks, M. F., & Tsatsakis, A. M., (2013). Linking pesticide exposure and dementia: what is the evidence? *Toxicology, 307*, 3–11.

Zahm, S. H., & Blair, A., (1992). Pesticides and non-Hodgkin's lymphoma. *Cancer Res., 52*, 5485s–5488s.

Zahm, S. H., Weisenburger, D. D., Babbitt, P. A., Saal, R. C., Vaught, J. B., Cantor, K. P., & Blair, A., (1990). A case-control study of non-Hodgkin's lymphoma and the herbicide 2,4- di-chlorophenoxy acetic acid (2,4-D) in eastern Nebraska. *Epidemiology, 1*, 349–356.

CHAPTER 7

Developmental Activities and Freshwater Biodiversity

RUQUIA GULZAR,[1] TABASUM HASSAN,[1] and AADIL GULZAR[2]

[1]*Department of Botany, University of Kashmir, Srinagar, Jammu and Kashmir – 190006, India*

[2]*Department of Environmental Science, University of Kashmir, Srinagar, Jammu and Kashmir – 190006, India, E-mail: ehaan900@gmail.com*

ABSTRACT

Freshwater bodies such as deltas, peatlands, swamps, fens, and springs—home to a significant portion of flora and fauna comprise about 0.01% of the entire world's water and occupies about 0.8% of earth's total surface. The freshwater habitats affect biogeochemical cycles and other ecological interactions and dynamics that determine ecosystem productivity and biodiversity. The endemism level and species richness among freshwater ecosystems are higher in several major groups, including crustaceans, decapod molluscs, and aquatic insects such as mayflies and caddis. Despite the benefits of freshwater resources, these freshwater bodies and the biodiversity associated with these water bodies face many threats.

7.1 INTRODUCTION

Freshwaters are amongst the world's varied or diverse habitats. Although rivers, lakes, and reservoirs cover only 2.3% of the earth's surface, about

Biodiversity of Freshwater Ecosystems: Threats, Protection, and Management. Suhaib A. Bandh, PhD, Basharat Mushtaq, PhD, Mohammad Farooq Mir, PhD, & Shahid A. Wani (Eds.)

9.5% of recognized animal species live in freshwater in addition to other life forms (Reid et al., 2018). This small proportion of water supports almost 100,000 species of a total of 1.8 million species and comprises nearly 6% of all the known species (Dudgeon et al., 2006). Despite its lesser proportion, the freshwater ecosystems support a significant proportion of biodiversity, including vertebrates and various fish species (~ 10,000) (Lundberg et al., 2000). In addition to providing a favorable environment and support to biodiversity, fresh water ecosystems also support billions of people, including vulnerable and impoverished communities, by providing ecosystem services (Lynch et al., 2016). The associated riparian habitats with the freshwater ecosystems are among the most diverse habitats and have immeasurable health, economic, cultural, educational, and scientific values. The different anthropogenic activities threaten freshwater biodiversity by inculcating various stressors like pollution, flow regulation, habitat degradation, alien species and overfishing. As a result, the aquatic biodiversity is dramatically declining, including many fresh water species that became threatened or near-extinct than their marine or terrestrial counterparts. Throughout the globe, wetlands are degrading three times more than forests, due to which freshwater biota has reduced more than twice compared to the marine or terrestrial populations. This decline indicates our unsustainable practices and absence of shifting baselines which causes underestimation of ancient drops. The rising pressures due to increasing human population and water resources consumption, intensified by global warming and other environment alterations, will lead to future freshwater extinctions. These losses can significantly influence the ecosystem functioning, goods, and services upon which human life depends.

There is insufficient knowledge of the relationship between biodiversity and ecosystem functioning in the present world, and this field is not so mature to allow a detailed prediction of the outcome of biodiversity loss. Therefore, there is a dire need for management schemes for preserving freshwater biodiversity in the current world. Freshwater bodies such as deltas, peatlands, swamps, fens, and springs are home to a more significant flora and fauna proportion. Despite their benefits, these freshwater bodies and the biodiversity associated with them are facing enormous threats. The present chapter mainly discusses the various developmental activities and their associated dangers for freshwater ecosystems.

A number of anthropogenic factors are associated with developmental activities that are responsible for the loss of freshwater biodiversity.

7.2 CONSTRUCTION OF DAMS OR HYDROELECTRIC PROJECTS

About 940 million people, particularly in low-income areas, do not have a reliable electricity supply (World Bank, 2018). Simultaneously, many industries are expanding in the developing world, and their demand for energy (electricity) is rising exponentially. Dam construction is a proven hydropower generation technology that is sustainable but not climate-neutral. Renewable energy currently supplies 22% of the electricity globally, 73% of which is hydroelectric. This greatly affects the global river systems because of their fragmentation or flow control (Grill et al., 2015). Presently there are about 3,700 large hydropower dams in developing countries, either in the planning or construction phase (Zarfl et al., 2015). Dam construction on the freshwater ecosystems amongst the worlds varied and diverse habitats is a major threat to their biodiversity. One of the largest threats to freshwater biodiversity is dam construction (Dudgeon et al., 2010). Dams, for instance, block migratory pathways for large fish such as sturgeons and giant catfish, decrease access to their breeding places, and cause habitat fragmentation and degradation in both upper and lower portions of the freshwater body (Winemiller et al., 2016). Dams also play a major role in the habitat contraction and population loss of many freshwater fauna species, like the Indus river dolphin, sturgeon, and many Chinese paddlefish (Grill et al., 2015). Dams cause fragmentation, accumulation of solutes, changes in water temperature, sedimentation, flow, and fish movement (Reidy et al., 2012; Pelicice, Pompeu, and Agostinho, 2015). As dams alter the natural flow and temperature systems and reduce river-floodplain mobility, aquatic production, and fish connectivity to breeding areas and nursery ecosystems greatly threaten freshwater biodiversity (Juracek, 2015). Although some hydropower projects possess structures for fish movement, such structures are inefficient and often act as environmental traps (Pelicice, 2012). Despite emerging opinions on the sustainable development of big hydropower plants (LHPs), assistance for the rampant establishment of small hydropower plants (SHPs) has increased significantly. Tens of thousands of SHPs are working or are under development. If all possible generation ability were to be built, this number is expected to triple (Couto and Olden, 2018). The development of SHPs has significantly outpaced existing ecological research, driven by tremendous political and financial support in recent times. Sediment fluctuations connected with dam activity and tributary supplies lead to reservoirs' aging through sedimentation, beach erosion, and stream depletion, which differ regionally (Juracek, 2015). Sedimentation also affects fish's health and reproduction, reduces the yield of fish, decreases primary production, and limits storage capacity. The capacity of reservoirs to meet other human requirements (e.g., flood protection, drinking water, and navigation) is also compromised by modified waterfront accessibility (Juracek, 2015). As many dams were constructed in the middle of the 20th century, when political and financial circumstances favored dam construction, thus many large-river reservoirs are approaching 50+ years old (Avakyan and Iakovleva, 1998). Taking the case of Amazon, where there are 154 mega hydropower dams, the establishment of all 277 planned dams would destroy all three free-flowing waterways and thereby endangering the biodiversity of fish, fish-based resources, and other food production (Pelicice et al., 2015). The lower portion of Mekong, the world's major inland catch fishery, is similarly hindered by dam building along the river's mainstream (Ziv et al., 2012; Winemiller et al., 2016). Furthermore, draining land to build reservoirs raises mercury methylation and transfers to living organisms, especially fish, thus undermining community food security (Harris et al., 2007). Reservoirs are the main causes of GHG emissions, especially in tropical and subtropical areas (Deemer et al., 2016). Hydropower provides a sustainable but not environmentally neutral source of energy since dams greatly affect the state of flow to which biodiversity is adapted. Dam construction also replaces the upper portion of the riverine with a reservoir. Animal movement, particularly of migratory species, is blocked, and materials like organic carbon, silicates, and other nutrients are build-up in the water bodies. It has been estimated that reservoirs store more than 10,000 km^3 of water, five times greater than the standing volume of global rivers, decreasing more than 25% of sediment flow to the marine environment (Vorosmarty et al., 2000). Around 48% of the water volume is moderate to strongly impaired by flow control or degradation, or by both (Grill et al., 2015), and only about 37% of rivers more than 1,000 kilometers long-stay free-flowing (Grill et al., 2019). Water compartmentalized for irrigation purposes and other requirements of human's leads to decreased downstream flows. Dams mostly act as drainage basins, change water flow, shift conditions between donor and receiver river systems, thus facilitate exchange between previously separated biota's (Shumilova et al., 201). These exchanges are mostly associated with negative impacts on the resident biota.

7.3 LAND USE CHANGES

The modification of existing catchment land use into agricultural land, urban areas, wastelands, and construction areas are the major threats to the diversity of freshwater taxons (Weijters et al., 2009). People have tended to live near lakes, rivers, and other freshwater bodies for centuries because it provides essential human survival resources. As a result, freshwater habitats and diversity have been under constant threat due to human use since the dawn of civilization. Freshwater habitats are inextricably related to their catchments because of water's down slope flow (Hornung and Reynolds, 1994). Living organisms, particularly in sediments, are highly vulnerable to land-use changes because of the direct effect of changes in adjacent terrestrial habitats. The biota that lives above and below the sediment is diverse and plentiful, and each one has a significant impact on ecosystem services, including primary production and mineral nutrition. Shifts in the connectivity between upper sediment and sediment biodiversity are predicted to be one of the immediate responses to rising levels of human disturbances, e.g., the increased decomposition of phytoplankton's and macrophytes in the eutrophic lakes, increases the concentration of phosphorus in the water. Under anaerobic conditions, this phosphorus further enhances the effects of eutrophication (Lampert and Sommer, 1997). Deforestation is often involved in modifying river catchments (Bouwman et al., 2005), resulting in complete or partial loss of natural vegetation and enhancing the runoff contributing to soil erosion and subsequent water body sedimentation. The substitution of natural flora by vegetation with different water requirements often alters the runoff from surfaces, enrichment of allochthonous organic materials like leaf litter, woody debris, etc. Agrochemicals and excess nutrients found in runoff from agricultural lands is more significant and faster than runoff from land with natural vegetation. Urbanization has some substantial consequences since the frequency and concentration of surface runoff are significantly enhanced by permeable surfaces (Revenga et al., 2005). Habitat modification is widely cited as the leading cause of biodiversity loss and has led to 73% of fishes' loss in North America through the 20th century (Sala et al., 2000). In some regions, the intensification of agriculture in the last 50 years has resulted in strong enrichment consequences on rivers and stream ecosystems. This has led to remarkable changes in species diversity with decreased susceptible species and an enhancement of more robust, often non-native species (Nijboer et al., 2004).

7.4 WATER POLLUTION

Water pollution is caused due to the dumping of different types of physical, chemical, and biological pollutants in water bodies.

7.4.1 PHYSICAL POLLUTANTS

Physical pollutants include materials like solid wastes, e.g., plastic materials, metal elements taken away by flowing water to the water bodies (Free et al., 2014). One of the essential concerns regarding water quality degradation is arsenic contamination by human activities (Fendorf et al., 2010; Rodríguez-Lado et al., 2013). The ongoing urbanization and industrialization have increased exposure of these metal elements in the aquatic flora and fauna, which can decline their number. Moreover, various types of physical pollutants are increasingly released into the water bodies because of some freshwater bodies' location in lower terrains and valleys.

Water channels carry some physical pollutants to central water bodies, and some are directly discarded in freshwater ecosystems (Gasperi et al., 2014). The disposal of these wastes into the water bodies have been carried over the past many years limited by modern legislation. From various sources, the plastic and other solid particles, when ultimately reached the freshwater bodies, cause physical, behavioral, and toxic effects on freshwater biodiversity, e.g., Physical pollutants can cause adverse effects on Algae and plants. *Chlorella pyrenoidosa* (Chick, 1903), when exposed to plastic material, reduces photosynthetic production (Wu et al., 2019) along with its effects on root development. Some algae and plants absorb the physical pollutant particles, which become available to secondary consumers (Chae et al., 2018), thereby affecting the food webs. Various other aquatic groups affected by these pollutants include crustaceans, fishes, invertebrates, amphibians, and some freshwater birds.

7.4.2 CHEMICAL POLLUTANTS

Chemical pollutants include chemical substances released from different sources through human activities, which ultimately led to an increase in the concentration of these pollutants in water bodies (Jürgens et al., 2014). Many aquatic plants and animals can take up these contaminants in their bodies which can lead to physiological changes in aquatic biodiversity or sometimes may prove lethal to the aquatic flora and fauna (Fu et al., 2013). The main groups affected by water pollution include fishes, plants, amphibians, aquatic mammals, birds, and invertebrates, leading to their death or even their extinction.

7.5 EUTROPHICATION

Eutrophication is caused due to excessive presence of nutrients in water bodies (Dodds and Smith, 2016), released into the water bodies through anthropogenic activities. Due to the enrichment of nutrients in freshwater bodies, algal blooms flourish and develop due to growth of more algal species. Invertebrates and algae are the most important base of the food web. Therefore, any change due to the accumulation of these nutrients can alter the function and structure of freshwater ecosystems (Cardinale 2011; Carpenter and Kitchell, 1993). Throughout the globe, eutrophication is considered a severe problem (Jeppesen et al., 2014), as it affects the growth of other aquatic biota and, ultimately, the decrease in aquatic biodiversity.

7.6 ACIDIFICATION

Acidification of freshwater bodies is caused due to precipitation containing nitrogen oxide and sulfur oxides released from motor vehicles and coal, and other petroleum products (Manca et al., 2016). These substances get mixed up with the water vapors producing acidic rain, affecting lake ecosystems and other water bodies. As a result of this acidity, many plant and animal species die except for some mosses and insects (Vrba et al., 2016).

7.7 CLIMATE CHANGE

This is one of the most severe issues in the modern era. Contemporary climate change is caused due to a large number of anthropogenic developmental activities—this change in climatic variables affects almost all regions of the globe, including freshwater ecosystems. Most of the hotspots are restricted geographically; as many species do not possess capability to migrate to other ecosystems (Fialkowski et al., 2003), they are more prone to changing climate. Hence any change in temperature and precipitation affects these freshwater ecosystems by disturbing the environment of aquatic flora and fauna (Portner and Knust, 2007), e.g., the increased temperature of freshwater bodies cannot be tolerated by aquatic fauna, therefore, led to their death.

According to the Intergovernmental Panel on Climate Change (IPCC, 2007), in Africa, inland freshwater ecosystems are vulnerable to changing climate due to water quality deterioration. Climate change impacts freshwater ecosystems through changes in the water cycle or various changes incorporated in terrestrial ecosystems. Timing in snowmelt and flow rates are other negative impacts that are caused due to climate change. The global climate has resulted in a 15% reduction in the flowing rivers' freshwater biodiversity (Kramer, 1997).

7.8 OVEREXPLOITATION

Overexploitation of various resources, e.g., sand excavation from the water bodies, can directly influence aquatic ecosystems with fishes and some amphibians as the most affected. For example, due to the forest clearance, a lot of silt and clay get deposited into the water bodies, thereby increasing sediments, which became the leading cause of alteration in aquatic life environments.

7.9 AGRICULTURAL WASTES

As far as agricultural practices are concerned, throughout the world, almost 3 million km^2 land is under cultivation, which constitutes about 22% of the total land. Agricultural activities degrade the fresh water's ecological status and quality due to discharge of large quantities of fertilizers, pesticides, and insecticides. Further, the expansion of agricultural land area elevates evapotranspiration (Destouni et al., 2013). It causes a decline in the runoff freshwater amount (Ceballos et al., 2015), thereby affecting the freshwater biota.

Overall, the anthropogenic activities due to developmental projects can severely affect aquatic communities, e.g., change in the composition of phytoplankton food webs, affecting fish and other aquatic organisms, which ultimately leads to their extinction. In boreal peatland, species composition can get changed due to change in hydrology and precipitation due to various anthropogenic activities, e.g., in Urmia Iran, large numbers of fresh water bodies are drying up, ultimately proving fatal to the aquatic biota. In some wetlands, precipitation changes led to a loss of biodiversity due to a lack of seasonal migration (Dudgeon, 2011). Although the effect of increased concentration of carbon dioxide differs according to the type of wetland, in most cases, it can lead to the production of gases like methane (Fialkowski, 2003), which ultimately deteriorates the aquatic ecosystems. Due to changing rainfall patterns, many bird species are affected in different wetlands (Roshier, 2004).

Due to various socio-economic developments, including economic growth and increasing human populations over the last 50 years (Crist et al., 2017), there has been a change in freshwater bodies' status. This increment in human population and developmental activities led to an increase in the concentration of pollutants in the water bodies, thereby degrading water quality and ecosystem services across different world areas (Destouni and Jarsjo, 2018). In the middle of the 20th century, the reduction in the river volume has tripled, thereby reducing the volume of river flow in the world's half rivers. All these processes negatively impact their aquatic biodiversity. As a result, there is a sharp decrease in freshwater biodiversity due to these developmental activities. Since 2000, the number of freshwater fishes and insects has sharply reduced in tropical and temperate zones (Wagner, 2019). From 1970 to 2012, there has been a sharp decline of megafauna by ~88%, including fishes, reptiles, mammals, and amphibians, with more reduction in mega fishes by ~94% (He et al., 2019).

Over the latter half of the century, human activities and their utilization of freshwater resources show an abrupt increase. The demand for freshwater resources begins to grow with expansion by 50–70% before 2050. If this utilization continues, the future generation would have to face scarcity of freshwater resources.

7.10 CONSERVATION STRATEGIES FOR FRESHWATER BIODIVERSITY

Freshwater biodiversity can still be preserved with much efforts at local, regional, national, and international levels (Ceballos et al., 2015). Many eco-friendly steps should be taken in temperate and tropical regions, which should be designed in their ecological and socio-economic contexts (Udall, 2017). Action should be taken at the local and community scales to maintain fresh water biodiversity and ecosystem services and avoid the scarcity and water stress of these freshwater ecosystems.

One of the most important steps is to motivate stakeholders, including municipalities, consumers, and farmers locally, to reduce the degradation of freshwater bodies and increase conservation of watersheds (Barbier, 2017). Steps should inform the general public about the changes in species diversity in freshwater ecosystems and how these changes are associated with loss of biodiversity. Actions must be taken to properly use water, energy, and biodiversity at the regional level.

There is an urgent need to limit the pollution and extraction of freshwater resources. International communities should take steps to reduce global warming to <2°C and return the globe towards the pre industrial era (UNFCCC, 2015). Some mitigation agreements, including pollution control and transboundary sharing of water resources, should be developed to reduce freshwater contamination, and sharing irrigation and waste treatment technologies with the developing world.

7.11 CONCLUSIONS

Freshwaters are hotspots of biodiversity throughout the world. The environmental factors that triggered the initial trend of species richness are also to blame for the vulnerability of biodiversity to human threats. The present series of threats to freshwater habitats is peculiar due to an abrupt increase in human populations and their rising water needs. To limit the associated dangers, mitigation of relative risks to freshwater habitats must be paired with more extensive global interventions. The former needs urgent intervention but gives only short-term benefits in biodiversity conservation without the latter's support. Even in the most promising case, where global threats are significantly reduced, failure to resolve proximate threats would limit freshwater habitats' strength and weaken their biota, making them more susceptible to species loss. At this scale, conservation, and recovery interventions can only be successful if they are focused on a detailed understanding of the mechanisms that influence freshwater habitats and biodiversity and the distinct challenges that they face, such as flow alteration and loss of connectivity. On the other hand, carefully developed portfolios of sustainable management measures that address the most pressing threats and factors will result in rapid changes in the state of freshwater ecosystems.

KEYWORDS

- **biodiversity**
- **biogeochemical cycles**
- **freshwater**
- **polychlorinated biphenyls**
- **polynuclear aromatic hydrocarbons**
- **species richness**

REFERENCES

Avakyan, A. B., & Iakovleva, V. B., (1998). Status of global reservoirs: The position in the late twentieth century. *Lakes & Reservoirs: Research & Management, 3*(1), 45–52.

Barbier, E. B., (2017). The economics of aquatic ecosystems: An introduction to the special issue. *Water Economics and Policy, 3*, 1202002.

Bernstein, L., Bosch, P., Canziani, O., Chen, Z., Christ, R., & Riahi, K. (2008). *IPCC, 2007: Climate Change 2007: Synthesis Report.* Geneva: IPCC. ISBN 2-9169-122-4

Bouwman, A. F., Van, D. G., Knoop, J. M., Beusen, A. H. W., & Meinardi, C. R., (2005). Exploring changes in river nitrogen export to the world's oceans. *Global Biogeochemical Cycles, 19*(1).

Cardinale, B. J., (2011). Biodiversity improves water quality through niche partitioning. *Nature, 472*, 86–89.

Carpenter, S. R., & Kitchell, J. F., (1993). *The Trophic Cascade in Lakes.* Cambridge University Press, Cambridge.

Ceballos, G., Ehrlich, P. R., Barnosky, A. D., Garcı´a, A., Pringle, R. M., & Palmer, T. M., (2015). Accelerated modern human-induced species losses: Entering the sixth mass extinction. *Science Advances, 1*, e1400253.

Chae, Y., Kim, D., Kim, S. W., & An, Y. J., (2018). Trophic transfer and individual impact of nano-sized polystyrene in a four species freshwater food chain. *Scientific Reports, 8*, 284.

Cheng, L., Crochetiere, H., *et al., (*2019*).* Mapping the world's free-flowing rivers. *Nature.*

Couto, T. B., & Olden, J. D., (2018). Global proliferation of small hydropower plants–science and policy. *Frontiers in Ecology and the Environment, 16*(2), 91–100.

Crist, E., Mora, C., & Engelman, R., (2017). The interaction of human population, food production, and biodiversity protection. *Science, 356*, 260–264.

Deemer, B. R., Harrison, J. A., Li, S., Beaulieu, J. J., DelSontro, T., Barros, N., &Vonk, J. A., (2016). Greenhouse gas emissions from reservoir water surfaces: a new global synthesis. *BioScience, 66*(11), 949–964.

Destouni, G., & Jarsjo, J., (2018). Zones of untreatable water pollution call for better appreciation of mitigation limits and opportunities. *Wiley Interdisciplinary Reviews: Water, 5*, e1312.

Destouni, G., Jaramillo, F., & Prieto, C., (2013). Hydroclimatic shifts driven by human water use for food and energy production. *Nature Climate Change, 3*, 213.

Dodds, W., & Smith, V. H., (2016). Nitrogen, phosphorus, and eutrophication in streams. *Inland Waters, 6*(2), 155–164.

Dudgeon, D., (2010). Prospects for sustaining freshwater biodiversity in the 21st century: linking ecosystem structure and function. *Current Opinion in Environmental Sustainability, 2*(5, 6), 422–430.

Dudgeon, D., (2011). Asian river fishes in the Anthropocene: threats and conservation challenges in an era of rapid environmental change. *Journal of Fish Biology, 79*(6), 1487–1524.

Dudgeon, D., Arthington, A. H., Gessner, M. O., Kawabata, Z. I., Knowler, D. J., Lévêque, C., & Sullivan, C. A., (2006). Freshwater biodiversity: importance, threats, status and conservation challenges. *Biological Reviews, 81*(2), 163–182.

Fendorf, S., Michael, H. A., & Van, G. A., (2010). Spatial and temporal variations of groundwater arsenic in South and Southeast Asia. *Science, 328*, 1123–1127.

Fialkowski, W., Rainbow, P. S., Smith, B. D., & Zmudzinski, L., (2003). Seasonal variation in trace metal concentrations in three talitrid amphipods from the Gulf of Gdansk, Poland. *Journal of Experimental Marine Biology and Ecology, 288*, 81–93.

Free, C. M., Jensen, O. P., Mason, S. A., Eriksen, M., Williamson, N. J., & Boldgiv, B., (2014). High-levels of microplastic pollution in a large, remote, mountain lake. *Marine Pollution Bulletin, 85*(1), 156–163.

Fu, J., Hu, X., Tao, X., Yu, H., & Zhang, X., (2013). Risk and toxicity assessments of heavy metals in sediments and fishes from the Yangtze River and Taihu Lake, China. *Chemosphere, 93*(9), 1887–1895.

Gasperi, J., Dris, R., Bonin, T., Rocher, V., & Tassin, B., (2014). Assessment of floating plastic debris in surface water along the Seine River. *Environmental Pollution, 195*, 163–166.

Grill, G., Lehner, B., Lumsdon, A. E., MacDonald, G. K., Zarfl, C., & Liermann, C. R., (2015). An index-based framework for assessing patterns and trends in river fragmentation and flow regulation by global dams at multiple scales. *Environmental Research Letters, 10*(1), 015001.

Grill, G., Lehner, B., Thieme, M., Geenen, B., Tickner, D., Antonelli, F., Babu, S., et al., (2007). Whole-ecosystem study shows rapid fish-mercury response to changes in mercury deposition. *Proceedings of the National Academy of Sciences, 104*(42), 16586–16591.

He, F., Zarfl, C., Bremerich, V., David, J. N., Hogan, Z., Kalinkat, G., Tockner, K., & Ja¨hnig, S. C., (2019). The global decline of freshwater megafauna. *Global Change Biology, 25*, 3883–3892.

Houghton, R. A., (1994). The worldwide extent of land-use change. *BioScience, 44*(5), 305–313.

Intergovernmental Panel on Climate Change (IPCC), (2007). *Climate Change 2007: Synthesis Report* http://www.ipcc.ch.

Jeppesen, E., Meerhoff, M., Davidson, T. A., Trolle, D., SondergaarD, M., Lauridsen, T. L., Beklioglu, M., et al., (2014). Climate change impacts on lakes: an integrated ecological perspective based on a multi-faceted approach, with special focus on shallow lakes. *Journal of Limnology, 73*(1), 88–111.

Juracek, K. E., (2015). The aging of America's reservoirs: In-reservoir and downstream physical changes and habitat implications. *JAWRA Journal of the American Water Resources Association, 51*(1), 168–184.

Jürgens, M. D., Johnson, A. C., Jones, K. C., Hughes, D., & Lawlor, A. J., (2013). The presence of EU priority substances mercury, hexachlorobenzene, hexachlorobutadiene and PBDEs in wild fish from four English rivers. *Science of the Total Environment, 461, 462*, 441–452.

Kramer, R. A., Richter, D. D., Pattanayak, S., & Sharma, N. P., (1997). Ecological and economic analysis of watershed protection in Eastern Madagascar. *Journal of Environmental Management, 49*(3), 277–295.

Lampert, W., & Sommer, U., (2007). *Limnoecology: The Ecology of Lakes and Streams.* Oxford university press.

Lundberg, J. G., Kottelat, M., Smith, G. R., Stiassny, M. L., & Gill, A. C., (2000). So many fishes, so little time: an overview of recent ichthyological discovery in continental waters. *Annals of the Missouri Botanical Garden,* 26–62.

Lynch, A. J., et al., (2016). The social, economic, and environmental importance of inland fish and fisheries. *Environmental Reviews, 24*, 115–121.

Manca, M. M., Bonacina, C., & Yan, N. D., (2016). What have we learned about ecological recovery from liming interventions of acid lakes in Canada and Italy? *Journal of Limnology, 75*(s2), 1–3.

Nijboer, R. C., Verdonschot, P. F. M., Johnson, R. K., Sommerhäuser, M., & Buffagni, A., (2004). Establishing reference conditions for European streams. In: *Integrated Assessment of Running Waters in Europe* (pp. 91–105). Springer, Dordrecht.

Pelicice, F. M., Pompeu, P. S., & Agostinho, A. A., (2015). Large reservoirs as ecological barriers to downstream movements of Neotropical migratory fish. *Fish and Fisheries, 16*(4), 697–715.

Portner, H. O., & Knust, R., (2007). Climate change affects marine fishes through the oxygen limitation of thermal tolerance. *Science, 315*(5808), 95–97.

Reid, A. J., et al., (2018). Emerging threats and persistent conservation challenges for freshwater biodiversity. *Biol. Rev., 94*, 849–873.

Reidy, L. C. A., Olden, J. D., Beechie, T. J., Kennard, M. J., Skidmore, P. B., Konrad, C. P., & Imaki, H., (2012). Hydrogeomorphic classification of Washington State rivers to support emerging environmental flow management strategies. *River Research and Applications, 28*(9), 1340–1358.

Revenga, C., Campbell, I., Abell, R., De Villiers, P., & Bryer, M., (2005). Prospects for monitoring freshwater ecosystems towards the 2010 targets. *Philosophical Transactions of the Royal Society B: Biological Sciences, 360*(1454), 397–413.

Rodríguez-Lado, L., et al., (2013). Groundwater arsenic contamination throughout China. Science 341, 866–868.

Roshier, D. A., & Rumbachs, R. M., (2004). Broad-scale mapping of temporary wetlands in arid Australia. *Journal of Arid Environments, 56*(2), 249–263.

Sala, O. E., Stuart, C. F., Juan, A. J., Eric, B., Janine, B., Rodolfo, D., Huber-Sanwald, E., et al., (2000). "Global biodiversity scenarios for the year 2100." *Science, 287*(5459), 1770–1774.

Shumilova, O., Tockner, K., Thieme, M., Koska, A., & Zarfl, C., (2018). Global water transfer megaprojects: A potential solution for the water-food-energy nexus? *Frontiers in Environmental Science, 6*, 150.

The World Bank, (2018). *Database World Development Indicators*. https://data.worldbank.org (accessed on 14 December 2021).

Udall, S. L., (2017). *Battle Against Extinction: Native Fish Management in the American West*. Tucson: University of Arizona Press.

UNFCCC, (2015). *United Nations Framework Convention on Climate Change*. Adoption of the Paris Agreement FCCC/CP/2015/L.9/Rev.1. http://www.unfccc.int/resource/docs/2015/cop21/eng/l09r01.pdf (accessed on 14 December 2021).

Vörösmarty, C. J., & Sahagian, D., (2000). Anthropogenic disturbance of the terrestrial water cycle. *Bioscience, 50*(9), 753–765.

Vrba, J., Bojková, J., Chvojka, P., Fott, J., Kopáček, J., Macek, M., Nedbalová, L., et al., (2016). Constraints on the biological recovery of the Bohemian Forest lakes from acid stress. *Freshwater Biology, 61*(4), 376–395.

Weijters, M. J., Janse, J. H., Alkemade, R., & Verhoeven, J. T., (2009). Quantifying the effect of catchment land use and water nutrient concentrations on freshwater river and stream biodiversity. *Aquatic Conservation: Marine and Freshwater Ecosystems, 19*(1), 104–112.

Winemiller, K. O., McIntyre, P. B., Castello, L., Fluet-Chouinard, E., Giarrizzo, T., Nam, S., &Sáenz, L., (2016). Balancing hydropower and biodiversity in the Amazon, Congo, and Mekong. *Science, 351*(6269), 128–129.

Wu, Y., Guo, P., Zhang, X., Zhang, Y., Xie, S., & Deng, J., (2019). Effect of microplastics exposure on the photosynthesis system of freshwater algae. *Journal of Hazardous Materials, 374*, 219–227.

Zarfl, C., Lumsdon, A. E., Berlekamp, J., Tydecks, L., & Tockner, K., (2015). A global boom in hydropower dam construction. *Aquatic Sciences, 77*(1), 161–170.

CHAPTER 8

Global Climate Change and Freshwater Biodiversity

SAMI ULLAH QADIR,[1] VASEEM RAJA,[2] NASEER AHMAD DAR,[3] and SAYAR YASEEN WANI[4]

[1]*Department of Environmental Sciences, Government Degree College, Kokernag, Kashmir – 192202, India, Phone: 9797831608; 7889310156; E-mail: samievs84@gmail.com*

[2]*Department of Botany, Government Degree for Women, Pulwama, Jammu and Kashmir – 192301, India*

[3]*Department of Environmental Sciences, Government Degree College, Charar-i-Sharief, Jammu and Kashmir – 191112, India*

[4]*Department of Environmental Sciences, Government Degree College, Shopian, Jammu and Kashmir – 192303, India*

ABSTRACT

Freshwater, comprising a tiny portion of water on land, is home to about 6% species, out of the total 1.8 million species documented globally. Inland freshwater resources and biodiversity constitute practical importance in scientific, cultural, esthetic, economic, and educational aspects. Therefore, the conservation of these resources is of critical importance for all humans inhabiting the earth and using these resources. But these precious resources are in a great predicament. Freshwaters are facing far more significant declines in biodiversity than most invaded terrestrial

Biodiversity of Freshwater Ecosystems: Threats, Protection, and Management. Suhaib A. Bandh, PhD, Basharat Mushtaq, PhD, Mohammad Farooq Mir, PhD, & Shahid A. Wani (Eds.)

ecosystems in the world. Thus, the conservation of this part of biodiversity and its associated resources is the most important for the survival of life on this planet.

Keeping in view the patterns of precipitation, annual mean temperatures, and evaporation rates, it is evident that the climate is changing. The consequences of this rapid environmental change on freshwater biodiversity needs a clear understanding that could be very dangerous if neglected. Among these factors' enhancement in water temperature, alteration in hydrological regimes, reduction in dissolved oxygen levels, drop in the availability of habitats, and proliferation of invasive species are the noted factors affected in freshwater ecosystems causing significant declines in biodiversity. Another cause that leads to the loss of biodiversity in freshwater habitats is the increasing intensity and frequency of droughts and river intermittence during summer seasons in various parts of the world. The species that are naturally adapted to cold perineal water sources suffer from reduced range distribution and species extinction. In contrast, the tolerant ones survive by enlarging their range of distribution. In this context, freshwater biodiversity has become the priority of overriding conservation during the "water for life" decade.

8.1 INTRODUCTION

Global Climate Change has witnessed several changes in the world as per the published scientific literature. It is one of the most critical factors affecting disaster risk as defined by the Intergovernmental Panel on Climate Change (IPCC). It occurs due to the unparalleled development and economic undertakings resulting in the release of Green House Gases (GHGs) into the global atmosphere, consequently leading to the rise of the earth's mean temperature due to the radiative properties of these gases. As per the projected reports of IPCC global warming is expected to continue with an increase of 5.8° C by 2100 and (Lama and Devkota, 2016) due to swelling temperatures, the changing climate may have a wide range of effects on environmental resources and habitats of biodiversity. The rapidly increasing warming trend has already shown visible and generally negative impacts on vital ecological resources such as land, animals, and water over the last few decades. Climate Change has already affected the seasonal patterns, glacial melting, and permafrost, which led to the range land shifts in wildlife migratory patterns at higher altitudes (Bhattarai, 2017).

Climate change affects biodiversity at all levels ranging from low lying areas to high mountain ranges and from tiny brooks to vast oceans. Climate Change in some ecosystems leads to the extinction of some species, while others become endangered or critically endangered (Sodhi et al., 2011). Thus, biodiversity loss is one of the noteworthy global climate change features (Martens and Rotmans, 2005). The millennium ecosystem assessment (MEA) showed that 60% of the ecosystem services deteriorate or are already overexploited (Mooney et al., 2009). This situation is further worsened by global climate change, triggering the extinction and immigration of the globe's biodiversity components. The current climate change data shows that more ecological processes will occur shortly (Bhattarai, 2017). In the present Anthropocene era, the threats responsible for the loss of biodiversity become more apparent, and the new unrecognized ones become more deceptive (Reid et al., 2018). The loss of freshwater biodiversity in this era becomes so rapid that it is acknowledged as a "hidden invisible" underwater tragedy (Richter et al., 1997) that attracts insufficient scientific, political, and social attention (Cooke et al., 2016). Human activities are among the most critical factors affecting negatively aquatic ecosystems, leading to severe biodiversity reduction. The noted anthropogenic factors negatively affecting freshwater ecosystems include climate change (Woodward et al., 2010), species extinction and invasion, flow modification, pollution, overexploitation, higher incidence of Ultra Violet radiations, thermal changes, and habitat modification/destruction (Benateau et al., 2019; Dodds et al., 2013; Reid et al., 2018), channelization, dam construction and, eutrophication (Green et al., 2015).

The resultant of global climate change is the alteration of aquatic ecosystems' physical characteristics, possibly leading to the formation of new environmental circumstances that subsequently affect the position and biological processes in these freshwater ecosystems (Woodward et al., 2010). Much attention has been paid to assess the effect of climate change on meteorological parameters. The resultant environmental consequences of climate change for water quality and the ecological position of freshwater ecosystems were somewhat given poor consideration. The present chapter aims to find the gaps by summarizing impacts of climate change knowledge on the direct and indirect processes by which changing climate patterns affect freshwater ecosystems and the intact biodiversity (Pachauri and Leo, 2014).

8.2 DISTRIBUTION OF BIODIVERSITY AND CLIMATE

Rainfall and air temperature generally tend to co-vary with increasing altitude and latitude and directly bear biodiversity distribution. Moreover, the well-documented association between altitude, climate, and distribution of different species have guided us to study the potential effects of climate change on the biodiversity of different taxa's (Knouft and Ficklin, 2017; Parmesan, 2006). Freshwater species follow general global biodiversity patterns exhibited by many terrestrial groups with much species richness found in warm, tropical regions and relatively more minor species richness in the cold high altitudes. However, these biodiversity patterns vary among the groups of organisms (Tisseuil et al., 2013).

In terms of climate importance, it has been reported that temperature and rainfall are the better interpreters of continental variation in the richness of freshwater fish species than latitudes (Knouft and Page, 2011). Additionally, climate acts as a reasonable indicator of species richness of macrophytes, invertebrates, and fishes at regional and spatial scales (Heino, 2001). The biodiversity richness of macroinvertebrates and fish species decreases with increasing temperature with associated elevation (Knouft and Ficklin, 2017). The climate variables are also linked with the variations in abundance across the geographical distribution of freshwater species (Knouft and Anthony, 2016).

8.3 FRESHWATER BIODIVERSITY STATUS

The impacts of climate change have already been evident in different realms of biodiversity such as terrestrial, marine, and freshwater. The ongoing predictions suggest a gruesome picture of future biodiversity in the natural habitats (Sala et al., 2000; Thomas et al., 2004, 2006). The declines observed in freshwater ecosystems is of grave concern as more and more reductions were observed in these ecosystems in comparison to their marine and terrestrial counterparts and are exceeding those estimated for the biodiversity of tropical forests called the biodiversity cradles (Heino et al., 2009; Ricciardi and Rasmussen, 1999; Xenopoulos et al., 2005). This rapid decline in freshwater biodiversity is because freshwater ecosystems possess more incredible biodiversity than other complexes of biodiversity (Balian et al., 2008; Dudgeon et al., 2006). Further, the expected increase in the levels of freshwater biodiversity is only possible when underground water sources (Gibert and Deharveng, 2002), particularly the microorganisms and different other species of invertebrates inhabiting tropical surface water resources, are studied more comprehensively (Balian et al., 2008; Dudgeon, 2019; Dudgeon et al., 2006; Heino et al., 2009).

Multiple anthropogenic factors are responsible for freshwater biodiversity declines, with the most effective changes in water temperature and hydrological regimes induced by climate change on a global-scale analysis (Xenopoulos et al., 2005). Regarded as a global phenomenon, its effects on global biodiversity are manifested at different levels of biological organization (genes, species, communities, and ecosystems) and scales of spatial distribution (habitat, regional, local, and continental).

8.4 FRESHWATER ECOSYSTEMS AND PATTERNS OF CLIMATE CHANGE

Projected climate change evidence has shown the significant impacts on biodiversity at regional, local, and global levels. Although the earth has witnessed a considerable change in climate events during the past few decades and soon, the projected climate patterns changes will accelerate at an unprecedented rate (Sala et al., 2000).

The observed results show that at least 15–37% of terrestrial biodiversity will become extinct, possibly due to present climate change impacts in the next five decades (Thomas et al., 2004). A similar tragedy has been predicted in the next few years for freshwater species (Xenopoulos et al., 2005). Despite the current research, there is still a lack of widespread understanding and predictive competency of climate change effects on biodiversity in various groups of organisms and their ecosystems.

Freshwater ecosystems are vulnerable to climate change like terrestrial and marine ecosystems (Heino et al., 2009). At global scale assessments, lotic, and lentic freshwater ecosystems are more sensitive to land-use alterations, the introduction of non-native species and climate change (Sala et al., 2000). The factors that lead to the change in biodiversity vary among regions and latitudes, with significant changes due to climate change on aquatic ecosystems and biodiversity at higher latitudes than other responsible factors.

In addition to these effects, climate change is likely to affect the global patterns of precipitation. It has been suggested that differences in rainfall amounts with both regional increases or decreases will occur due to changes in global temperatures (Pachauri and Leo, 2014). This is likely to affect the characteristics of freshwater ecosystems directly or indirectly and the organisms harboring these ecosystems (Poff et al., 2002).

Effects of climate change vary among ecosystem types. It has been observed that small streams are likely to get mainly influenced than rivers due to the close relationship between air and water temperatures of these small ecosystems (Poff et al., 2002). These tiny ecosystems (streams) are also highly vulnerable to low flow regimes and flash flooding, arising from the changes in precipitation patterns; in the same way, small ponds are likely to be more affected than large lakes. Increased water temperature also affects the different species residing in cold, shallow lakes.

8.5 FRESHWATER BIODIVERSITY: THE EXPANDING PREDICAMENT

Of the total water resources, freshwater comprises 0.1%, with 2.3% occupied by lakes, reservoirs, and rivers and 5.4–6.8% occupied by the freshwater wetlands (Reid et al., 2018). In the last two decades, the global freshwater biodiversity in lakes, reservoirs, streams, wetlands, and other ecosystems has seriously deepened due to various stress factors. Covering only 2.3% of the global surface area, these ecosystems harbor at least 9.5% of the world's described animal species (Reid et al., 2018), with one-third of the vertebrate population. They are considered local or global hot spots of biodiversity (Benateau et al., 2019), providing numerous ecological services such as provisioning, cultural, recreational, etc. (Green et al., 2015). Despite covering a large area and total productivity of marine counterparts, the species richness of saline and freshwater fishes is slightly similar, with all saline species derived from freshwater descendants (Vega and Wiens, 2012).

Drastic reduction in population size and large-scale extinction is going at an alarming rate in organisms harboring global freshwater ecosystems, as reported from various indicators. As suggested from the World Wide

Fund for natures living planet index (LPI) documented observations, the population of freshwater vertebrates has declined by about 81% relative to index declines of 38 and 36% for land and sea by 2014, respectively (Collen et al., 2009). For freshwater biodiversity, the LPI value shows an average declined of 84%, equivalent to 4% annually since 1970, which is four times bigger than that of terrestrial dwellers (WWF, 2020). The reports from the International Union for Conservation of Nature (IUCN) "Red List" also depicts that a large proportion of threatened species of vertebrates are associated with freshwater. It was also reported that about 40% of freshwater fishes are at risk in European and North American countries (Jelks et al., 2008). Vertebrates of freshwater ecosystems are also facing worse situation than those dwelling in terrestrial ecosystems (Clausnitzer et al., 2009; Cumberlidge et al., 2009; Reid et al., 2018; Taylor et al., 2007).

Despite a declining trend of the freshwater taxa's, the literature regarding the conservation of organisms harboring terrestrial ecosystems is poorly estimated, with approximately 20% of the scientific papers dealing with aquatic organisms (Di Marco et al., 2017). The three most important reasons for this might be:

- The terrestrial inhabiting organisms are the poor substitutes for freshwater dwellers (Darwall et al., 2011);
- The primary conservation problem solutions for freshwater depends on land-fresh water interface management; and
- The terrestrial conservation efforts for freshwater biodiversity require large-scale spatial extents such as channel, riparian, and catchment scales (Darwall et al., 2011).

For instance, the vast majority of freshwater threatened megafauna (~84%) fall outside the existing network of protected areas (Carrizo et al., 2017). These freshwater ecosystems represent the hotspots of endangered species due to the convergence between species richness and different forms of human freshwater species exploitation, not only driven by land-based activities. The projected reports suggest that the rate of freshwater extinction risk will remain higher in the coming decades regardless of the actions taken at present, owing to an incurred "debt" arising from lower population availabilities resulting in declining to extinction (Strayer and Dudgeon, 2015).

8.6 CLIMATE CHANGE: INFLUENCES ON FRESHWATER BIOTA

Scientific literature is bestowed with good observational records and scientific evidence regarding the changing climate at global levels. At present, it is changing at an accelerated pace with more profound consequences on mean temperature, precipitation, and evaporation. However, the results arising from this rapid climate change on the world's freshwater species are still ambiguous, but the predicted consequences could be very catastrophic. The main effects of the changing environmental conditions we can individuate, are the increase in surface water temperatures, especially for the poikilothermic organisms (Fenoglio et al., 2010). Alteration in the hydrological regimes (Figure 8.1) is another essential characteristic of climate change coupled with a rapid increase in intensity and frequency of extreme events such as droughts and floods (Dallas and Rivers-Moore, 2014) and groundwater recharge rates. This scenario could severely affect freshwater biodiversity, particularly in mid-latitude areas. Shifts in phenology (Dallas and Rivers-Moore, 2014), life cycles and distributional ranges of organisms are likely to get affected for many of the species along with the extinction of sensitive ones. In particular, species inhabiting and adapted to cold and perennial water bodies suffer due to their reduction in the ranges of distribution and local extinctions. In contrast, the more tolerant ones may enlarge their distribution ranges (Heino et al., 2009).

The rapid climate change is expected to bring some drastic and dramatic changes in freshwater biodiversity of large regions concerning the presence and distribution of several species of organisms, ranging from declines in taxonomic richness to the entire disappearance of ecosystems (Svenning et al., 2009). There is already a plentiful of information available that climate change has terrible effects on our planet's biodiversity. Recent evidence reported that there is a close correspondence between increased temperatures and changes in fish diversity of Atlantic ocean (Fenoglio et al., 2010), death, and bleaching of coral reefs in Australian and Caribbean regions, changes in the life cycle patterns of some species of insects (Fenoglio et al., 2010; Gomi et al., 2007) and growing risks of extinction and the disappearance or extinction of Golden Toad (*Bufo periglenes*) of Costa Rica has been linked to global climate change.

Keeping in view the vulnerability of freshwater ecosystems, the inhabiting biota could expect some dramatic changes. Some recent reports have presented glaring evidence of a decline in freshwater biodiversity at a higher rate than marine and terrestrial ecosystems from the last 30 years (Jenkins, 2003). This was the possible reason for the consequences of climate change on inland water ecosystems and freshwater biota.

It is evident that freshwater bodies present on the earth's surface are closely attached to weather and atmospheric temperature variations, so changes in climate patterns may have substantial direct and indirect consequences on the global freshwater biota. Among different factors affecting the biodiversity of freshwater jurisdictions, changing temperatures of inland water bodies, the distribution in hydrological cycles, alterations in the patterns of precipitation and snow cover in different regions of the world are some of the noted factors. In general, climatic conditions could have some indirect effects, including landscape alteration, river morphology amendments, upsurge in dam constructions associated with river disintegration, enhancement in irrigational needs for agricultural purpose, poor water quality and allochthonous invasion of non-native species (Table 8.1).

8.6.1 FRESHWATER BIODIVERSITY AND INCREASE IN WATER TEMPERATURE

The air temperature strictly determines the temperature of inland water streams. The other factors responsible for this change are the shading effect from nearby vegetation, changes in the watershed, storage in wetlands, land use land cover change (LULCC) and groundwater networks. Water temperature has a profound impact on the biodiversity of freshwaters (Dallas and Rivers-Moore, 2014). For instance, the solubility of oxygen in water decreases with an increase in temperature. Most of the freshwater organisms are included in the poikilothermic group with internal temperatures fluctuating with surrounding environments, thereby affecting the basic physiological and other metabolism activities, which in turn strongly affects the survival, longevity, and reproductive processes (Fenoglio et al., 2010) of these animals (Figure 8.1). In particular, the growth of organisms inhabiting freshwater aquatic environments is dependent on temperature since elevations in temperature is known to increase the feeding activity and rate of digestion in freshwater organisms.

Recent evidence from North Italian rivers revealed that the population of mayflies (*Oligoneuriella rhenana*) and *Potha-mantus luteus* has faster

TABLE 8.1 Effect of Global Climate Change on Some Freshwater Organisms (Biodiversity)

TAXA	Effect of Global Climate Change	Type of Effect	References
Mollusca			
Dreissena polymorpha (Pallas, 1771)	+	Population increase	Mouthon and Daufrresne (2006)
Pisidium amnicum (Müller, 1774)	–	Population Collapse	Mouthon and Daufrresne (2006)
Pisidium casertanum (Poli, 1791)	–	Population Collapse	Mouthon and Daufrresne (2006)
Pisidium henslowanum (Sheppard, 1823)	–	Population Collapse	Mouthon and Daufrresne (2006)
Pisidium moitessierianum (Paladilhe, 1866)	–	Population Collapse	Mouthon and Daufrresne (2006)
Pisidium nitidum (Jenyns, 1832)	–	Population Collapse	Mouthon and Daufrresne (2006)
Pisidium subtruncatum (Malm, 1855)	–	Population Collapse	Mouthon and Daufrresne (2006)
Crustacea			
Thermocyclops oithonoide (Sars, 1863)	+	Change in phenology and abundance increase	Adrian et al. (2006)
Hesperodiaptomus arcticus (Marsh, 1920)	–	Population reduction	Holzapfel and Vinebrooke (2015)
Eudiaptomus intermedius (Steuer, 1897)	–	Modification of life history strategies	Fenoglio et al. (2010)
Insecta			
Gomphus vulgatissimus (L.)	–	Change in phenology and emergence	Richter et al. (2008)
Chloroperla sp.	–	Habitat reduction	Daufresne et al. (2003)
Protonemura sp.	–	Habitat reduction	Daufresne et al. (2003)
Nemoura sp.	–	Habitat reduction	Daufresne et al. (2003)
Ecdyonurus sp.	–	Habitat reduction	Daufresne et al. (2003)
Serratella ignita (Poda, 1761)	+	Habitat expansion	López-Rodríguez et al. (2009)
Dinocras cephalotes (Curtis, 1827)	–	Habitat reduction	Daufresne et al. (2003)
Cloeon dipterum L.	0	No direct or immediate effect	Buisson et al. (2008)

TABLE 8.1 *(Continued)*

TAXA	Effect of Global Climate Change	Type of Effect	References
Vertebrata			
Salmo trutta fario (Linnaeus, 1758)	–	Habitat reduction	Buisson et al. (2008); Hari et al. (2006)
Cottus gobio (Linnaeus, 1758)	–	Habitat reduction	Buisson et al. (2008)
Lampetra planeri (Bloch, 1784)	–	Habitat reduction	Buisson et al. (2008)
Gobio (Goc)	+	Habitat expansion	Buisson et al. (2008)
Leuciscus cephalus (Lec)	+	Habitat expansion	Buisson et al. (2008)

developments in body size and warmer habitats (Fenoglio et al., 2008, 2005). In addition to this, some researcher's observed recently that thermal characteristics are the chief factors responsible for observed differences in life cycle patterns of Baetidae in the Alps of Italy (Fenoglio et al., 2010). Hence it can be postulated that with projected enhancement in water temperatures, the phenology and life cycle patterns of many freshwater organisms change. Coupled with these changes, increased water temperatures are likely to decrease the average body size, disturbance in sex ratios and reduced fecundity (Hogg and Williams, 1996).

In running waters (lotic ecosystems), the species distribution along longitudes is strictly associated with water temperature. Keeping in view the sedentary nature and confined habitat ranges, the freshwater organisms are therefore incapable of rapid reactions to environmental variations by a progressive increase in temperatures. For instance, in this response, drastic alterations were reported in the distribution of fishes in response to climate change because of the physiological inability to regulate the body temperature (Grant et al., 2007). It is also predicted hypothetically from the recent studies that fish assemblages in streams will change in the coming future due to climate change patterns throughout Europe. The warm ones such as *Barbus meridionalis*, *Gobio gobio* and *Leuciscus cephalus* possibly increase in their range of distribution, while the cold counterparts like those of *Salmo trutta fario* and *Cottus gobio* expected to lose many of the previously occupied habitats (Buisson et al., 2008). These impacts are likely to worsen in the coming future because of excessive water withdrawal. In a paper authored by Xenopoulos et al. (2005), the author predicted a model related to fish extinction in freshwaters based on two scenarios of climate change suggested by IPCC. Both the models predicted a drastic reduction in freshwater fish richness in two lotic ecosystems of Italy. In one scenario, the worst loss of 7–13% fish species was reported, and in the second, the decline was 3–6%. The expected extinction range was 10–15% in the first case and 5% in the second case.

Similar results are also expected in lentic ecosystems because most mountainous lakes are characterized by a long period of cold water with abysmal productivity (primary and secondary). The increased temperature results in strong thermal stratification enhanced nutrient re-suspension resulting in the enhanced production in these lakes. Subsequently, a zone shift in productivity occurs from benthic to pelagic. In turn, the taxonomic composition could be powerfully transformed (Fenoglio et al., 2010) because different taxa prefer different responses under warming conditions in environmental gradients. Long term declines in the population of *Daphnia* and trophic linkage disturbances were observed in large temperate lakes because of increasing temperatures. Early blooming in Phytoplanktons' growth was also observed that creates a mismatch with the presence of consumers of zooplanktons and may have severe repercussions for resource flow to higher-order trophic levels (Winder and Schindler, 2004). It indicates that the last part of the 20th century has witnessed a severe mismatch in the timing conditions of algal-herbivore interaction, resulting in severe consequences in ecosystem interactions, especially when involving keystone species. In pelagic ecosystems, the energy required at higher trophic levels is produced by algae-zooplankton interactions. This interaction's decoupling may be communicated to all trophic levels, causing catastrophic ecological and economic concerns (Platt et al., 2003).

8.6.2 CHANGES IN HYDROLOGICAL CYCLE

The most significant global warming changes include the modification of hydrological cycles and subsequent alteration in streamflow regimes (Huntington, 2006). It is an evident fact that climate change is one of the prominent factors transforming the timing and quantity of river flows, which are the most critical parameters for flora and fauna of world river systems. In recent years, considerable changes were observed in precipitation and hydrological regimes patterns on a global scale (Mauget, 2003). There is increasing evidence suggesting that there is a significant loss in ice cover and snow melt due to climate change. These regimes in future will be characterized by an increase due to the rise in temperatures, droughts, and flooding, which will lead to the shifting of permanent water bodies to temporary ones, especially in mid-latitudes (Boulton and Lake, 2008). This hydrological change brought about by climate change will strongly affect the biota of freshwater bodies (Figure 8.1). These extreme climate events, such as droughts, are likely to affect directly due to decreased water flow, hydrological connectivity, and habitat alterations in world rivers and indirectly due to increased competition, predation, and change in food resources of organisms (Miller and Golladay, 1996). The fresh water biota of discontinuous environments show adaptation strategies to survive dry periods, but inferior information is available about the drought impact of perennial water bodies. While analyzing the drought impact on macroinvertebrate assemblage in different reaches of the Po river, the abundance and taxonomic richness of these macroinvertebrates were found to be inversely linked to the length of dry periods, coupled with the disturbances in benthic community functional organization due to disturbances in flow perpetuity (Fenoglio et al., 2010).

Moreover, in other areas marked with water's discontinued presence, the colonization of small, fast-growing plurivoltine organisms was reported. Simultaneously, the substratum traps showed poor taxa richness, representing that only a few taxa could survive drought conditions by colonizing interstitial and hyporheic habitats (Fenoglio et al., 2006). This study demonstrated that in natural perennial pre-alpine rivers, the drought conditions would cause general depletion of freshwater biodiversity. Remarkable plasticity in the life cycle, secondary production and feeding behavior was observed in *Serratella ignita* mayfly in two Mediterranean streams. This species shows high fitness in both these streams and supports that such species will cope with the global climate change scenario by maintaining or even increasing their population at the expense of more stenosis ones (López-Rodríguez et al., 2009).

There is increasing evidence that the flooding phenomenon is likely to increase in frequency and intensity because of increased precipitation rates (Brunetti et al., 2004). Severe flooding causes "Pulse disturbances" (Lake, 2000), thereby affecting aquatic organisms by an increase in water velocity, alterations in habitats and physical scouring (Suren and Jowett, 2006). Generally, it has been observed that inhabiting river communities react to flooding by a reduction in density and taxonomic richness, shifting community composition mainly by affecting physiological and ecological characteristics of different taxa. It was also reported by Fenoglio et al. (2010) that flooding was responsible for halving the density of three species of *Hydropsyche* with no evident effects on the abundance of *Ephoron leucan*.

8.6.3 CLIMATE CHANGE AND ACIDIFICATION OF WATERBODIES

Climate change and acidification of water bodies have a complex interaction with each other. Due to climate warming, water bodies' acidification may increase rapidly, thereby negatively affecting their recovery (Heino et al., 2009). However, the opposite trend of alkalization of lakes was also observed (Winder and Schindler, 2004; Schindler, 2001). It was also observed that the rapid increase in acidification generally leads to the decline and overall insolvency in the biodiversity of freshwaters (Fenoglio et al., 2010; Heino et al., 2009). While a decrease in acidification rate generally has contrasting effects on freshwater biodiversity, climate change influences through acidification on freshwater biodiversity are challenging to predict. The expected scenario shows that fresh water ecosystems that are subjected to acidification show exacerbated negative consequences on biodiversity Climate change is likely to have opposite effects on temporal changes in the biodiversity of acidified and neutral streams (Durance and Ormerod, 2007).

The sensitive groups of organisms like fishes, molluscs, and cyanobacteria are likely to disappear with inland water bodies' increasing acidity. The diversity and species richness of Phytoplanktons, zooplanktons, and benthic invertebrates also decrease to very low levels. The population of herbivorous insects gets reduced, whereas carnivorous organisms are likely to increase numbers. Some groups of organisms increase in their abundance, whereas some of them decrease. The fungal diversity and richness are likely to increase; sulfate reducing bacteria in aerobic zones is also expected to outgrow acidified environmental conditions. *Sphagnum* moss or filamentous green algae will shift from mat-like structures to masses. The food chains will become more straightforward in acid precipitated lakes. The metabolism and productivity rates will be declined to lower levels, with the result accumulation of detritus and deadwood debris will get accumulated. In lakes, where pH was reported to be less than 5.7, most fish species were reported to disappear from La Cloche Mountains, while no fish species survived when pH was said to be < 4.3. keeping in view the acidity in the freshwater ecosystem, the fish diversity in Norway affected about 25%, and by 1975 southern Norway reported a decline of about 50% in brown trout fish species. This damage to native biota was due to the combination of metal toxicity, acidity, sedimentation, and salinization (Geller and Schultze, 2009).

8.6.4 CLIMATE CHANGE AND EUTROPHICATION

Climate change is directly linked to both the increase and decrease of nutrients that are likely to enter freshwater ecosystems. It has been reported that climate change leads to a general decline in the concentration of phosphorous in lakes that leads to changes in the growth of phytoplankton. The general decrease in chlorophyll concentration but an increase in phytoplankton biomass and diversity to some degree is also reported (Heino et al., 2009).

Increasing levels of nutrients with a concomitant increase in precipitation and runoff was also expected, and enough of the rise may lead to the reduced phytoplankton diversity in freshwater lakes. However, it should be kept in mind that nutrient increase in freshwater bodies depends on these ecosystems' present and natural state. As evident from the fact that most of the lentic and lotic ecosystems of northern boreal regions are oligotrophic, the increased levels of nutrients will lead to the general increase in biodiversity levels. In contrast, the southern boreal areas are characterized by eutrophic water bodies where high nutrient levels may decrease biodiversity (Knouft and Ficklin, 2017; Reid et al., 2018).

In artic regions, climate change will lead to the thawing of upper permafrost layers, leading to increasing phosphorous levels in arctic lakes and streams, thereby leading to an increase in the production of diatoms (Heino, 2001; Heino et al., 2009). The increase in lower trophic levels may lead to the subsequent increase at higher levels, leading to increased freshwater fish biomass. The excessive load of nutrients may lead to the dominance of moss diversity as primary producers, leading to nutrient depletion in these aquatic ecosystems. Such changes in key organisms providing habitat to other microorganisms may also have severe impacts, and this role is attributed to the moss diversity in streams (Dudgeon et al., 2006; Heino et al., 2009). This increase in moss abundance leads to the consequent rise in benthic macro-invertebrate organisms' abundance and diversity (Heino et al., 2009).

Arctic lakes have shown to increase algal production and biomass with nutrient input by climate change. Keeping in view the relationship between altitude, nutrient input (phosphorous), and algal biomass, climate change is predicted to increase arctic lakes' productivity if the projected association, increase in nutrients, and climate change holds (Eissa and Zaki, 2011; Flanagan et al., 2003). An increase in productivity and biomass at lower trophic levels may significantly impact higher order trophic levels in food chains. For instance, deficient levels of productivity support fish population in arctic lakes, but this increased production leads to colonization of these systems by predatory fish species (Flanagan et al., 2003). In the absence of fishes grazing, zooplanktons may control the state of increased production in these systems (Dallas and Rivers-Moore, 2014).

8.6.5 CLIMATE CHANGE AND INVASION OF NON-NATIVE SPECIES

Aquatic systems become more and more vulnerable when the rate of climate change accelerates or becomes rapid soon (Reid et al., 2018). Temperature and flow regimes are the two principal factors connected with the success and introduction of non-native species in rivers. Considering the disaggregation in aquatic ecological communities due to the changes brought about by climate change and the resulting increase of temperatures and flow regimes. This will lead to the shift in community equilibrium that makes the communities more vulnerable to invasion by non-native species, especially in species-poor and with more open niche space ecosystems (Dallas and Rivers-Moore, 2014). These non-native species may outcompete native or non-indigenous species (Fenoglio et al., 2010).

Following climate change, many freshwater ecosystems may become suitable breeding grounds for populations of exotic species with more detrimental effects on native species, biotic communities, and ecosystem processes (Rahel and Olden, 2008). These effects arise from competition, predation, and development and the spread of diseases to freshwater habitats to which native species are not adapted (Reist et al., 2006). The negative effects of exotic species are very calamitous and rapid if they:

- come in direct contact with keystone species;
- lead to a general decline in biodiversity;
- change trophic interactions in ecosystems to which they are introduced (Heino et al., 2009).

The invasion process and impacts on biodiversity of freshwater aquatic organisms by aquatic invasive species have been closely associated and anticipated with other factors, including climate change (Rahel, 2002; Rahel and Olden, 2008; Ward and Ricciardi, 2007).

In nutshell, we can conclude that number of exotic species is likely to increase with the increase in global temperatures in freshwater ecosystems. Keeping in view the context-dependent nature of invasive species' effect, it is difficult to provide any generalized statement of predictions about the impacts of biodiversity's non-native introduction on biodiversity. There are many examples of the introduction of invasive species in inland waters, mostly because in many of the freshwater ecosystems, the important human activities such as recreation, sport, fishing, aquaculture practices, pet trade (Fenoglio et al., 2010), e-commerce (Reid et al., 2018) that have led to the allochthonous invasive species import and the eradication of these is very difficult (Heino et al., 2009).

8.7 LEVELS OF ORGANIZATION: INDIVIDUALS POPULATIONS AND SPECIES

The impacts of climate change are pervasive due to their non-uniform nature. The reported evidence indicates that climate change responses change as a function of relative vulnerability because of variabilities in exposure time, organisms' sensitivity, and adaptive capacity (Beever et al., 2017; Kovach et al., 2019; Weiskopf et al., 2020). A brief description is provided on the major impacts of climate change on organization levels in subsections.

8.7.1 MORPHOLOGY AND BEHAVIOR

One of the most critical species' strategies to adapt and cope with environmental changes is changing behavior or morphology. Behavior changes in response to climate change due to global temperatures and manifest before changes at the levels of species and population, such as distribution changes and population declines (Beever et al., 2017; Weiskopf et al., 2020). Among behavioral responses, the most noted ones include the refuge or shade seeking, alteration in feeding times, changing sites, and shifting of circadian or circannual rhythms (e.g., migration, hibernation) (Beever et al., 2017; McCann et al., 2017; Weiskopf et al., 2020). Common morphological responses include body size changes (Weeks et al., 2019; Weiskopf et al., 2020). In ectothermic organisms increase in temperature leads to faster growth rates and ultimately small body size. The direct impact of the changing the temperature has been observed on the growth of American lobster (*Homarus americanus*) and Atlantic cod (*Gadus morhua*) in response to recent warming in the Northwest Atlantic.

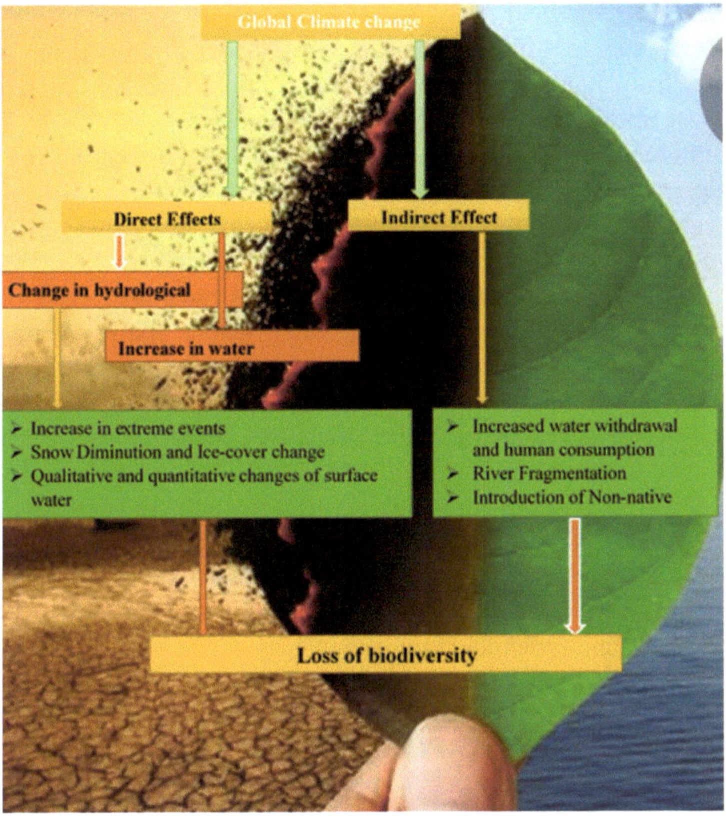

FIGURE 8.1 Climate change factors linked to change of biodiversity in aquatic ecosystems.

8.7.2 CLIMATE CHANGE AND PHENOLOGY

The critical and most crucial part of the ecological relationship is the time-specific seasonal and recurring biological events called phenology, one of the essential indicators of species' responses to global environmental change (Rudolf, 2019). Across the world, broad changes in seasonality are evident by the early arrival of springs compared to the 20th century. Changes in the phenology of organisms are well documented (Weiskopf et al., 2020). Keeping in view the variability in climatic events, organisms' phenological responses vary across different habitats (Pearson, 2019; Weiskopf et al., 2020).

These phenological shifts are significantly less evident in aquatic organisms due to difficulty in detecting and tracking. But the clear evidence from aquatic ecosystems includes the directional shifts in abiotic drivers in timing and seasonality, earlier changes from winter to spring temperatures (Post, 2016; Weiskopf et al., 2020). Phytoplanktons are the only taxa that respond quickly to these changes resulting in an alteration of phytoplankton blooms (Wasmund et al., 2019), leading to mismatch second-order consumers and thereby changing food web structures (Post, 2016; Weiskopf et al., 2020). These changes have also been observed in freshwater ecosystems, including early winter spawning phenology of Coho (*Oncorhynchus kisutch*) and Chun salmon (*O. keta*). These phenology shifts in interacting organisms lead to reproductive declines through reproductive alterations or increased predation and competition (Visser and Gienapp, 2019; Wann et al., 2019; Zimova et al., 2016).

Asynchronous shifts in phenology have the potential ability to disturb the functioning, persistence, and resilience of population dynamics and the service they provide (Visser and Gienapp, 2019; Weiskopf et al., 2020). Overall, the most common observed changes in phenology are advances of emergence and rapid development.

8.7.3 SHIFTS IN GEOGRAPHICAL RANGES

Large scale shifts in distribution, abundance, and reorganization have been observed in species of terrestrial and aquatic ecosystems. The range shifts are generally regarded as the most essential and primary mechanisms by which species adapt to climate change effectively (Weiskopf et al., 2020). As a result of changes in habitat stability, the spatial position or altitudinal and latitudinal limits for fish species are expected to change. The most likely responses are the shifting of fishes to higher latitudes or altitudes, especially for cold water species. The poleward shifts in the northern range have been observed in 15 stream fish species over the last 25 years of Great Britain's climate change. This pattern has also been reported for Salmonid species in several northern hemispheres (Comte et al., 2013; Hickling et al., 2006).

The declines in the population of brown trout (*Salmo trutta*) at the vulnerable southern fringe of its range has also been recently related to the loss of its thermal habitat (Almodovar et al., 2011). There have also been several pieces of evidence of pacific salmon (*Oncorhynchus spp.)* located outside their previously known distribution area, which parallel an observed increase in water temperature (Babaluk et al., 2000). Other studies have quantified the potential altitudinal shifts of several trout species under climate change scenarios.

8.8 EFFECT OF CLIMATE CHANGE ON SPECIES, SPECIFIC GROUPS, AND BIODIVERSITY

Some of the most immediate effects of climate change on freshwater ecosystems' ecological status are those experienced by some specific species. The underlying reasons have two-fold effects in nature.

In the first instance, the individual organisms, communities, and biodiversity of ecosystems are the backbones of stability, resilience, and functionality (Urban et al., 2016). Any change resulting from climate change alters the ecological system by shifting species occurrence and abundance, affecting these ecosystems' essential aspects. The well-defined ecological niches of organisms are likely to get disturbed by climate-induced changes in abiotic components in ecosystems, especially those in water temperatures, drought, and runoff regimes. Land-use changes directly associated with the basic species elements of the "ecological niche" affect the species occurrence. These changes will be directly observed in those species or groups of highly naked species and directly associated with humans, such as fish. Many of the climate-induced changes and their adverse effects on aquatic ecosystems are best indicated by some individual species. For example, when changes in mean local temperature or stratification in lakes are caused by heat waves that lead to local extinction of fishes like those of brown trout (*Salmo trutta*) or harmful algal blooms in lakes, hence, can be used as best substitutions to illustrate the effect of change in freshwater ecosystems.

Second, the status of individuals, species, and biodiversity, in particular, is one of the most important elements that determines the ecological status of ecosystems and any change in these determinants are the best indicators for monitoring freshwater ecosystems. If the occurrence and abundance of these elements get changed, it is reflected in these ecosystems' structure and dynamics. Loss of biodiversity and change in species composition due to the introduction of non-native species are reflected in fresh water than other ecosystems that are mostly climate-induced (Vorosmarty et al., 2010). Thus, any change occurring in biodiversity and species occurrence must be linked to multiple anthropogenic drivers leading to proper forecasting of biodiversity changes in the context of climate change (Urban et al., 2016).

Many past studies on the ecological status and diversity of freshwater ecosystems concentrated on local environmental or human factors. In aquatic ecosystems, diversity patterns are strongly caused by a network with higher levels of diversity in down streams and lower in upstreams. This pattern has higher and substantial implications in the process of extinction driven by climate change. For instance, when system structure network changes due to damming, desiccation, or river fragmentation, or when climate change temperatures make these systems more vulnerable to foreign species invasion. Thus, an integrated approach is needed to study and evaluate climate change's effects on richness and biodiversity.

8.8.1 EFFECT ON SPECIES RICHNESS AND COMPOSITION

Climate change will manifest its general effects on diversity patterns by the homogenizing composition of communities (Urban et al., 2016). Depending upon the local context, climate change may lead to an increase or decrease in species richness; keeping in mind the ecosystems, a local group of organisms will have different effects for ecosystems lying at lower or higher elevations (Brown et al., 2007). The expected richness in these areas may be slower and take a longer time to adjust to new habitats and conditions arising from climate change (Menendez et al., 2006). However, a rapid shift in richness and composition of aquatic invertebrates has been reported in alpine streams with increased numbers of species and overall similarity in communities due to loss of specialist species (Robinson et al., 2014). In contrast, extreme habitats' disappearance is predicted to decrease habitat heterogeneity, thereby reducing β-diversity (Brown et al., 2007; Jacobsen et al., 2012).

Climate-induced change in hydrological systems affect the biodiversity at regional (α diversity), local (β-diversity) and between ecosystems (ϒ-diversity). The receding water of glaciers is expected to increase the connectivity in flood plains that are likely to increase the general richness of some taxa but subsequent shrinking, leading to the decrease in α-diversity. A different effect of climate change is observed on biodiversity at lower elevations from small to medium-sized rivers to higher expected intermittence in summers due to decreasing rainfall and less melting of glaciers. Such hydro-period changes in these rivers are likely to reduce species richness at local scales (Soria et al., 2017). Thus, the intermittent rivers, in general, have a lower average α-diversity than perineal water sources. But in streams and lakes, the changes in diversity are unclear, e.g., an increase in the variety of phytoplankton was observed in perialpine low land lakes in Switzerland, north Italy, arising from the combination of warming and global environmental change (Monchamp et al., 2017).

8.8.2 EFFECT ON FISH SPECIES

The structural diversity of fishes is well studied in lakes and streams. Besides climate change, essential factors responsible for the fish decline include habitat modification, damming, hydropower generation, eutrophication, and non-native species.

In Switzerland, brown trout (*Salmo trutta*) is one of the most important species of freshwater fishes and is a true indicator of climate change, used as the surrogate species because the effects of climate change on this fish species are likely to be equally relevant for other cold inhabiting oxygen-rich streams (Hari et al., 2006). This species' population has experienced severe declines in the last 25 years (Borsuk et al., 2006). It has been observed that warmer waters are detrimental to the reproduction and survival of this fish species. Temperature profoundly affected spawning onset, sex determination, prey availability, fry development, and disease dynamics (Carraroa et al., 2017). Climate change also affects aquatic organisms through changes in host-pathogen interactions (Borsuk et al., 2006), and it has been observed that fish mortality increases with increasing temperature.

8.8.3 EFFECT ON MACROZOOBENTHOS

The diversity of macroinvertebrates is highly linked to topography, water temperature, land-use, chemical pollution, and discharges (Fuente et al., 2018). Climate change is expected to drive environmental conditions towards a more homogeneous habitat, leading to the modification in macroinvertebrates' diversity patterns to more uniform species distribution. These species of macroinvertebrates respond quickly to global environmental changes and has drawn strong impacts on macroinvertebrate communities. Further substantial changes are expected in macroinvertebrates' diversity and composition (Jacobsen et al., 2012). According to a study conducted by Rosset and Oertli (2011), the warming of lentic waters in Switzerland resulted in a risk of extinction to 11% of Coleoptera and 33% of Odonata species thermal preferences. Out of which 8–10 least resilient Odonata species were at risk due to their low resilience to temperature alterations.

In future, warmer temperatures may increase pressure on native macroinvertebrates. Still, those non-native macroinvertebrates will strengthen their pressure on native species through resource competition and habitat, thereby altering food web structures (Gallardo et al., 2016). Climate change is expected to increase the frequency and intensity of extreme events leading to flooding, thereby resulting in modification of seasonality in these species towards earlier flood seasons that will likely impact the macroinvertebrates by altering emergence timing, etc. (Lytle, 2014).

8.8.4 EFFECT ON MACROPHYTES

Climate change has a profound and counterintuitive effect on growth and an overall decrease in macrophytes' population throughout the world (Zhang et al., 2017). The development of macrophytes like that of other organisms also interacts with other anthropogenic activities like land-use, aquaculture, and eutrophication. With the increased intensity and frequency of extreme events, the growth of macrophytes will be impaired strongly. Floods and storms decrease the availability of light by increasing turbidity levels, resulting in the loss of submerged and emergent macrophytes (Havens et al., 2004).

The extreme events transport nutrients by precipitation and lead to an increase in eutrophication conditions with potential impacts on the growth of phytoplanktons, which compete for light and results in depletion of macrophytic cover (Short et al., 2016).

8.8.5 EFFECT ON DIATOMS

Diatoms are a group of very common unicellular microalgae, important producers, and indicators for reconstructing past climate change events and water quality. Because of climate change's effect on physical lake conditions (stratification), the peak of diatom seasonality has changed enormously in global lakes (Yankova et al., 2017). The reduced mixing in different layers of lakes results in the lowering of phosphorous (P) in the epilimnion, which results in higher silicon to phosphorous ratio (Si:P), critical for the species of diatoms as they are associated with silicon shells. As a result, the diatom dominance in these water bodies changes. The centric group of diatoms' density has shown a recent decline in lake Zurich because of a shift in the density towards summer (Yankova et al., 2017). The pennate group of diatoms is more adapted to elevated Si:P ratios and are therefore less affected by changes in lake nutrient dynamics. The changes resulting from climate change alter these diatoms' structure and diversity, thus affecting their use as water quality indicators. Therefore, it may be essential to consider how the effect of climate change influences the interpretability of the diatom index (Visco et al., 2015).

8.8.6 EFFECTS ON PHYTOPLANKTONS

Climate-induced eutrophication has enormous effects on lake phytoplankton dynamics due to enhanced nutrient enrichment, which changed the community dominance from eukaryotic phytoplanktons cyanobacteria (Sukenik et al., 2015). Cyanobacterial phytoplanktons are likely to benefit from climate warming, leading to macrophytes' recession by increasing resource competition, e.g., light, and nutrients (Short et al., 2016). It has been observed that increased rates of eutrophy, oligotrophy, and climate change affect lake phytoplankton communities strongly. Climate change primarily affects phytoplanktons by strong stratification, increased water column stability and gradient of environmental conditions (Longhi and Beisner, 2014).

The blooms of cyanobacteria are detrimental to other species because they alter environmental conditions by reducing light and nutrients for phytoplankton species, thereby impacting higher organisms at different trophic levels, zooplankton, and fishes by creating anoxic conditions (Sukenik et al., 2015). The bacterial toxicity is also of great importance because these blooms contain harmful cyanobacterial species, a growing phenomenon throughout the world (Salmaso et al., 2017). This may be one of the most prominent and obvious impacts of climate change on freshwater ecosystems, directly relevant to the general public.

8.9 EFFECT ON GENETIC DIVERSITY

In the past, most of the research on the effect of climate change on ecosystems focused on changes in richness and composition of species. Recently it was reviewed that the impact on genetic diversity is of good relevance (Fourtune et al., 2016; Paz-vinas et al., 2015), particularly in the context of adaptability to new environments in the events of climate change.

In mountainous freshwaters, the high degree of isolation results in the limited flow of genes and the genetic variation will be very high (Hotaling et al., 2017). A decrease in habitat heterogeneity with the shrinking of glacial water, lake homogeneity, and temperature dynamics results in genetic diversity loss (Hotaling et al., 2017). Loss in genetic diversity was higher than expected when morphologically different species were used to study climate change's effect on diverse populations' genetic variability. The studies have revealed that 67% of species were predicted to resist under two different emission scenarios of CO_2. While only 35% of the diversity survived when genetic metrics were employed. This loss in genetic diversity resulted in the loss of phenotypic plasticity (Bálint et al., 2011).

While analyzing the effect of glaciation catchment on genetic diversity of macroinvertebrates compared to catchment without glaciers, it was observed that the β-diversity was very high in highly glaciated areas, the highest turnover of diversity between mid and high glacial catchments. These patterns are most probably driven by the prevailing environmental conditions with low temperature, increased instability, and isolation. Therefore, it can be concluded that a reduction in glaciation is responsible for reduced genetic diversity.

8.10 CONCLUSIONS

Over the last few decades, rapid increasing warming trends have already shown visible and generally negative impacts on critical environmental resources such as land, animal, and water. Climate Change has negatively affected seasonal climate patterns, glacial melting, and permafrost, leading to the rangeland shifts in the migratory patterns of biodiversity components in various habitats. The threats responsible for the loss of biodiversity includes exploitation, overuse of water resources, destruction of habitats, modification in flow regimes, and proliferation of exotic species. Their combined and interacting effects on the biodiversity of freshwaters are recognized throughout the globe. The consequences are exacerbated by the ongoing rapid changes in climate patterns at the global scale coupled with the deposition of nitrogen and accumulation of CO_2 in our environment. The information regarding these threats is increasing among conservationists and ecologists but is poorly incorporated with global water resources' ongoing development, thus requiring extensive broadcasting and emphasis.

An amalgam of strategies is required that are essential for the preservation of freshwater biodiversity in the long run. It must include all the areas of protected and conservation reserves that protect these biodiversity-rich freshwater bodies, particularly those with essential species radiations and their associated areas of the catchment. It must be coupled with habitat and species centered plans that resolve the uses of biodiversity conservation and public service of water resources in the perspective of human-altered ecosystems. In the same way, biodiversity scientists must more effectively communicate the importance and value of freshwater biodiversity to Govt. policymakers to ensure that all assimilated information regarding freshwater biodiversity is effectively taken into consideration to ensure the conservation of these vital resources of life.

KEYWORDS

- **biodiversity**
- **extinction**
- **fresh water**
- **global climate change**
- **invasive species**
- **macrophytes**
- **species richness**
- **zoobenthos**

REFERENCES

Adams, G., Pichler, D. E., Cox, E. J., Gorman, E. J. O., Seeney, A., Woodward, G., & Reuman, D. C., (2013). Diatoms can be an important exception to temperature-size rules at species diatoms can be an important exception to temperature – size rules at species and community levels of organization. *Global Change Biology, 19*, 3540–3552. doi: 10.1111/gcb.12285.

Adrian, R., Wilhelm, S., & Gerten, D., (2006). Life-history traits of lake plankton species may govern their phenological response to climate warming. *Global Change Biology, 12*, 652–661. doi: 10.1111/j.1365-2486.2006.01125.x.

Almodovar, A., Nicola, G. G., Ayllo, D., & Elivira, B., (2011). Global warming threatens the persistence of Mediterranean brown trout. *Global Change Biology, 13*, 1–12. doi: 10.1111/j.1365-2486.2011.02608.x.

Atwood, T. B., Hammill, E., Kratina, P., Greig, H. S., Shurin, J. B., & Richardson, J. S., (2015). Warming alters food web-driven changes in the CO_2 flux of experimental pond ecosystems. *Biology Letters, 11*, 1–5.

Babaluk, J. A., Reist, J. D., Johnson, J. D., & Johnson, L., (2000). First Records of Sockeye (*Oncorhynchus nerka*) and pink salmon (*O. gorbuscha*) from banks island and other records of Pacific Salmon in Northwest Territories, Canada. *Arctic, 53*, 161–164. doi: 10.14430/arctic846.

Balian, E. V., Segers, H. H., & Martens, K., (2008). An introduction to the freshwater animal diversity assessment (FADA) project an introduction to the freshwater animal diversity assessment (FADA) project. *Hydrobilogia, 595*, 3–8. doi: 10.1007/s10750-007-9235-6.

Bálint, M., Domisch, S., Engelhardt, C. H. M., Haase, P., Lehrian, S., Sauer, J., Theissinger, K., et al., (2011). Cryptic biodiversity loss linked to global climate change. *Nature Climate Change, 1*, 313–318. doi: 10.1038/NCLIMATE1191.

Beever, E. A., Leary, J. O., Mengelt, C., West, J. M., Julius, S., Green, N., Schuurman, G. W., et al., (2017). Improving conservation outcomes with a new paradigm for understanding species' fundamental and realized adaptive capacity. *Conservation Letters*, *9*(2), 131–137. doi: 10.1111/conl.12190.

Benateau, S., Gaudard, A., Stamm, C., & Altermatt, F., (2019). *Climate Change and Freshwater Ecosystems Impacts on Water Quality and Ecological Status*. Hydro-CH 2018 project. Federal office of environment (FOEN). Bern, Switzerland. doi: 10.5167/uzh-169641.

Bhattarai, U., (2017). Impacts of climate change on biodiversity and ecosystem services: direction for future research. *Hydrology Nepal,* 41–48.

Borsuk, M. E., Reichert, P., Peter, A., Schager, E., & Burkhardt-holm, P., (2006). Assessing the decline of brown trout (Salmo trutta) in Swiss rivers using a Bayesian probability network. *Ecological Modeling, 192*, 224–244. doi: 10.1016/j.ecolmodel.2005.07.006.

Boulton, A. J., & Lake, P. S., (2008). Effects of drought on stream insects and its ecological consequences. in: Lancaster, J., & Briers, R. A., (eds.), *Aquatic Insects Challenges to Populations Proceedings of the Royal Entomological Society's 24th Symposium* (pp. 81–102). CAB International CABI, Cambridge USA.

BROWN, J. H., Gillooly, J. F., Allen, A. P., Savage, V. M., & West, G. B., (2004). Perspectives. *Ecology, 85*, 1771–1789.

Brown, L. E., Hannah, D. I. D. M., Lner, A. E. X. R. M., (2007). Vulnerability of alpine stream biodiversity to shrinking glaciers and snowpacks. *Global Change Biology, 13*, 958–966. doi: 10.1111/j.1365-2486.2007.01341.x.

Brunetti, M., Buffoni, L., Mangianti, F., Maugeri, M., & Nanni, T., (2004). Temperature, precipitation and extreme events during the last century in Italy. *Global and Planetary Change, 40*, 141–149. doi: 10.1016/S0921-8181(03)00104-8.

Buisson, L., Thuiller, W., Lek, S., Lim, P., & Grenouillet, G., (2008). Climate change hastens the turnover of stream fish assemblages. *Global Change Biology, 14*, 2232–2248. doi: 10.1111/j.1365-2486.2008.01657.x.

Carraroa, L., Bertuzzob, E., Maric, L., Fontesd, I., Hartikainend, H., Strepparavaf, N., Schmidt-Posthausf, H., et al., (2017). Integrated field, laboratory, and theoretical study of PKD spread in a Swiss prealpine river. *Proceedings of Natural Academy of Sciences*, 1–6. doi: 10.1073/pnas.1713691114.

Carrizo, S. F., Jahnig, S. C., Bremerich, V., Freyhof, J., Harrison, I., HE, F., Langhans, S. D., et al., (2017). Freshwater megafauna: Flagships for freshwater biodiversity under. *BioScience, 67*, 919–927. doi: 10.1093/biosci/bix099.

Clausnitzer, V., Kalkman, V. J., Ram, M., Collen, B., Baillie, J. E. M., Darwall, W. R. T., Dijkstra, K. B., et al., (2009). Odonata enter the biodiversity crisis debate: The first global assessment of an insect group. *Biological Conservation Journal, 142*, 1864–1869. doi: 10.1016/j.biocon.2009.03.028.

Collen, B. E. N., Loh, J., Whitmee, S., McRae, L., Amin, R., & Baillie, J. E. M., (2009). Monitoring change in vertebrate abundanc: the living planet index. *Conservation Biology, 23*, 317–327. doi: 10.1111/j.1523-1739.2008.01117.x.

Comte, L., Buisson, L., Daufresne, M., & Grenouillet, G., (2013). Climate-induced changes in the distribution of freshwater fish: observed and predicted trends. *Fresh water Biology, 58*, 625–639. doi: 10.1111/fwb.12081.

Cooke, S. J., Allison, E. H., Beard, T. D., Arlinghaus, R., Arthington, A. H., Bartley, D. M., Cowx, I. G., et al., (2016). On the sustainability of inland fisheries: Finding a future for the forgotten. *Ambio*, 753–764. doi: 10.1007/s13280-016-0787-4.

Cumberlidge, N., Ng, P. K. L., Yeo, D. C. J., Magalhães, C., Campos, M. R., Alvarez, F., Naruse, T., et al., (2009). Freshwater crabs and the biodiversity crisis: Importance, threats, status, and conservation challenges. *Biological Conservation, 142*, 1665–1673. doi: 10.1016/j.biocon.2009.02.038.

Dallas, H. F., & Rivers-Moore, N., (2014). Ecological consequences of global climate change for freshwater ecosystems in South Africa. *Climate Change and Freshwater Ecosystems, 110*, 1–11.

Darwall, W. R. T., Holland, R. A., Smith, K. G., Allen, D., Brooks, E. G. E., Katarya, V., Pollock, C. M., et al., (2011). Implications of bias in conservation research and investment for freshwater species. *Conservation Letters, 4*, 474–482. doi: 10.1111/j.1755-263X.2011.00202.x.

Daufresne, M., Roger, M., Capra, H., & Lamouroux, N., (2003). Long-term changes within the invertebrate and fish ˆ ne river: effects of communities of the Upper Rho climatic factors. *Global Change Biology, 10*, 124–140. doi: 10.1046/j.1529-8817.2003.00720.x.

Di Marco, M., Chapman, S., Althor, G., Kearney, S., Besancon, C., Butt, N., Maina, J. M., et al., (2017). Changing trends and persisting biases in three decades of conservation science. *Global Ecology and Conservation, 10*, 32–42. doi: 10.1016/j.gecco.2017.01.008.

Dodds, W. K., Perkin, J. S., & Gerken, J. E., (2013). Human impact on freshwater ecosystem services: A global perspective. *Environmental Science and Technology, 47*, 9061–9068.
Dudgeon, D., (2019). Multiple threats imperil freshwater biodiversity in the Anthropocene. *Current Biology, 29,* R960–R967. doi: 10.1016/j.cub.2019.08.002.
Dudgeon, D., Arthington, A. H., Gessner, M. O., Kawabata, Z., Knowler, D. J., Leveque, C., Naiman, R. J., et al., (2006). Freshwater biodiversity: importance, threats, status and conservation challenges. *Biological Reviews, 81*, 163–182. doi:10.1017/S1464793105006950.
Durance, I., & Ormerod, S. J., (2007). Climate change effects on upland stream macroinvertebrates over a 25-year period. *Global Change Biology, 13*, 942–957. doi: 10.1111/j.1365-2486.2007.01340.x.
Eissa, A. E., & Zaki, M. M., (2011). The impact of global climatic changes on the aquatic environment. *Procedia Environmental Sciences*, 4. doi: 10.1016/j.proenv.2011.03.030.
Fenoglio, S., Bo, T., Cucco, M., Mercalli, L., & Malacarne, G., (2010). Effects of global climate change on freshwater biota: A review with special emphasis on the Italian situation. *Italian Journal of Zoology*, 77. doi:10.1080/11250000903176497.
Fenoglio, S., Tiziano, B., & Bosi, G., (2006). Deep interstitial habitat as a refuge for *Agabus paludosus* (Fabricius) (Coleoptera: Dytiscidae) during summer droughts. *The Coleopterists Bulletin, 1*, 37–41. doi: 10.1649/842.1.
Fenoglio, S., Tiziano, B., Battegazzore, M., & Morisi, A., (2005). Growth of Oligoneuriella Rhenana (Imhoff, 1852) (Ephemeroptera: Oligoneuriidae) in Two Rivers with Contrasting Temperatures in NW Italy. *Zoological Studies, 44*(2), 271–274.
Fenoglio, S., Tiziano, B., Figueroa, J. M. T. De., & Cucco, M., (2008). Nymphal growth, life cycle, and feeding habits of *Potamanthus luteus* (Linnaeus, 1767) (*Insecta*: *Ephemeroptera*) in the Bormida River, Northwestern Italy. *Zoological Studies, 47*, 185–190.
Flanagan, K. M., Mccauley, E., Wrona, F., & Prowse, T., (2003). Climate change: the potential for latitudinal effects on algal biomass in aquatic ecosystems. *Canadian Journal of Fisheries Aquatic Sciences, 60*, 635–639. doi: 10.1139/F03-062.
Flato, G. M., & Boer, G. J., (2001). Asymmetry in Climate Change Simulations. Geophysical research letters, 28, 195–198.
Fourtune, L., Paz-vinas, I., Loot, G., Prunier, J. G., & Blanchet, S., (2016). Lessons from the fish: a multi-species analysis reveals common processes underlying similar species-genetic diversity correlations. *Fresh Water Biology, 8*, 1–17. doi: 10.1111/fwb.12826.
Fuente, M. D. L., Bonada, N., Beche, L., Dahm, C. N., Mendez, P. K., Tockner, K., Uehlinger, U., & Vicenc, A., (2018). Evolutionary responses of aquatic macroinvertebrates to two contrasting flow regimes. *Hydrobiologia, 808*, 353–370. doi: 10.1007/s10750-017-3437-3.
Gallardo, B., Clavero, M., Sanchez, M. I., & Vila, M., (2016). Global ecological impacts of invasive species in aquatic ecosystems. *Global Change Biology, 22*, 151–163. doi: 10.1111/gcb.13004.
Geller, W., & Schultze, M., (2009). *Pollution and Remediation.* Encyclopedia of Inland waters.
Gibert, J., & Deharveng, L., (2002). Subterranean ecosystems: A truncated functional biodiversity. *BioScience, 52*, 473–481.

Gomi, T., Nagasaka, M., Fukuda, T., & Hagihara, H., (2007). Shifting of the life cycle and life-history traits of the fall webworm in relation to climate change. *Entomologia Experimentalis et Applicata, 125*, 179–184. doi: 10.1111/j.1570-7458.2007.00616.x.

Grant, E. H. C., Lowe, W. H., & Fagan, W. F., (2007). Review and living in the branches: population dynamics and ecological processes in dendritic networks. *Ecology Letters, 10*, 165–175. doi: 10.1111/j.1461-0248.2006.01007.x.

Green, P. A., Vörösmarty, C. J., Harrison, I., Farrell, T., Sáenz, L., & Fekete, B. M., (2015). Freshwater ecosystem services supporting humans: Pivoting from water crisis to water solutions. *Global Environmental Change, 34*, 108–118. doi: 10.1016/j.gloenvcha.2015.06.007.

Hari, R. E., Livingstone, D. M., Sibber, R, Holm, P. B., & Guttinger, H., (2006). Consequences of climatic change for water temperature and brown trout populations in Alpine rivers and streams. Global Change Biology 12, 10–26. doi:10.1111/j.1365-2486.2005.01051.x

Havens, K. E., Sharfstein, B., Brady, M. A., East, T. L., Harwell, M. C., Maki, R. P., & Rodusky, A. J., (2004). Recovery of submerged plants from high water stress in a large subtropical lake in Florida, USA. *Aquatic Botany, 78*, 67–82. doi: 10.1016/j.aquabot.2003.09.005.

Heino, J., (2001). Regional gradient analysis of freshwater biota : do similar biogeographic patterns exist among multiple taxonomic groups ? *Journal of Biogeography*, 69–76.

Heino, J., Virkkala, R., & Toivonen, H., (2009). Climate change and freshwater biodiversity: detected patterns, future trends and adaptations in northern regions. *Biological Reviews, 84*, 39–54. doi: 10.1111/j.1469-185X.2008.00060.x.

Hickling, R., Roy, D. B., Janek, H., Fox, R., & Thomas, C., (2006). The distributions of a wide range of taxonomic groups are expanding Polewards. *Global Change Biology, 12*, 450–455. doi: 10.1111/j.1365-2486.2006.01116.x.

Hogg, I. D., & Williams, D. D., (1996). Response of stream invertebrates to a global-warming thermal regime: an ecosystem-level manipulation. *Ecology, 77*, 395–407.

Holzapfel, A. M., & Vinebrooke, R. D., (2015). Environmental warming increases invasion potential of alpine lake communities by imported species. *Global Change Biology, 11*, 2009–2015. doi: 10.1111/j.1365-2486.2005.01057.x.

Hotaling, S., Finn, D. S., Giersch, J. J., Weisrock, D. W., & Jacobsen, D., (2017). Climate change and alpine stream biology: progress, challenges, and opportunities for the future. *Biological Reviews, 10*, 1–21. doi: 10.1111/brv.12319.

Huntington, T. G., (2006). Evidence for intensification of the global water cycle: Review and synthesis. *Journal of Hydrology, 319*, 83–95. doi: 10.1016/j.jhydrol.2005.07.003.

Jacobsen, D., Milner, A. M., Brown, L. E., & Dangles, O., (2012). Biodiversity under threat in glacier-fed river systems. *Nature Climate Change, 2*, 361–364. doi: 10.1038/nclimate1435.

Jelks, H. L., Walsh, S. J., Burkhead, N. M., Contreras-balderas, S., Díaz-pardo, E., Hendrickson, D. A., Lyons, J., et al., (2008). Conservation status of imperiled North American freshwater and diadromous fishes. *Fisheries, 33*, 372–407.

Jenkins, M., (2003). *Prospects for Biodiversity. Science, 302*, 1775–1777. doi: 10.1126/science.1088666.

Judson, W. S., & Nelson, C. R., (2011). Diversity, phenology, and elevational distribution of *Ephemeroptera, Plecoptera*, and *Trichoptera* in American Fork Canyon, Utah American fork canyon, utah. *BioOne, 70*, 526–540.

Knouft, J. H., & Anthony, M. M., (2016). Climate and local abundance in freshwater fishes subject category: Subject Areas: Author for correspondence. *Royal Society Open Science, 3*, 1–9.

Knouft, J. H., & Ficklin, D. L., (2017). The potential impacts of climate change on biodiversity in flowing freshwater systems. *Annual Review of Ecology, Evolution, and Systematics*, 111–133.

Knouft, J. H., & Page, L. M., (2011). Assessment of the relationships of geographic variation in species richness to climate and landscape variables within and among lineages of North American freshwater fishes. *Journal of Biogeography, 38*, 2259–2269. doi: 10.1111/j.1365-2699.2011.02567.x.

Kovach, R. P., Dunham, J. B., Al-Chokhachy, R., Snyder, C. D., Letcher, B. H., Young, J. A., Beever, E. A., et al., (2019). An integrated framework for ecological drought across riverscapes of North America. *BioScience, XX*, 1–14. doi: 10.1093/biosci/biz040.

Lake, P., (2000). Disturbance, patchiness, and diversity in streams. *Journal of the North American Benthological Society, 19*, 573–592. doi:10.2307/1468118.

Lama, S., & Devkota, B., (2016). Vulnerability of mountain communities to climate change and adaptation strategies. *Journal of Food Agriculture and Environment, 10*, 65–71. doi: 10.3126/aej.v10i0.2133.

Longhi, M. L., & Beisner, B., (2014). Environmental factors controlling the vertical distribution of phytoplankton in lakes. *Journal of Plankton Research, 31*, 1195–1207. doi:10.1093/plankt/fbp065.

López-rodríguez, Figueroa, J. M. T. D., & Alba-tercedor, J., (2009). The life history of *Serratella ignita* (Poda, 1761) (*Insecta: Ephemeroptera*) in a temporary and permanent Mediterranean stream. *Quatic Science, 71*, 179–188. doi: 10.1007/s00027-009-9211-5.

Lytle, D. A., (2014). Flash floods and aquatic insect life-history evolution: Evaluation of multiple models flash floods and aquatic insect life-history evolution. *Ecology, 83*, 370–385.

Magnuson, J. J., Webster, K. E., Assel, Y. R. A., Bowser, C. J., Dillon, P. J., Eaton, J. G., Evans, H. E., et al., (1997). Potential effects of climate changes on aquatic systems: Laurentian great lakes and Precambrian shield region. *Hydrological Processes*, 11.

Martens, P., & Rotmans, J., (2005). Transitions in a globalizing world. *Futures, 37*, 1133–1144. doi: 10.1016/j.futures.2005.02.010.

Mauget, S. A., (2003). Multidecadal regime shifts in U. S. streamflow, precipitation, and temperature at the end of the twentieth century. *Journal of Climate, 16*, 3905–3916.

McCann, E. L., Johnson, N. S., & Pangle, K. L., (2017). Corresponding long-term shifts in stream temperature and invasive fish migration. *Canadian Journal of Fisheries Aquatic Sciences Canadian, 2*, 1–25.

Mekonnen, M. M., & Hoekstra, A. Y., (2016). Four billion people facing severe water scarcity. *Science Advances, 2*(2), 1–7.

Menendez, R., Megias, G. A., Hill, J. K., Braschler, B., Willis, S. G., Collingham, Y., Fox, R., et al., (2006). Species richness changes lag behind climate change. *Proc. R. Soc. B 273*, 1465–1470. doi: 10.1098/rspb.2006.3484.

Miller, A. M., & Golladay, S. A., (1996). Effects of spates and drying on macroinvertebrate assemblages of an intermittent and a perennial prairie stream. *Journal of the North American Benthological Society.* doi: 10.2307/1467815.

Monchamp, M., Spaak, P., Domaizon, I., Dubois, N., Bouffard, D., & Pomati, F., (2017). Homogenization of lake cyanobacterial communities over a century of climate change and eutrophication. *Nature Ecology & Evolution, 2*, 317–324. doi: 10.1038/s41559-017-0407-0.

Mooney, H., Larigauderie, A., Cesario, M., Elmquist, T., Hoegh-guldberg, O., Lavorel, S., Mace, G. M., et al., (2009). Biodiversity, climate change, and ecosystem services. *Current Opinion in Environmental Sustainability, 1*, 46–54. doi: 10.1016/j.cosust.2009.07.006.

Mouthon, J., & Daufrresne, M., (2006). Effects of the 2003 heatwave and climatic warming on ˆ ne: A large lowland river mollusc communities of the Sao and of its two main tributaries (France). *Global Change Biology, 12*, 441–449. doi: 10.1111/j.1365-2486.2006.01095.x.

Nelson, D., Benstead, J. P., Huryn, A. D., Cross, W. F., Hood, J. M., Johnson, P. W., Junker, J. R., Gíslason, G. M., & Ólafsson, J. S., (2017). Experimental whole-stream warming alters community size structure. *Global Change Biology, 23*, 2618–2628. doi: 10.1111/gcb.13574.

O'Gorman, E. J., Zhao, L., Pichler, D. E., Adams, G., Friberg, N., Rall, B. C., Seeney, A., et al., (2017). Unexpected changes in community size structure in a natural warming experiment. *Nature Climate Change Letters, 38*, 1–7. doi:10.1038/nclimate3368.

Pachauri, R. K., Leo, M., (2014). Climate Change 2014 *Synthesis Report*. Contribution of working groups I, II and III to the Fifth Assessment Report of the Intergovernmental Panel on Climate Change Geneva, Switzerland, 1–51.

Parmesan, C., (2006). Ecological and evolutionary responses to recent climate change. *Annu. Rev. Ecol. Evol. Syst., 37*, 637–671. doi: 10.1146/annurev.ecolsys.37.091305.110100.

Paz-vinas, I., LOOT, G., Stevens, V. M., & Blanchet, (2015). Evolutionary processes driving spatial patterns of intraspecific genetic evolutionary processes driving spatial patterns of intraspecific genetic diversity in river ecosystems. *Molecular Biology, 24*, 4586–4604. doi: 10.1111/mec.13345.

Pearson, K. D., (2019). Spring- and fall-flowering species show diverging phenological responses to climate in the Southeast USA. *International Journal of Biometeorology, 2*, 1–10.

Platt, T., Fuentes-Yaco, C., & Frank, K. T., (2003). Spring algal bloom and larval fish survival. *Nature, 423*, 398–399.

Poff, N. L., Brinson, M. M., & Day, J. W., Jr. (2002). Aquatic ecosytems & global climate change: Potential impacts on inland freshwater and coastal wetland ecosystems in the United States. Retrieved from https://www.pewtrusts.org/-/media/ legac y/uploadedfi les/ wwwpewtrustsorg/reports/protecting_ocean_life/envclimate aquaticecosystemspdf.pdf.

Post, E., (2016). Implications of earlier sea ice melt for phenological cascades in arctic marine food webs. *FOOWEB, 13*, 60–66. doi: 10.1016/j.fooweb.2016.11.002.

Rahel, F. J., & Olden, J. D., (2008). Assessing the effects of climate change on aquatic invasive species. *Climate Change and Aquatic Invasive Species, 22*, 521–533. doi: 10.1111/j.1523-1739.2008.00950.x.

Rahel, F. J., (2002). *Omogenization of Annual Reviews of Ecological Systems, 33*, 291–315. doi: 10.1146/annurev.ecolsys.33.010802.150429.

Reid, A. J., Carlson, A. K., Creed, I. F., Eliason, E. J., Gell, P. A., Johnson, P. T. J., Kidd, K. A., et al., (2018). Emerging threats and persistent conservation challenges for freshwater biodiversity. *Biological Reviews*, 1–18. doi:10.1111/brv.12480.

Reist, J. D., Wrona, F. J., Prowse, T. D., Power, M., Dempson, J. B., Beamish, R. J., King, J. R., et al., (2006). General Effects of Climate Change on Arctic Fishes and Fish Populations. *Ambio,* 35. doi:10.1579/0044-7447(2006)35.

Ricciardi, A., & Rasmussen, J. B., (1999). Extinction rates of North American freshwater Fauna. *Conservation Biology, 13*, 1220–1222.

Richter, B. D., Braun, D. P., Mendelson, M. A., & Master, L. L., (1997). Threats to imperiled freshwater Fauna. *Conservation Biology, 11*, 1081–1093.

Richter, O., Suhling, F., Muller, O., & Dietrich, K., (2008). A model for predicting the emergence of dragonflies in a changing climate. *Fresh water Biology, 53*, 1868–1880. doi: 10.1111/j.1365-2427.2008.02012.x.

Robinson, C. T., Thompson, C., & Freestone, M., (2014). rapid glacial recession. *Fundam. Appl. Limnol., 18*, 235–246. doi: 10.1127/fal/2014/0667.

Rosset, V., & Oertli, B., (2011). Freshwater biodiversity under climate warming pressure : Identifying the winners and losers in temperate standing waterbodies. *Biological Conservation, 144*, 2311–2319. doi: 10.1016/j.biocon.2011.06.009.

Rudolf, V. H. W., (2019). The role of seasonal timing and phenological shifts for species coexistence. *Conservation Letters, 22*, 1324–1338. doi: 10.1111/ele.13277.

Sala, O. E., Iii, F. S. C., Armesto, J. J., Berlow, E., Bloomfield, J., Dirzo, R., Huber-sanwald, E., et al., (2000). Global Biodiversity Scenarios for the Year 2100. *Science*, 1770–1774. doi: 10.1126/science.287.5459.1770.

Salmaso, N., Boscaini, A., Capelli, C., & Cerasino, L., (2017). Ongoing ecological shifts in a large lake are driven by climate change and eutrophication: evidences from a three-decade study in Lake Garda. *Hydrobiologia, 4*, 1–19. doi:10.1007/s10750-017-3402-1.

Schindler, D. W., (2001). The cumulative effects of climate warming and other human stresses on Canadian freshwaters in the new millennium. In: Bendell-young, L., & Gallaugher, P., (eds.), *Waters in Peril* (pp. 18–29). Springer, Boston, MA.

Short, F. T., Sarian, K., Morgan, P. A., Malone, S., & Moore, G. E., (2016). Impacts of climate change on submerged and emergent wetland plants. *Aquatic Botany, 135*, 3–17. doi: 10.1016/j.aquabot.2016.06.006.

Sodhi, N. S., Sekercioglu, C. H., Jos, B., & Robinson, S. K., (2011). Conservation of Tropical Birds, (I[st] edn.). Wiley-Blackwell, West Sussex, UK.

Soria, M., Leigh, C., Datry, T., Bini, L. M., & Bonada, N., (2017). Biodiversity in perennial and intermittent rivers: a meta-analysis. *Okios, 126*, 1078–1089.

Strayer, D. L., & Dudgeon, D., (2015). Freshwater biodiversity conservation: Recent progress and future challenges Freshwater biodiversity conservation : recent progress and future challenges. *Journal of the North American Benthological Society*, 29. doi: 10.1899/08-171.1.

Sukenik, A., Quesada, A., & Salmaso, A., (2015). Global expansion of toxic and non-toxic cyanobacteria: effect on ecosystem functioning. *Biodivers Conserv, 24*, 889–908. doi:10.1007/s10531-015-0905-9.

Suren, A. M., & Jowett, I. G., (2006). Effects of floods versus low flows on invertebrates in a new Zealand gravel- effects of floods versus low flows on invertebrates in a New Zealand gravel-bed river. *Freshwater Biology, 51*, 2207–2227. doi: 10.1111/j.1365-2427.2006.01646.x.

Svenning, J., Kerr, J., & Rahbek, C., (2009). Predicting future shifts in species diversity. *Ecography, 32*, 4, 5. doi: 10.1111/j.1600-0587.2009.06024.x.

Taylor, C. A., Schuster, G. A., Cooper, J. E., Distefano, R. J., Eversole, A. G., Hamr, P., Hobbs, H. H., Robison, H. W., Skelton, C. R., & Thoma, R. F., (2007). A Reassessment of the Conservation Status of Crayfishes of the United States and Canada agter 10 + years of Increased Awareness Fisheries 32:8, 372–389. doi:10.1577/1548-8446(2007)32.

Thomas, C. D., Cameron, A., Green, R. E., Bakkenes, M., Beaumont, L. J., Collingham, Y. C., & Erasmus, B. F. N., (2004). Extinction risk from climate change. *Letters of Nature, 247*, 145–148.

Thomas, C. D., Franco, A. M. A., & Hill, J. K., (2006). Range retractions and extinction in the face of climate warming. *Trends in Ecology and Evolution, 21*, 7–8. doi: 10.1016/j.tree.2006.05.012.

Tisseuil, C., Beauchard, O., Brosse, S., Holland, R., Hugueny, B., Tedesco, P. A., & Oberdorff, T., (2013). Global diversity patterns and cross-taxa convergence in freshwater systems. *Journal of Animal Ecology, 82*, 365–376. doi: 10.1111/1365-2656.12018.

Urban, M. C., Bocedi, G., Hendry, A. P., Mihoub, J. B., Pe'er, G., Singer, A., Bridle, J. R., et al., (2016). Improving the forecast for biodiversity under climate change. *Science, 353*, 1113–1120. doi: 10.1126/science.aad8466.

Vega, G. C., & Wiens, J. J., (2012). Why are there so few fish in the sea ? *Proceedings of the Royal Society of London B: Biological Sciences, 279*, 2323–2329. doi:10.1098/rspb.2012.0075.

Visco, J. A., Laure, A., Cordonier, A., Esling, P., Pillet, L., & Pawlowski, J., (2015). Environmental monitoring: inferring diatom index from next-generation sequencing data. *Environmental Science and Technology, 49*, 7597–7605. doi: 10.1021/es506158m.

Visser, M. E., & Gienapp, P., (2019). Evolutionary and demographic consequences of phenological mismatches. *Nature Ecology & Evolution, 12*, 1–8. doi: 10.1038/s41559-019-0880-8.

Vorosmarty, C. J., McIntyre, P. B., Gessner, M. O., Dudgeon, D., Prusevich, A., Green, P., Glidden, S., et al., (2010). and river biodiversity. *Nature, 467*. doi:10.1038/nature09440.

Wann, G. T., Aldridge, C. L., Seglund, A. E., Mccance, S. J. O., Kondratieff, B. C., & Braun, C. E., (2019). Mismatches between breeding phenology and resource abundance of resident alpine ptarmigan negatively affect chick survival. *Ecology and Evolution, 9*, 7200–7212. doi:10.1002/ece3.5290.

Ward, J. M., & Ricciardi, A., (2007). Impacts of *Dreissena* invasions on benthic macroinvertebrate communities: a meta- analysis. *Diversity and Distributions, 13*, 155–165. doi:10.1111/j.1472-4642.2007.00336.x.

Wasmund, N., Nausch, G., Gerth, M., Busch, S., Burmeister, C., Hansen, R., & Sadkowiak, B., (2019). Extension of the growing season of phytoplankton in the western Baltic Sea in response to climate change. *Marine Ecology Progress Series, 622*, 1–16. doi: 10.3354/meps12994.

Weeks, B. C., Zimova, M., Ellis, A. A., Witynski, M. L., Hennen, M., & Winge, B. M., (2019). Shared morphological consequences of global warming in North American migratory birds. *Ecology Letters*, 1–10. doi:10.1111/ele.13434.

Weiskopf, S. R., Rubenstein, M. A., Crozier, L. G., Gaichas, S., Grif, R., Halofsky, J. E., Hyde, K. J. W., et al., (2020). Climate change effects on biodiversity, ecosystems, ecosystem services, and natural resource management in the United States. *Science of the Total Environment, 9*, 1–18. doi: 10.1016/j.scitotenv.2020.137782.

Winder, M., & Schindler, D. E., (2004). Climate change uncouples trophic interactions in an aquatic ecosystems. *Ecology, 85*, 2100–2106.

Woodward, G., Perkins, D. M., & Brown, L. E., (2010). Climate change and freshwater ecosystems : impacts across multiple levels of organization Climate change and freshwater ecosystems : impacts across multiple levels of organization. *Philosophical Transactions, 365*, 2093–2106. doi:10.1098/rstb.2010.0055.

Wrona, F. J., Prowse, T. D., Reist, J. D., Hobbie, J. E., & Le, L. M. J., (2006). Climate change effects on aquatic biota, ecosystem structure and Function. *AMBIO A Journal of the Human Environment, 35*, 359–369. doi:10.1579/0044-7447(2006)35.

WWF, (2020). *Living Planet Report 2020 Bending the Curve of Biodiversity Loss*. WWF International, Gland, Switzerland.

Xenopoulos, M. A., Lodge, D. I. D. M., Alcamo, J. O., Rker, M. M., Schulze, K. R., & Vuuren, D. T. L. E. F. P., (2005). Scenarios of freshwater fish extinctions from climate change and water withdrawal. *Global Change Biology, 11*, 1557–1564. doi: 10.1111/j.1365-2486.2005.01008.x.

Yankova, Y., Neuenschwander, S., Köster, O., & Posch, T., (2017). Abrupt stop of deep water turnover with lake warming: Drastic consequences for algal primary producers. *Scientific Reports, 7*, 1–10. doi: 10.1038/s41598-017-13159-9.

Zhang, Y., Jeppesen, E., Liu, X., Qin, B., Shi, K., Zhou, Y., Thomaze, S. M., & Denga, J., (2017). Global loss of aquatic vegetation in lakes earth-science reviews global loss of aquatic vegetation in lakes. *Earth-Science Reviews, 173*, 259–265. doi: 10.1016/j.earscirev.2017.08.013.

Zimova, M., Mills, L. S., & Nowak, J. J., (2016). High fitness costs of climate change-induced camouflage mismatch. *Ecology Letters, 19*, 299–307. doi: 10.1111/ele.12568.

CHAPTER 9

Phytoremediation Potential of *Spirodela polyrhiza* for Cd and Cr Under Hydroponic Culture

RIFAT ARA WANI,[1] BASHIR AHMAD GANAIE,[1] and MANZOOR AHMAD SHAH[2]

[1]*Center of Research for Development, University of Kashmir, Jammu and Kashmir, India, E-mail: rifatevs@gmail.com (R. A. Wani)*

[2]*Department of Botany, University of Kashmir, Jammu and Kashmir, India*

ABSTRACT

The present investigation examined the phytoremediation potential of *Spirodela polyrhiza*, a free-floating aquatic plant, for two heavy metals, including Cadmium (Cd) and Chromium (Cr), from the contaminated waters. During the investigation, the plant was exposed to different concentration ranges (2.0–6.0 ppm) of the two metals and harvested after 21 days for metal accumulation. Results revealed that *S. polyrhiza* could remove 57–85% Cadmium and 67–79% Chromium, inferring its higher accumulation potential for Chromium over Cadmium. The plant was further found to be a moderate accumulator of heavy metals. The Pigment contents were seen decreasing with increasing concentration of both Cd and Cr, which indicates inhibition of photosynthesis due to higher metallic concentration.

Biodiversity of Freshwater Ecosystems: Threats, Protection, and Management. Suhaib A. Bandh, PhD, Basharat Mushtaq, PhD, Mohammad Farooq Mir, PhD, & Shahid A. Wani (Eds.)

9.1 INTRODUCTION

Pollution of the aquatic systems due to different human activities like growing industrialization, rapid urbanization, overuse of pesticides, detergents, and agricultural chemicals, excessive generation of liquid and solid wastes has become a widespread problem (Bashir et al., 2020). This problem is mainly manifested through eutrophication and heavy metal pollution of the natural resources of water. Consequently, the pollutants generated from different sources are of prime concern, not only from the viewpoint of human health but also their impacts on the ecosystem integrity and loss of biodiversity (Duruibe et al., 2007). Immediate action is therefore needed to remediate such polluted systems and to improve their deteriorating health. Over the past decades, many conventional and novel methods like alkaline precipitation, ion exchange columns, electrochemical removal, filtration, and membrane technologies (Kamel, 2013) have been applied to treat the negative impacts of heavy metal contamination and the subsequent eutrophication. However, none of them have shown promise. Phytoremediation, therefore, has become a preferred method because of its cost-effectiveness, eco-friendly nature, and simple maintenance (Klink, 2017; Newete and Byrne, 2016).

Further, it is considered a green technique that enjoys popularity among the general public (Manorama et al., 2020). In the 12th International Phyto-technology conference held in Manhattan, Kansas, in 2015, phytoremediation was considered an essential technological component for achieving the sustainable development goals (SDGs) (Erickson, 2017). This technique is used not only to remediate the remediation of common organic and inorganic pollutants but are also to detoxify the lethal heavy metals from the contaminated sites. Besides these commonly occurring pollutants, the water bodies are infected by heavy metals with diverse negative impacts on these systems. Cd and Cr are the two extremely harmful heavy metals in terms of their detrimental effect. They do not undergo biodegradation and accumulate in the living tissues resulting in many diseases and disorders even at relatively lower concentrations (Godwill et al., 2019). Cadmium, the most dangerous metal ion, characterized by high stability and toxicity, is also known to bind with essential respiratory enzymes (Nies, 2003), causing oxidative stress and cancer (Banjerdkji et al., 2005), while chromium (Cr) the 17th most abundant element in the

Earth's mantle presents different chemical, toxicological, and epidemiological characteristics (Kimbrough et al., 1999).

The Dal Lake is an ecologically and socio-economically important water body in the heart of the historical city of Srinagar. The water quality of this lake has been a significant concern for many years which is worsening due to rapid and unplanned urbanization, large quantities of raw sewage and other wastes that are discharged into the lake (Pandit, 1993; Qadri and Yousuf, 2008). Studies have shown that the lake supports a huge macrophytic diversity (Zutshi and Vass, 1982; Pandit, 1984), dominated by the species belonging to *Myriophyllum*, *Typha*, *Polygonum*, *Nelumbo*, *Trapa*, *Ceratophyllum*, *Potamogeton*, and *Spirodela*, etc., which have the potential of being used effectively in the process of phytoremediation. Accumulation of metals by these aquatic plants has opened novel gateways for phytoremediation and bio-indication (Favas et al., 2012). Over the last few decades or so, different workers have carried out plenty of work on the various ecological aspects of the lakes (Yaqoob and Pandit, 2009; Saba and Wanganeo, 2010; Bandh et al., 2016). However, very few research reports have been published regarding the phytoremediation potential of macrophytes found in the lake. In this context, the current study was taken up to assess the heavy metal uptake potential of *Spirodela polyrhiza*, commonly known *as* green duckweed, common duckweed, and duck meat. It is a perennial, mat-forming, floating aquatic plant growing in dense and abundant colonies in the lakes (Zimmo, 2003). Due to its different characteristics like fast growth (Wang et al., 2014) and easy harvesting, it is considered an ideal bioremediation and biofuel generation system. It is also considered among the most standardized test organisms in aquatic ecotoxicology (EPA, 1996; Eberius, 2001; OECD, 2002).

9.2 MATERIAL AND METHODS

9.2.1 HYDROPONIC CULTURE CONDITIONS

Before using the lake water to prepare hydroponics, its chemical composition was analyzed (Table 9.1). The desired quantities of Cadmium chloride ($CdCl_2 . 2H_2O$) and Potassium dichromate ($K_2Cr_2O_4$) salts were dissolved in distilled water separately to achieve the appropriate contamination levels of Cd and Cr. Working metal and standard solutions with varying concentrations of each metal (2, 4, and 6 ppm) were formulated by diluting the corresponding stock solution (1,000 mg L^{-1}) using deionized water.

TABLE 9.1 Chemical Composition Analysis of Lake Water

Parameters	Results of Chemical Analysis
pH	7.2
Total dissolved solids, mg/L	73.0
Dissolved oxygen mg/L	7.9
Total hardness (as $CaCO_3$), mg/L	105.0
Ca, mg/L	27.0
Mg, mg/L	9.2
Na mg/L	8.0
Potassium (K)	1.5
Chloride (Cl)	17.7
Ammonical nitrogen µg/L	54.0
Nitrate nitrogen µg/L	215.0
Total-phosphorus µg/L	241.0
Cadmium (Cd) mg/L	BDL
Chromium (Cr) mg/L	BDL

Note: BDL: Below detection level.

All the listed parameters were analyzed as per APHA (2005) methods.

9.2.2 PLANT SAMPLING

Samples of *Spirodela polyrhiza* were collected from the lake in polythene bags, washed several times with tap water and deionized water to remove the adhering soil, and transferred to the laboratory. For experimental setup, carried out in replicates, uniform size plants were selected and acclimatized for 15 days in acid cleaned plastic vessels containing lake water as culture media. The required quantity of acclimatized *S. polyrhiza* was transferred to different plastic containers containing Cd and Cr (2.0, 4.0, and 6.0 ppm) spiked lake water under natural conditions. A control setup was also used where the same quantity of plants was cultured in lake water without heavy metals. The experiments that continued for around 21 days were carried out at an average 11 hour day length and an average temperature of about 27±2°C. In the end, the plant samples were harvested from each enclosure and washed three times with deionized water. The washed samples were oven-dried at 60°C, weighed, and grounded to pass a 40 mesh screen using a Wiley mill.

9.2.3 HEAVY METAL ANALYSIS

The powdered plant samples were acid digested using an acid mixture of HNO_3: $HClO_4$: H_2SO_4 (10:4:1) at about 80°C. The acid digested samples were analyzed for Cd and Cr using Atomic Absorption Spectrophotometer (Perkin Elmer Model Analyst 800). The water samples collected from each enclosure were also digested with HNO_3 at 95°C for 2 hours and analyzed for Cd and Cr using AAS (Gupta, 2004).

9.2.4 CHLOROPHYLL ESTIMATION

Chlorophyll (Chl. a, Chl. b and Total Chlorophyll) content was estimated Spectrophotometrically in acetone extracts prepared from fresh samples (Lichtenthaler and Wellburn, 1983). Acetone extract of the plants was prepared by macerating 1 g of fresh weight of the plant leaves with 80% acetone using a mortar and pestle. The decanted suspension was centrifuged for 3 minutes at 1,320 rpm, and the upper green clear solution was decanted from the colorless residue. The volume was raised up to 10 ml with 80% acetone in a test tube, and the solution was subjected to centrifugation at 10,000 rpm for 10 min. The absorbance of the solution was determined using a spectrophotometer at wavelengths of 665 and 649 nm, respectively, and the results were expressed in μg/ml as follows:

$$\text{Chlorophyll a } (\mu g/ml) = 11.63\,(A_{665}) - 2.39\,(A_{649})$$
$$\text{Chlorophyll b } (\mu g/ml) = 20.11\,(A_{649}) - 5.18\,(A_{665})$$
$$\text{Total chlorophyll} = 6.45\,(A_{665}) + 17.72\,(A_{649})$$

9.2.5 DATA ANALYSIS

Data were presented as mean ± standard error of all the replicates. Removal of metals (Cd and Cr) was calculated on a percentage basis using the following formula (Chaudhuri et al., 2011; Tanhan et al., 2007).

$$R\% = \left[\frac{(Co - Ct)}{Co}\right] \times 100$$

where; C_o and C_t are initial and final concentrations of metal in medium (ppm).

9.2.6 BIOCONCENTRATION FACTOR

The bio-concentration factor (BCF) used to evaluate sample plants potentiality to accumulate metals with respect to metal concentration in the surrounding media was calculated by using the following formula (Zayed et al., 1998):

$$BCF = \frac{Concentration\,of\ metal\ in\ plant\ tissue\,(mg\,/\,kg)}{Initial\ metal\ concentraion\ in\,external\ solution\,(\mathrm{mg\,/\,L})}$$

9.2.7 UPTAKE INDEX

The metal uptake index was calculated as per the following formula (Chorom et al., 2012):

$$Uptake\ index = \frac{Metal\ in\,biomass\,(mg\,/\,g)}{Plant\,biomass\,(\mathrm{g})}$$

9.2.8 ENRICHMENT FACTOR (EF)

Enrichment factor (EF), measuring the degree of enrichment, was evaluated by the ratio of the metal in the plant (μg/g) to the metal in the water (mg/L) at the end of the experiment (Ahmad et al., 2014) using the following formula:

$$Enrichment\ Factor = \frac{Metal\ concentration\,in\ plant\,(\mathrm{\mu g\,/\,g})}{Metal\ concentration\,in\,external\ solution\,at\ the\,end\ of\ exp.\,(\mathrm{mg\,/\,L})}$$

9.3 RESULTS AND DISCUSSION

9.3.1 *HEAVY METAL REMOVAL EFFICIENCY OF S. POLYRHIZA*

The percentage removal of Cd and Cr by *S. polyrhiza* from different initial concentrations (Table 9.2) was seen to be influenced by their initial concentrations in the culture media, as it removed 85%, 68% and 56.33% of Cd and 77%, 72.25%, and 66.66 of Cr at their initial concentrations of 2.0, 4.0 and 6.0 ppm respectively. Results suggested that the test plant's overall removal efficiency was better for Cr compared to that of Cd except at two ppm. The amount of Cd and Cr removal by the test plant was higher at two ppm concentrations (85% for Cd and 77% for Cr) and decreased after that with the increase in metal concentration, which may be attributed to the toxicity of metal concentration itself. Similar observations were made by Kara and Kara (2005) while treating Cd from wastewater by *Lemna minor* and by Chaudhuri et al. (2014) while treating Cd spiked pond water by *Lemna minor* and *Spirodela polyrhiza.* The work carried on other plants like *Hydrilla verticillata* (Phukan et al., 2015), *Eichhornia crassipes* (Lu et al., 2005) corroborates with our results. Cadmium, although considered as non-essential for the everyday life of living organisms, has been recognized for its toxic effects on plants (Boddi et al., 1995). Likewise, Chromium interferes with several plant metabolic processes causing phytotoxicity such as reduced growth, chlorosis, and ultrastructural effects on organelles. It leads to plants' death (WHO, 1995).

9.3.2 *CADMIUM AND CHROMIUM ACCUMULATION IN PLANT BIOMASS*

In the test plant, the concentration of Cd and Cr was found to increase sharply with an increase in their concentration in the treated culture media. The metal amount accumulated in *S. polyrhiza* was 1.00 (±0.053) mg/g, 1.66 (±0.02) mg/g, and 2.03 (±0.015) mg/g when the medium was supplied with 2.0, 4.0, and 6.0 ppm of Cd and 0.94 (±0.04) mg/g, 1.72 (±0.02) mg/g and 2.52 (±0.02) mg/g when 2.0, 4.0 and 6.0 ppm of Cr was added to the medium (Figure 9.1). The metal uptake index, calculated for quantifying metal accumulation in the plant biomass (Figure 9.2) showed that the Cd uptake value for *S. polyrhiza* was between 0.59–1.24 mg, and Cr uptake

TABLE 9.2 Percentage Removal (R %) of Cd and Cr from Culture Media by *Spirodela polyrhiza**

Initial Metal Concentration	**Cd**			**Cr**		
Initial metal concentration in solution (ppm)	2	4	6	2	4	6
Final metal concentration in solution (ppm)	0.3 ± 0.061	1.28 ± 0.006	2.62 ± 0.021	0.46 ± 0.055	1.11 ± 0.026	2.0 ± 0.030
Percentage of Removal	85	68	56.33	77	72.25	66.66

*Observed values are the mean of triplicates. Metal levels in control groups have been subtracted in the calculation of R%.

value was between 0.58–1.62 mg. In the case of all the treatments, the highest value was attained at 6.0 ppm metal concentration. Results indicated considerable metal uptake values by *S. polyrhiza* both for Cd and Cr. Further, a strong inverse relationship was seen to exist between the two metals (Cd; $r = -0.969$; $p \leq 0.05$ and Cr; $r = -0.997$; $p \leq 0.05$) accumulated in *S. polyrhiza* and their residual concentration in culture media. The increase in accumulation was probably due to the availability of an increased number of binding sites for the complexation of heavy metal ions, which lead to the increased absorption by the test plant, as previously suggested by Rolli et al. (2013). The initial concentration of heavy metals in a given system is an essential factor influencing the plant accumulation capacity, which is often more potent in a higher metal concentration than in a lower metal concentration (Wang et al., 2014). In *S. polyrhiza*, the accumulation of Cd and Cr was found to be directly proportional to concentrations, and similar observations were reported by Garg and Chandra (1994) for Cd and Cr accumulation by a duckweed *Wolfia globose* and by Chaudhuri et al. (2014) for Cd accumulation by *Spirodela polyrhiza.* Further, reports for Cd accumulation by other aquatic plants like *Ceratophyllum demersum*, *Elodea canadensis* (Sun et al., 2011), *Pistia stratiotes* (Rolli et al., 2013) and for Cr accumulation by *Nymphaea alba* (Vajpayee et al., 2000) also supports our findings.

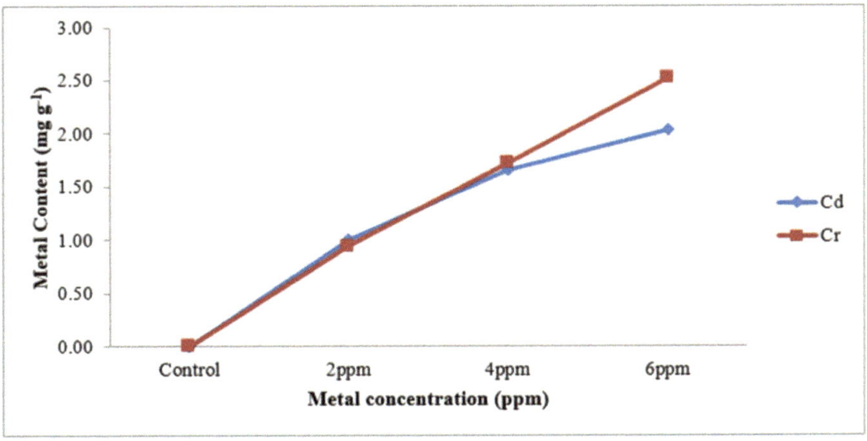

FIGURE 9.1 Metal content in *Spirodela polyrhiza* at different metal concentrations.

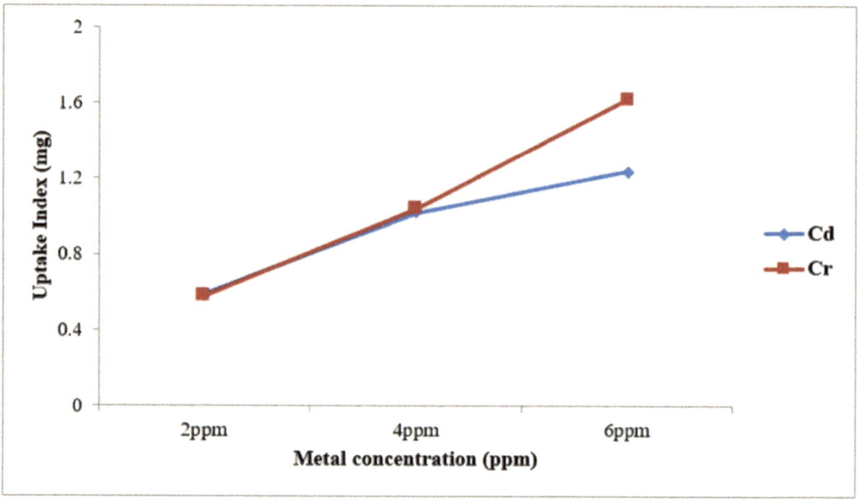

FIGURE 9.2 Cd and Cr uptake index in *Spirodela polyrhiza.*

The metal uptake depends on various factors, including the concentration of a contaminant, plant's ability to accumulate, metal availability, and interaction between plant habitat and climatic condition (Misra and Tripathi, 2008). To quantify metal accumulation in plant biomass, bioconcentration factor (BCF) is a more useful parameter than the amount accumulated in plants. It provides an index of the plants' ability to collect metals with respect to metal concentration in the surrounding media (Khellaf and Zerdaoui, 2009). The BCF calculation for Cd in *S. polyrhiza* was found to be 549 and 810, and Cr was found to be between 655.18 and 708.40, with the maximum value attainment at 2 ppm for both the metals (Figure 9.3). Similar observations were made by (Uysal and Taner, 2010; Chaudhuri et al., 2014), suggesting that the decrease of BCF value at higher Cd concentration was possible due to inhibition of growth of *Spirodela polyrhiza.* According to Razinger et al. (2007), a plant species that is considered a good accumulator must have a BCF over 1,000, indicating that *S. polyrhiza* can be regarded as a moderate accumulator in the case of the two metal under observation.

EF is another essential feature used to indicate the potential usefulness of the species for its metal sequestration from contaminated waters. The EF for Cd and Cr follows the same trend as that for BCF (Figure 9.4). The EF that must be greater than 1 to qualify to be called a better phytoremediator (Wei et al., 2006) was found high for both Cd and Cr in the test species allowing us to conclude that *S. polyrhiza* exhibits a better accumulation potential for Cr over Cd at all concentrations except at 2.0 ppm. Statistical analysis of the data revealed that the bioconcentration factor, EF, and removal percentage of heavy metals are positively related to each other but exhibit negative correlation with metal content in the plant and residual metal content in the culture media.

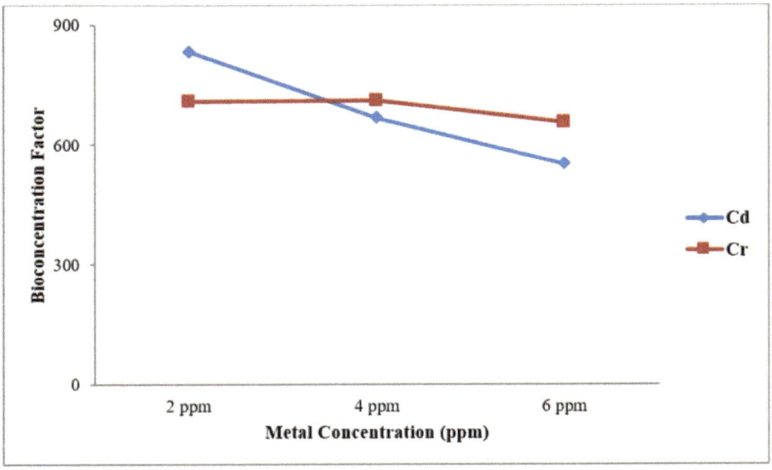

FIGURE 9.3 BCF of heavy metals in *Spirodela polyrhiza.*

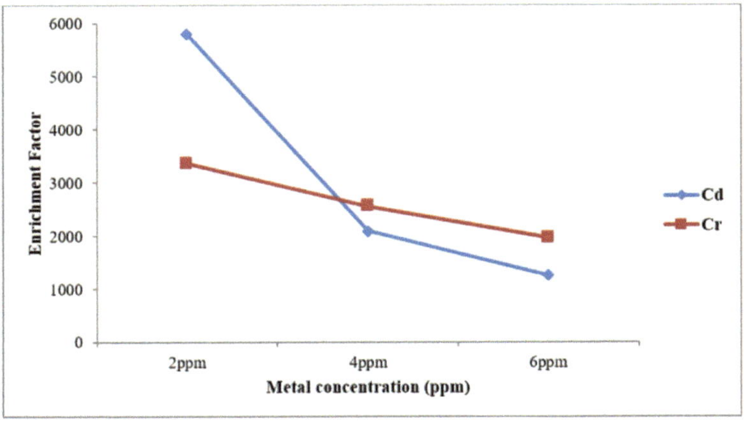

FIGURE 9.4 Enrichment factor in *Spirodela polyrhiza.*

9.3.3 *CHANGE IN CHLOROPHYLL CONTENT DUE TO CD AND CR TOXICITY*

The total chlorophyll content of the test plant revealed a gradual decrease with the increase in Cd and Cr concentration, as their toxicity caused a severe effect on its chlorophyll content. In the case of Cd, chlorophyll 'a' content of the plant sample reduced to 7.01 (±0.41) μgml^{-1} at 2 ppm, 5.79 (±0.28) μgml^{-1} at 4 ppm, and 4.4 (±0.15) μgml^{-1} at 6 ppm from 7.49 (±0.99) μgml^{-1}, the chlorophyll 'a' content of the control plant. Similarly, chlorophyll 'b' content decreased from 4.14 (±0.37) μgml^{-1} of the control plant to 3.01(±0.11) μgml^{-1} at 2 ppm, 2.67 (±0.01) μgml^{-1} at 4 ppm and 2.59 (±0.006) μgml^{-1} at 6 ppm, and the total chlorophyll content reduced to 10.03 (±0.95) μgml^{-1} at 2 ppm, 8.46 (±1.61) μgml^{-1} at 4 ppm and 6.99 (±0.98) μgml^{-1} at 6 ppm from 11.63 (±2.91) μgml^{-1}, the total chlorophyll content of the control plant (Figure 9.5). The trend shown by the test plant for Cr regarding the chlorophyll parameter was same as that for Cd (Figure 9.5). The decline in Chlorophyll content due to Cr and Cd toxicity may be attributed to the distortion of chloroplast ultrastructure, photosynthetic pigment biosynthesis inhibition, impaired δ-aminolevulinic acid dehydrates and protochlorophyllide reductase associated with chlorophyll biosynthesis and also to various abiotic

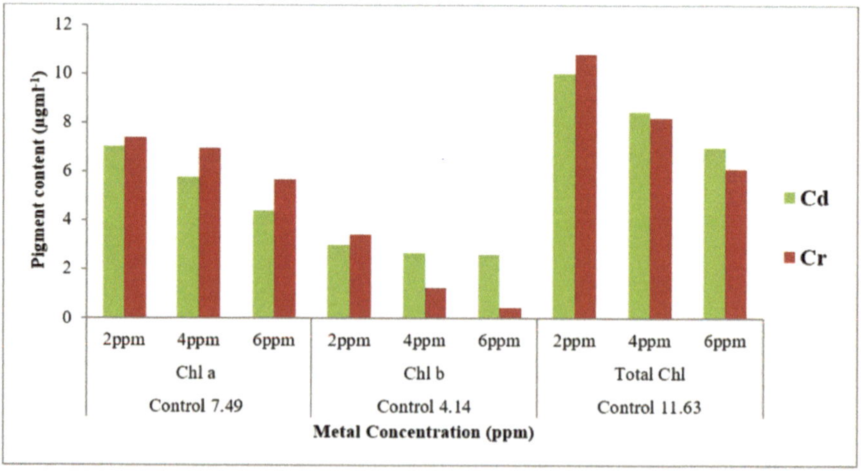

FIGURE 9.5 Change in chlorophyll content of *Spirodela polyrhiza* due to Cd and Cr.

stresses (Phukan et al., 2015; Vajpayee et al., 2000; Ouzounidou, 1995). Supporting our results (Gupta et al., 2011) observed that the *Vallisneria spiralis* and *Hydrilla verticillata* followed the same trend for chlorophyll content when exposed to different concentrations of treated tannery effluent collected from UASB, Jajmau, Kanpur. Similar observations were recorded by (Phetsombat et al., 2006) while studying Cd toxicity in *Salvinia cucullata*, who observed a significant decrease in total chlorophyll content with the increase exposure concentration. Reduction in chlorophyll content in plants treated with different metallic solutions has also been reported by several other authors (John et al., 2009; Wei et al., 2009; Hassan et al., 2007; Phukan et al., 2015).

9.4 CONCLUSIONS

Contamination of aquatic medium with toxic and undesirable heavy metals is a serious environmental problem that may be sustainably solved by phytoremediation approaches such as the one used in the current study wherein *S. polyrhiza* demonstrated a considerably high Cd and Cr removal efficiency, especially at lower initial concentrations. Excellent removal efficiency, high Cd, and Cr accumulation capacity, BCF values more than 500, and high EF values make *S. polyrhiza* a superb choice for phytoremediation of these heavy metals.

ACKNOWLEDGMENT

The authors acknowledge the Department of Environmental Science, the University of Kashmir, for the facilities provided to carry out the research work.

CONFLICT OF INTEREST

We declare that we don't have any conflict of interest.

KEYWORDS

- **bio-concentration factor**
- **cadmium**
- **chromium**
- **greenhouse gases**
- **phytoremediation**
- ***Spirodela polyrhiza***

REFERENCES

Ahmad, S. S., Reshi, Z. A., Shah, M. A., Rashid, I., Ara, R., & Andrabi, S. M. A., (2014). Phytoremediation potential of Phragmites australis in Hokersar wetland—A Ramsar site of Kashmir Himalaya. *Int. J. of Phytoremediation, 16*, 1183- 1191.

APHA, (2005). *Standard Methods for Examination of Water and Waste Water* (20th edn.). American Public Health Association, Washington DC USA.

Bandh, S. A., Kamili, A. N., Ganai, B. A., & Lone, B. A., (2016). Opportunistic fungi in lake water and fungal infections in associated human population in Dal Lake, Kashmir. *Microbial Pathogenesis, 93*, 105–110.

Banjerdkji, P., Vattanaviboon, P., & Mongkolsuk, S., (2005). Exposure to cadmium elevates expression of genes in the oxy R and Ohr R regulons and induces cross-resistance to peroxide killing treatment in *Xanthomona scampestris*. *Appl. Environ. Microbiol, 71*(4), 1843–1849.

Bashir, I., Lone, F. A., Bhat, R. A., Mir, S. A., Dar, Z. A., & Dar, S. A., (2020). Concerns and threats of contamination on aquatic ecosystems. *Bioremediation and Biotechnology: Sustainable Approaches to Pollution Degradation,* 1–26. https://doi.org/10.1007/978-3-030-35691-0_1.

Boddi, B., Oravez, A. R., & Lehozki, E., (1995). Effect of cadmium organization and photoreduction of photochlorophyllide in dark grown leaves and etioplast inner membrane preparation of wheat. *Photosynthetica, 3*, 411–420.

Chaudhuri, D., Goswami, C., Chatterjee, S., Bandyopadhyay, K., Majumder, A., & Misra, A. K., (2011). Phytoremediation of cadmium and nickel by *Spirodela polyrhiza. Indian J. Environ. Prot., 31*(9), 751–757.

Chaudhuri, D., Majumder, A., Misra, A. K., & Bandyopadhyay, K., (2014). Cadmium removal by *Lemna minor* and *Spirodela polyrhiza. Int. J. of Phytoremediation, 16*, 1119–1132. doi: 10.1080/15226514.2013.821446.

Chorom, M., Parnian, A., & Jaafarzadeh, N., (2012). Nickel removal by the aquatic plant (*Ceratophyllum Demersum L.*). *Inter. J. of Environ. Science and Development, 3*(4), 372–375.

Duruibe, J. O., Ogwuegbu, M. O. C., & Egwurugwu, J. N., (2007). Heavy metal pollution and human biotoxic effects. *Intl. J. of Physical Sciences, 2*(5), 112–118.

Eberius, M., (2001). *Assesment of Inhibition Values and Comprehensive Analysis of Biotests*. LemnaTec GmbH, Wurselen, online http://www.lemnatec.com/Literatur/LT003.pdf (accessed on 14 December 2021).

EPA, (1996). *Aquatic Plant Toxicity Test Using Lemna Spp: Tiers I and II- OPPTS 850.4400* (p. 9). United states Environmental protection Agency Prevention, pesticides and Toxic substances Unit, New York.

Erickson, L. E., (2017). Summary of the 12th international phytotechnology conference: Phytotechnologies for sustainable development, Manhattan, Kansas. *Int. J. of Phytoremediation, 19*(1), 1, 2.

Favas, P. J. C., Pratas, J., & Prasad, M. V., (2012). Accumulation of arsenic by aquatic plants in large-scale field conditions: Opportunities for phytoremediation and bioindication. *Science of the Total Environment, 433*, 390–397.

Garg, P., & Chandra, P., (1994). The duckweed *Wolffia globosa* as an indicator of heavy metal pollution: sensitivity to Cr and Cd. *Environ Monit Assess, 29*, 89–95.

Godwill, A. E. G. A., Paschaline, U. F. P. U., Friday, N. N. F. N., & Marian, N. U., M. N., (2019). *Mechanism and Health Effects of Heavy Metal Toxicity in Humans, Poisoning in the Modern World - New Tricks for an Old Dog?*, Ozgur Karcioglu and Banu Arslan, IntechOpen, doi: 10.5772/intechopen.82511.

Gupta, K., Gaumat, S., & Mishra, K. K., (2011). Chromium accumulation in submerged aquatic plants treated with tannery effluent at Kanpur, India. *J. of Environ. Biol., 32*(5), 591–597.

Gupta, P. K., (2004). *Soil, Plant, Water and Fertilizer Analysis*. Agro Botanica Vyas Nagar Bikaner, India.

Hassan, S. H., Talat, M., & Rai, S., (2007). Absorption of Zinc and Cd in aqueous solution by water hyacinth. *Bioresure. Technology, 98*, 918–928.

John, R., Amhad, P., Gadgil, K., & Sharma, S., (2009). Heavy metal toxicity: Effect on plant growth, biochemical parameters and metal accumulation by *Brassica juncea* L. *Int. J. Plant Prod., 3*(3), 161–165.

Kamel, K. A., (2013). Phytoremediation potentiality of aquatic macrophytes in heavy metal contaminated water of El-Temsah Lake, Ismailia, Egypt. *Middle-East J. of Scientific Research, 14*, 1555–1568.

Kara, Y., & Kara, I., (2005). Removal of Cadmium from water using duckweed (*Lemna trisulca* L.). *Int. J. Agr. Biol.*, 660–662.

Khellaf, N., & Zerdaoui, M., (2009). Phytoaccumulation of zinc by the aquatic plant, *Lemna gibba* L. *Bioresource Technology, 100*, 6137–6140.

Kimbrough, D. E., Cohen, Y., Winer, A. M., Creelman, L., & Mabuni, C., (1999). A critical assessment of chromium in the environment. *Critical Reviews in Environmental Science and Technology, 29*(1), 1–46.

Klink, A., (2017). A comparison of trace metal bioaccumulation and distribution in *Typha latifolia* and *Phragmites australis*: implication for phytoremediation. *Environ. Sci. Pollut. Res., 24*, 3843–3852.

Kundangar, M. R. D., & Abubakar, A., (2004). Thirty years of ecological research on Dal Lake, Kashmir. *J. Res. Dev., 4*, 45–57.

Lichtenthaler, H. K., & Wellburn, A. R., (1983). Determination of total carotenoids and chlorophyll 'a' and 'b' of leaf extracts in different solvents. *Biochem. Soc. Trans., 11*, 591–592.

Lu, X., Kruatrachus, M., Pokethitiyook, P., & Homyak, K., (2004). Removal of cadmium and zinc by water hyacinth. *Science Asia, 30*, 93–103.

Manorama, T. K. C., Beena, V. I., Meera, A. V., & Ajayan, A. S., (2020). Phytoremediation of metals by aquatic macrophytes. In: Shmaefsky, B., (ed.), *Phytoremediation. Concepts and Strategies in Plant Sciences*. Springer, Cham.

Misra, K. V., & Tripathi, B. D., (2008). Concurrent removal and accumulation of heavy metals by the three aquatic macrophytes. *Bioresource Technol., 99*, 7091–7097.

Newete, S. W., & Byrne, M. J., (2016). The capacity of aquatic macrophytes for phytoremediation and their disposal with specific reference to water hyacinth. *Environ. Sci. Pollut. Res., 23*, 10630–10643.

Nies, D. H., (2003). Efflux mediated heavy metal resistance in prokaryotes. *FEMS Microbiol. Rev., 27*(2, 3), 313–339.

OECD, (2002). *Guidelines for the Testing of Chemicals, Lemna Sp.* Growth Inhibition Test, Draft guideline, 221.

Ouzounidou, G., (1995). Cu-ions mediated changes in growth, chlorophyll and other ion contents in a Cu tolerant Koeleria splendens. *Biological Plant, 7*, 71–79.

Pandit, A. K., (1984). Role of macrophytes in aquatic ecosystems and management of freshwater ecosystems. *J. Environ. Manage., 18*, 73–88.

Pandit, A. K., (1993). Dal lake ecosystem in Kashmir Himalaya: Ecology and management. In: Mishra, P. C., & Trivedy, R. K., (eds.), *Ecology and Pollution of Indian Lakes and Reservoirs* (pp. 131–202). Ashish Publishing House, New Delhi, India.

Phetsombat, S., Kruatrachue, M., Pokethitiyook, P., & Upatham, S., (2006). Toxicity and bioaccumulation of cadmium and lead in *Salvinia cucullata*. *J. of Environl. Biol., 27*(4), 645–652.

Phukan, P., Phukan, R., & Phukan, S. N., (2015). Heavy metal uptake capacity of *Hydrilla verticillata:* A commonly available Aquatic plant. *Int. Res. J. Environment Sci, 4*(3), 35–40.

Qadri, H., & Yousuf, A. R., (2008). What is ailing the Dal Lake? In: Ranade P. S., (ed.), *Lakes Conservation and Management* (pp. 95–103). The ICFAI University Press, Hyderabad, India.

Razinger, J., Dermastia, M., Drinovec, L., Drobne, D., Zrimec, A., & Koce, J. D., (2007). Antioxidative responses of duckweed (*Lemna minor* L.) to short-term copper exposure. *Environ. Sci. Pollut. Res., 14*, 194–201.

Rolli, N. M., Suvarna, K. S. S., Gadi, S. B., Mulagund, G. S., & Taranath, T. C., (2013). Effect of cadmium toxicity on aquatic macrophyte *Pistia* stratiotes (L.). *J. Environ. Anal. Toxicol., 4*, 201. doi: 10.4172/2161-0525.1000201.

Saba, S., & Wanganeo, A., (2010). Excessive phosphorus loading to Dal Lake, India: implications for managing shallow eutrophic lakes in urbanized watersheds. *Int. Rev. Hydrobiol., 93*(2), 148–166.

Sun, Q., Liu, W. B., & Wang, C., (2011). Different response of phytochelatins in two aquatic macrophytes exposed to cadmium at environmentally relevant concentrations. *Afri. J. Biotech., 10*(33), 6292–6299.

Tanhan, P., Kruatrachue, M., Pokethitiyook, P., & Chaiyarat, R., (2007). Uptake and accumulation of cadmium, lead and zinc by Siam weed. *Chemosphere, 68*, 323–329.

Uysal, Y., & Taner, F., (2010). Bioremoval of cadmium by *Lemna minor* in different aquatic conditions. Wiley inter science. *Clean – Soil, Air, Water, 38*(4), 370–377. www.clean-journal.com (accessed on 14 December 2021).

Vajpayee, P., Tripathi, R. D., & Rai, U. N., (2000). Chromium (VI) accumulation reduces chlorophyll biosynthesis, nitrate reductase activity and protein content in *Nymphaea alba* L. *J. Chemosphere, 41*(7), 1075–1082.

Wang, C., Zheng, S., Wang, P., & Qian, J., (2014). Effects of vegetations on the removal of contaminants in aquatic environments: A review. *J. of Hydrodynamics, 26*(4), 497–511.

Wei, S. H., Zhou, Q. X., & Koval, P. V., (2006). Flowering stage characteristics of cadmium hyperaccumulator *Solanum nigrum* L. & their significance to phytoremediation. *Sci. Total Environ., 369*, 441–446.

Wei, S., Zhou, Q., Srivastava, M., Xiao, H., Yanga, C. Z., & Zheng, Q., (2009). *Kallimeris integrifolia* turcz.Ex.Dc: An accumulation of Cd. *J. Hazardous Mat., 162*, 1571–1573.

WHO, (1995). *Guidelines for Drinking Water Quality* (Vol. 3). Geneva.

Yaqoob, K. U., & Pandit, A. K., (2009). Distribution and abundance of macrozoobenthos in Dal Lake of Kashmir Valley. *J. Res. Dev., 9*, 20–29.

Zayed, A., Gowthaman, S., & Terry, N., (1998). Phytoaccumulation of trace elements by wetland plants: L Duckweed. *J. Environ. Qual., 27*, 715–721.

Zimmo, O., (2003). *Nitrogen Transformations and Removal Mechanisms in Algal and Duckweed Waste Stabilization Ponds*. Ph.D Thesis, International Institute for Infrastructural, Hydraulic and Environmental Engineering, Delft, the Netherlands.

Zutshi, D. P., & Vass, K. K., (1982). Limnological studies on Dal lake III. Biological features. *Pro Indian Natn. Sci. Academy, 48*(2), 234–241.

CHAPTER 10

Factors Affecting Habitat Selection of Freshwater Fishes of Sri Lanka: A Review

SUDESH BATUWITA,[1] SAMPATH UDUGAMPALA,[2] and UDENI EDIRISINGHE[3]

[1] *The Society for the Biodiversity Conservation, Kandy, Sri Lanka*

[2] *Wildlife Conservation Society-Galle, Hiyare Reservoir, Hiyare, Sri Lanka*

[3] *Postgraduate Institute of Agriculture, University of Peradeniya, Sri Lanka*

ABSTRACT

Habitat selection of fish is a recurrently active process and the models being used to evaluate the fish habitat selection is evolving rapidly. Using these rapidly evolving models, it has been seen that in freshwater ecosystems, the most important factors shaping the living conditions include temperature, ionic content of water and flow rate. Other important factors, that also influence the presence of these fishes in the aquatic ecosystems include substratum type, oxygen concentration, turbidity and light or pH. To a large extent these factors also depend on the above-mentioned conditions. Biotic factors like the type of predator, amount of prey, competitors, foraging, and predator abundance also contribute to the habitat selection by fish**es.** Like the external factors, habitat selection by a fish is also

Biodiversity of Freshwater Ecosystems: Threats, Protection, and Management. Suhaib A. Bandh, PhD, Basharat Mushtaq, PhD, Mohammad Farooq Mir, PhD, & Shahid A. Wani (Eds.)

governed by the intensity of its searching behavior *and* by the interaction of internal drives for needed resources.

10.1 INTRODUCTION

10.1.1 HABITAT SELECTION

Differences in morphology and behavior among species have become essential in promoting their coexistence, often through partitioning of resources in their habitats (Berumen and Pratchett, 2006). However, differences among individuals within single species populations can be just as significant as the differences between species and can have similar ecological consequences (MacArthur and Pianka, 1966; Berumen and Pratchett, 2006). Habitat selection of fish appears to be based on several factors but can be summarized broadly into: (i) morphology; (ii) abiotic factors; and (iii) niche dimensions, including biotic interactions. Considering the morphology, fish have enormous morphological diversity because currently, about 34,300 fishes have been described (FishBase, 2020), and almost all of them are differently classified due to their morphological differences.

Biotic interactions such as predation (i.e., types, amount of prey, competitors, foraging, predator abundance) also contributed to the habitat selection by fish (Southwood, 1977). The community's ecology depends on relative amounts of the different habitats in the ecosystem as well (Morris, 1988; Oksanen, 1990; Pulliam and Danielson, 1991). The strength and magnitude of biotic interactions may change populations' distribution among different habitats (Svärdson, 1949; Werner et al., 1983b; Gilliam and Fraser, 1987; Rosenzweig, 1987; Brown, 1988; Morris, 1988). Thus, density-dependent habitat selection occurs (Figure 10.1), i.e., individuals will move from a more crowded habitat to a less crowded one (Fretwell and Lucas, 1970).

Natural selection will favor individuals that maximize fitness when choosing between habitats (Cerri and Fraser, 1983; Gilliam and Fraser, 1987). Individuals will, therefore, try to maximize energy gain and minimize predation risk (Cerri and Fraser, 1983; Gilliam and Fraser, 1987). The capacities to forage and escape predators are size-dependent (Persson, 1988; Polis, 1988). Thus, in size-structured populations, the distribution among habitats is governed by size-dependent trade-offs between growth and mortality (Werner et al., 1983a; Werner and Gilliam, 1983b; Fraser and Gilliam, 1987; Gilliam and Fraser, 1987). Since fish continue to grow throughout their life, biotic interactions in fish communities depend on the ontogenetic niche-shifts (i.e., changes in feed sources and/or habitats, from the origin, throughout development during their life span; e.g., eels, gobies, and marine water fish, which breed in brackish water). Many large-sized fish species are omnivorous, and thus, they are predators of more than one trophic level (Polis, 1991). Therefore, interactions within and among size-structured populations with ontogenetic niche shifts will affect the whole fish community (Ebenman and Persson, 1988).

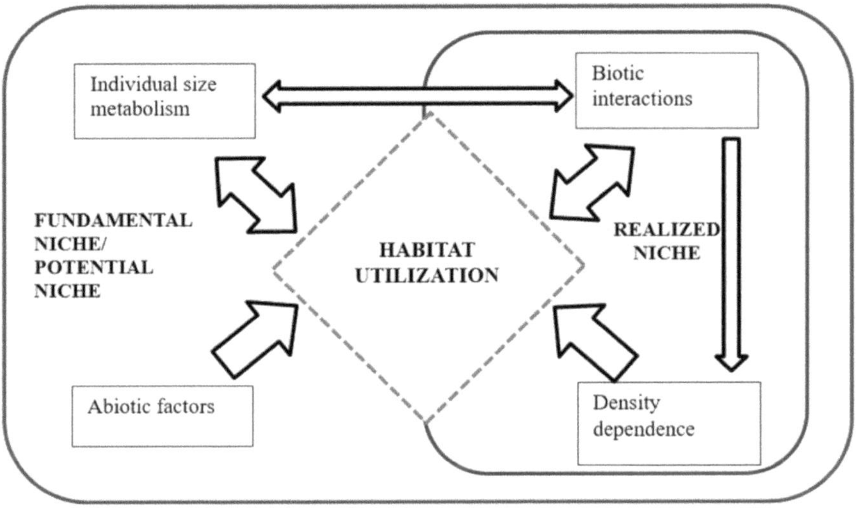

FIGURE 10.1 Factors that affect habitat selection of fish.

Source: Adapted from Beier, 2016.

Altered competition and predation situations may affect habitat selection and change species' distribution or size groups within species and between habitats (Werner et al., 1983a; Gilliam and Fraser, 1987; Brown, 1988). Thus, direct (predation) and indirect (competition for the same resource) interactions may be mediated by altered habitat use and may also affect the habitat use of stream fishes (Fausch and White, 1981; Werner et al., 1983a; Persson, 1988; Strauss, 1991; Grossman et al., 1982). Predation can be a significant determinant of the abundance and distribution of species (Paine, 1966). In the absence of some Spatio-temporal heterogeneity, predator-prey systems may be unstable and terminated, e.g., laboratory experiments; both prey and predator go extinct (Gause, 1934).

According to Sales (1969a) model, habitat selection of fish is a recurrently active process, governed by the intensity of searching behavior *via* a negative feedback loop and with exploratory behavior governed by the interaction of internal drives (motivation) for needed resources (mostly abiotic factors). On the other hand, foraging strategies that enable a species to respond to the changing environment also benefits their survival and reproduction (Berumen and Pratchett, 2006). Optimal foraging theory (MacArthur and Pianka, 1969) described that, when an optimal resource is abundant, prey would be exclusively consumed by the forager and avoid substandard prey items (Berumen and Pratchett, 2008), but when preferred prey is not sufficiently available, shifts in prey selectivity occur, and broader diets of less-preferred prey would be consumed (Charnov, 1976; Brooker et al., 2013a; Chandler et al., 2016).

Habitat selection can be modified by other individuals' presence, including competitors (theory based on ideal free distributions) and predators (theory based on trade-offs of growth *vs*. predation risk). However, most current models use game theory to describe habitat selection dynamics for prey species and their mobile predators (Weinstein and Kreeger, 2000). Predatory fish's functional morphology primarily governs the type and size of prey consumed (Marshal, 1965; Okamura, 1970; Geistdoerfer, 1973, 1978; McLellan, 1977).

10.1.2 FRESHWATER ECOSYSTEMS

Water that contains less than 1,000 parts per million (ppm) of dissolved salts (USGS, 2020) is known as freshwater. It can change due to other sources, which may increase to slightly saline water: from 1,000 ppm to 3,000 ppm (USGS, 2020). Freshwater in the Earth is just 2.5%, with most of it locked up in ice or deep underground aquifers (WWF, 2020). Only a small fraction is found in rivers and lakes, but that is the portion that supports Earth's enormous life. Freshwater ecosystems also contain about 41% of the world's known fish species (Daily, 1997).

There are three basic types of freshwater ecosystems:

1. **Lentic:** Slow-moving water, including natural and artificial pools, ponds, and lakes and reservoirs.

2. **Lotic:** Faster-moving water, e.g., streams, and rivers.
3. **Wetlands:** Areas where the soil is saturated or inundated for at least some months.

The zone, which bridges groundwater and lotic systems, is also known as the hyporheic zone (Bencala, 2000). This zone underlies many larger rivers and contains considerably more water than is seen in the open channels. It may also be in direct contact with the groundwater.

Freshwater ecosystems of Sri Lanka are mainly divided into three major groups: rivers, flood plain lakes, and groundwater (De Silva, 1988). Drainage system (pattern) is the type of ecosystem formed in a basin due to the first group, streams, rivers, and lakes. Geomorphologists and hydrologists often view streams as parts of drainage basins (Pidwirny, 2006). A drainage basin is a topographic region from which a stream receives runoff and groundwater flow (Pidwirny, 2006). Drainage basins are divided by topographic barriers called watersheds, representing all the stream tributaries that flow to some location along the stream channel (Pidwirny, 2006).

Dendritic systems are the most common form of drainage systems found in Sri Lanka. There are many contributing streams (analogous to the twigs of a tree) joined together into the main river's tributaries (the branches and the trunk of the tree, respectively). They develop where the river channel follows the slope of the terrain (David, 1998). Dendritic systems form V-shaped valleys, and as a result, the rock types must be impervious and non-porous (David, 1998). Sri Lanka has an extensive network of rivers (with Mahaweli River as the largest on the island) and streams that drain a total of 103 distinct natural river basins (De Silva, 1988).

The other freshwater bodies, like wetlands, are very diverse. Although large natural lakes are not found in Sri Lanka, several flood plain lakes called 'Villu' cover 12,500 ha. Handapana Villu and Pendiya Villu (796 ha) are the largest of the entire Mahaweli Villu System (De Silva, 1988). Examples of freshwater marshes are Muthurajawela in the Colombo suburbs and Walauwawatta-Wathurana swamp forest (12 ha) in the Kalu River basin (De Silva, 1988). Salt-water wetlands (e.g., lagoons, estuaries, mangroves, seagrass beds, and coral reefs) are also abundant in Sri Lanka. Apart from the above categories, three major types of human-made wetlands also exist in Sri Lanka, i.e., reservoirs or 'tanks,' rice fields, and salt pans (De Silva, 1988).

10.1.3 FRESHWATER FISHES OF SRI LANKA

Sri Lanka is a relatively large island (~65,610 km^2), which during the more significant part of its geological history, was a part of the Indian subcontinent (Erdelen, 1989). It is geologically differentiated into the Precambrian Highland Series, Cambrian Vijayan Series, South-western Group, and the Miocene limestones (Cooray, 1967). Sri Lanka is also a continental island in the Indian Ocean (Somasekaram, 1997). Owing to its position in the tropical zone and due to its geodiversity (*sensu* Domrös, 1998), a wide variety of habitats are found in the island nation. Sri Lanka also possesses a highly diverse endemic fauna and flora (Myers et al., 2000; Bossuyt et al., 2004). Furthermore, Sri Lanka is a biodiversity hotspot in the revised hotspots of the world and India's Western Ghats (Mittermeier et al., 1998). Moreover, recently, the Central Highland of Sri Lanka (Horton Plains, Knuckles Range, and Peak Wilderness) was declared a World Heritage Site (UNESCO, 2010).

Among the enormous vertebrate diversity of Sri Lanka, freshwater fishes appear to be one of the well-studied groups through some morphological and molecular studies (e.g., Silva et al., 2010; Pethiyagoda et al., 2012; Batuwita et al., 2013, 2017; Sudasinghe et al., 2020). Until these contemporary studies, the literature available about the freshwater fishes of Sri Lanka was largely based on the reviews published during the colonial period (Günther, 1868; Day, 1888, 1889), owing to the comprehensive classic works of P. E. P. Deraniyagala and I. S. R. Munro (Deraniyagala, 1952; Munro, 1955). Subsequently, a few color guides of the freshwater fishes of Sri Lanka were published (Pethiyagoda, 1991; Goonatilake, 2007; De Silva et al., 2015), which provided almost all up to date data on the freshwater species.

Aspects of freshwater fish ecology are very scanty in Sri Lanka. In this review, however, several essential studies done by some distinguished ecologists, Moyle and Senanayake (1984); Kortmulder (1982, 1987); and Wickramanayake and Moyle (1989); Wickramanayake (1990); and Piet (1996), as well as some recent invaluable studies especially on the fish feeding, abundance, and niche overlaps have been discussed. Also, a recent survey of Batuwita (2016) is also added. In this review, old species names have been replaced with the new ones following the most recent taxonomic studies.

10.2 RESULTS

10.2.1 FISH MORPHOLOGY

Inter-specific and intra-specific interactions always usher in mutual survival. However, competition may eliminate one or more species. Under natural conditions (e.g., absence of invasive species and pollution), species avoid such interactions by resource partitioning. It is proven that the phenotypic variation of a species depends on its genetic variation and the environmental effects. The ecological factors influence the species morphology and thereby adaptations to a particular environment. Thus, when factors that affect fishes are considered, it is essential to study their internal (e.g., digestive system, respiratory system morphology) and external morphology (e.g., mouth gape size, mouth orientation, caudal fin morphology).

Based on the morphology, Moyle and Senanayake (1984) classified the tropical stream fishes into five general types of fishes including: (i) elongate fishes; (ii) deep-bodied fishes and round to square in cross-section; (iii) generalized fishes (mid-water spp.); (iv) Snakeheads; and (v) bottom-oriented species. According to them, conspicuously different fish forms were with elongate bodies (depth/standard length ratio, <0.15), e.g., *Schistura notostigma, Mastacembelus armatus,* and *Xenentodon cancila.* Also, *Xenentodon cancila* has a long snout, sharp teeth, and pike-like body (a predator), whereas the other two have small mouths and eel-like bodies (Moyle and Senanayake, 1984). Due to the flattened head, snakeheads (*Channa* spp.) can be separated readily from the other species. They are the potential predators, e.g., *Channa orientalis* has more dorsally placed eyes and a more upward-pointing and more massive mouth than *Channa kelaartii,* indicating an orientation towards the water surface. Differences in mouth and gill structure indicate that *Dawkinsia singhala* is more bottom oriented and omnivorous than *Devario micronema. In contrast, Puntius titteya* and *Systomus pleurotaenia* have sub-terminal mouths, indicative of bottom-dwelling habits, while *Rasbora dandia* has a terminal mouth expressive of a mid-water feeder (Moyle and Senanayake, 1984).

However, a more rational division is based on depth/vertical distribution (Moyle and Senanayake, 1984). As indicated earlier, freshwater ecosystems, classified as lotic and lentic, have more or less similar zonation, e.g.: (i) surface; (ii) midwater (hereafter referred to as column zone);

and (iii) bottom. Thus, fishes were classified according to their place in those zones. In each zone, fishes show some unique adaptations to survive. To gain the ecological need, fishes show various morphological transformations, and these adaptations are well pronounced in those species inhabiting the surface and the bottom zones. Although the column zone fishes do not show greater morphological adaptations, they have specific morphological adaptations to survive under certain factors like variations in the flow rate (i.e., either slow-flowing or fast-flowing) or dietary habits (e.g., omnivorous). The lentic ecosystems also have the same major vertical zones (surface, column, and bottom). Besides, horizontal zonation of broader and deeper water bodies is also classified as inshore and offshore/pelagic zones or littoral and limnetic zones.

Surface inhabiting species have diverse adaptations to survive in a particular zone. Most of them have an elongated body with a superior mouth. Depending on the flow rate, the body length may also vary (e.g., *Aplocheilus, Xenentodon*). Bottom dwellers have more significant modifications like sub-terminal or inferior mouth, dorso-ventrally flattened anterior half of the body (e.g., *Glossogobius*), and even some unique features like sucking disc of *Garra* spp. and united pelvic fins of the members of the family Gobiidae (e.g., *Glossogobius, Sicyopterus*). Species, which live in the column zone, explore a broad range compared to the other two groups. Also, unlike surface and bottom zones species, they (e.g., *Belontia, Devario, Pethia, Puntius*) have the least modifications to the mouth's morphology. Still, they have well-modified features to explore the column zone, e.g., laterally compressed body, well-developed fins (dorsal, caudal, pectoral, and anal fins) lateral line. However, a single species in the genus *Devario, D. monticola* showed a unique fin length reduction for the pectoral fin (Batuwita et al., 2017). This condition is mystifying because all other congeners have relatively long pectoral fins and prefer to live in runs and riffles. It may be due to the absence of potential (fish) predators in its range. It is the only fish found in its type locality (Batuwita et al., 2017). It is restricted to hill streams of Sri Lanka, where the species richness is very low.

Piet (1996) published an interesting and important study on the reservoir-freshwater fishes of Sri Lanka. Based on size-class research, he demonstrated how the body/parts change through their ontogenetic development and concluded that the species' whole dimensions of their morphological characters increase with increased size (Piet, 1996; Figure

10.1). He has also mentioned several species in the species assemblage of Tissawewa (in the Southern Province of Sri Lanka), which are having relatively large adult sizes, e.g., *Glossogobius giuris, Systomus sarana, Mystus* spp. and *Oreochromis mossambicus* and some other species with relatively smaller sizes, i.e., *Amblypharyngodon melettinus, Rasbora dandia,* and *Hyporhamphus limbatus*. Interestingly, in most fish species, their morphological characters' increase was not proportional to the increase in body length (Piet, 1996). He further demonstrated that both mouth height and width (in the larger size-classes) increased proportionally with the length for most species. In contrast, for the smallest size-class(es), the mouth opening was disproportionally small. However, conversely, the *Mystus* spp. and *Amblypharyngodon melettinus* showed a relative decrease in the mouth opening with increased body size (Piet, 1996).

Piet (1996) found two species with very long relative gut length, *Amblypharyngodon melettinus* and *Oreochromis mossambicus,* while most barbs, *Puntius thermalis, P. dorsalis* and *Systomus sarana* had an intermediate relative gut length. According to the same source, throughout their ontogeny, each species' relative body depth appears to be related to the relative gut length except for higher relative body depths of barbs and *Mystus* spp. Interestingly, the snout length remains proportional to the smaller species' head length, such as *Amblypharyngodon melettinus, Hyporhamphus limbatus*, and *Rasbora dandia.* In contrast, the increase was well pronounced in the remaining species, with a more considerable maximum length (Piet, 1996).

Accordingly, the mouth's orientation and barbels' presence were the morphological characters, which differed between species. They did not change during ontogeny, while fish with barbels preferred benthic food and fish with a dorsally oriented mouth preferred zooplankton. It was further mentioned that the fish species with both barbels and a ventrally oriented mouth would forage closer to the bottom. In contrast, fish without barbels will feed in the column or on the surface, depending on the mouth's position.

Moreover, species with an upward oriented, relatively small mouth, fed chiefly on zooplankton and terrestrial insects had significant positive correlations ($p<0.01$) between the size of the food and the size of the mouth opening (both absolute and relative), head length, standard length, and relative gut length (Piet, 1996). It was also revealed that a long gut length was associated with a high proportion of residues in the diet and fish with a longer gut or with a higher relative gut length. Consequently, higher relative body depth had a higher proportion of plant material in their diet. However, in some species (e.g., *Amblypharyngodon melettinus* and *Systomus sarana*), the food type was better correlated with the absolute gut length during ontogeny.

10.2.2 ABIOTIC FACTORS

Several abiotic factors, of which velocity, substrate, and water column depth, are the well-documented factors tested on the fish distribution in Sri Lanka. Also, water temperature, pH, and turbidity were also used to understand the fishes' resource partition. However, some other factors need to be studied in the future. A study on the rainforest stream fishes (Moyle and Senanayake, 1984) explained some 88.0% of the variance for the following ecological parameters: two velocity variables (velocity at fish and mean water column velocity), two depth variables (depth at fish and water column depth) and the dominant substrate.

10.2.2.1 VELOCITY (FLOW RATE)

Moyle and Senanayake (1984) grouped fishes according to the following stream velocities: (i) low gradient streams with flow rates of less than 2 m^3/minute (= 0.01 ms^{-1}); (ii) streams with flow rates greater than 15 m^3/min (= 0.09 ms^{-1}); (iii) large pools of low gradient streams; and (iv) fast waters. Also, some authors described certain species, which represented three habitat generalists distributed in forest streams of 2–15 m^3/min (= 0.01–0.09 ms^{-1}): found in quiet small streams, *viz., Aplocheilus dayi, Channa orientalis* and *Puntius titteya.* The latter was recorded almost entirely in relatively undisturbed streams surrounded by thick forests (Moyle and Senanayake, 1984). As Moyle and Senanayake (1984) stated, *Puntius titteya* was the only fish present in such streams. The two low-gradient pool specialists were *Rasboroides pallidus* and *Xenentodon cancila,* of which the latter was a predator on the former.

According to them, fast water specialists were *Devario malabaricus, Garra ceylonensis* and *Sicyopterus lagocephalus,* where all were most abundant in sections of the more massive streams (5–20 m^3/min [= 0.03–0.12 ms^{-1}]) with high gradients. Jayaratne and Surasinghe (2010) also reported *Garra ceylonensis* and *Schistura notostigma* from a wide range of habitats. Moyle and Senannayake (1984) stated that *Schistura notostigma* was found in a wide variety of other habitats and was less tolerant of high gradients. In contrast, *Belontia signata* and *Rasbora dandia* were habitat generalists. Jayaratne and Surasinghe (2010) reported that *Devario malabaricus, Glossogobius giuris, Lepidocephalichthys thermalis, Mystus nanus, Rasbora dandia,* and *Xiphophorus maculatus* (an invasive species) were confined to runs.

Schistura notostigma, Belontia signata, and *Rasbora dandia*, together with *Pethia nigrofasciata, Puntius dorsalis, Systomus pleurotaenia* and *Awaous melanocephala* formed the typical diurnal fish assemblage in the rainforest streams with average flows of 2–15 m^3/min (= 0.01–0.09 ms^{-1}). In contrast, *Mastacembelus armatus* and *Channa kelaartii* were probably mostly nocturnal (Moyle and Senanayake, 1984). *Puntius bimaculatus, Hypostomus* sp. (an invasive species), and *Tor khudree* were present only in pools (Jayaratne and Surasinghe, 2010). Still, *Tor khudree* juveniles were observed together with *Sicyopus jonklaasi* in a riffle near the Sinharaja Pitadeiya camp in the Gin river (Sudesh Batuwita, pers. obs., 1997). However, water velocity at each fish is considered the poorest discriminator among species, and the only species that was consistently found in fast water was *Devario micronema.* In contrast, other fast water species were usually seen in crevices, behind rocks, or in depressions in which current velocities were low (Moyle and Senanayake, 1984).

It is known that the biology and ecology of riverine species, which complete their entire life cycle in riverine habitats, are closely linked with the annual hydrological regimes and flood patterns (Poff et al., 1997; Welcomme and Halls, 2001). With the onset of monsoons, stream flow rates increase in the rainforests. Batuwita (2016) reported that the sampling points' stream flow rate fluctuated with monsoons' start (Figure 10.2). The flow rate significantly increased in Homadola (stream) area, indicating the stream's highest disturbance, possibly due to encroachments. All the other sites did not show such a sudden increase in flow rates. Mathur and Sajwan (1978) described rainfall interception by dense canopy cover. High infiltration due to thick litter in the forest has reduced surface runoff and peak flow rate indicating the least streamflow fluctuation. Thus, the effects of encroachment for the destruction of these ecosystems are evidenced. Batuwita (2016) also pointed out that an endemic threatened fish, *Rasboroides pallidus* prefers to live in slow-flowing waters (0.02–0.17

ms^{-1}) at Kanneliya and Kottawa, which conformed with the data provided by Wikramanayake and Moyle (1989); and Ediriweera et al. (2003).

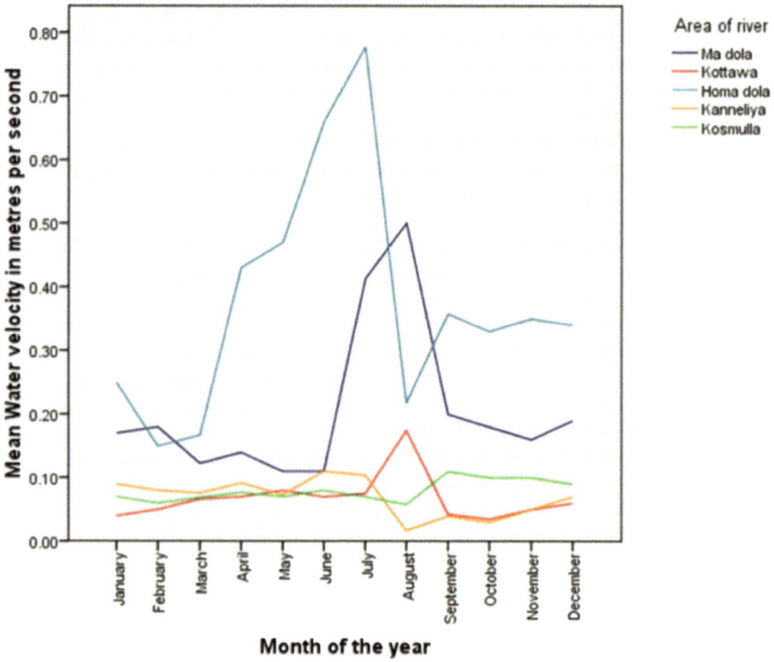

FIGURE 10.2 Variations in the mean monthly velocity in different sampling sites in Gin and Polathumodera rivers throughout the study period (January–December).
Source: Reprinted from Batuwita (2016).

10.2.2.2 DEPTH OF WATER COLUMN

Relative depth (which is a measure of fish's position in the column) showed the most differences among species (Moyle and Senanayake, 1984). According to them, column depth measurements at each fish indicated that most species occupied the full range of depths available to them. However, a few (four) species preferred the shallow water, either at the stream edge or in small streams. Still, *Puntius dorsalis* and *P. titteya* were both bottom-oriented and were frequently found at 0.02–0.04 m above the substrate. However, *Awaous melanocephala* was observed on sand and *Sicyopterus lagocephalus* on boulders.

Wickramanayake and Moyle (1989) reported that the upper-column species, such as *Devario malabaricus* was associated with moderate water velocities. In contrast, *Rasboroides pallidus, Belontia signata,* and *Rasbora dandia* were found in slow waters. According to Batuwita (2016), the water column's depths in all the sites gradually increased with the rainfall. Thus, the two abiotic factors of the sites, velocity (flow rate), and water depth, simultaneously increased with the onset of rains from April to July (Figure 10.3). *Rasboroides pallidus* was abundant at 0.4–0.7 m depths at Kanneliya and Kottawa. Wikramanayake and Moyle (1989) had given a similar depth (~0.6 m) as the optimum for *Rasboroides pallidus.*

Batuwita (2016) reported a shallow stream depth at a sampling point in the fragmented forest patch (Kottawa) during the dry season (April), that had been comparatively very small (only the Arboretum area; on the left side of the Galle to Udugama road has an approximate about area ~14 ha). The drying up of streams may be due to the fragmentation (due to the Galle-Udugama road) of this small forest patch because the arboretum side has comparatively less width and shallow streams.

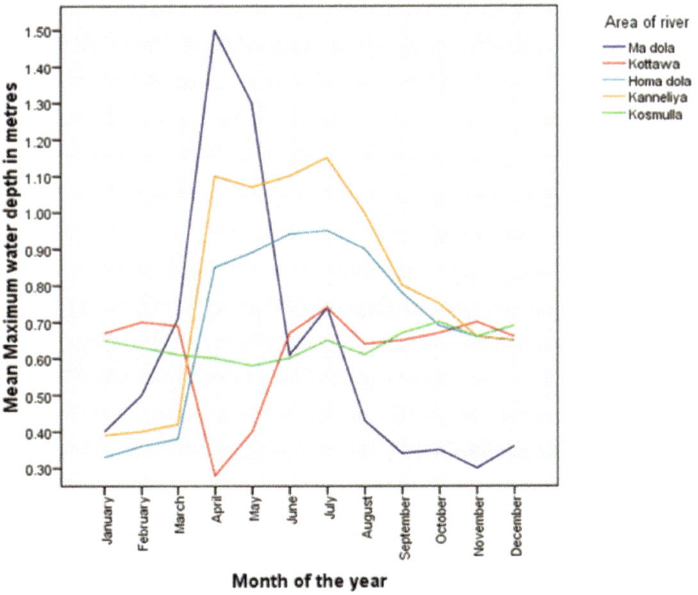

FIGURE 10.3 Fluctuations of mean monthly water depth in the selected sites in Gin river (January–December).
Source: Batuwita (2016).

Wickramanayake and Moyle (1989) revealed that there were fast water-inhabitant species in the lower-column zone like *Garra ceylonensis* and *Tor khudree*. *Pethia nigrofasciata* and *Puntius bimaculatus* were found in moderate velocity and lower-column zone, while *Lepidocephalichthyus thermalis, Pethia cumingii,* and *Puntius titteya* were observed as the slow water-bottom. A few species reported by Wickramanayake and Moyle (1989) showed wider-habitat utilization: *Tor khudree* (fast water to slow water), *Devario malabaricus* (moderate to fast), and *Schistura notostigma* (slow to fast) as well.

Based on the Wickramanayake and Moyles (1989) study, *Rasbora pallidus* was consistently found in the upper-column zone, *Devario malabaricus* lower-column zone (below the former species, but above *Rasbora dandia*). The latter species showed a considerable variability concerning the relative depths, probably due to its wider-habitat utilization, including *Belontia signata,* a deep-water inhabitant. Wickramanayake and Moyle (1989) considered the above four species as column species with a vertical stratification among them, while *Channa kelaartii* and *C. orientalis* were benthic.

Pethia nigrofasciata's wider-habitat utilization was possibly due to its breeding behavior (found in the upper-column zone), and probably both *Puntius bimaculatus,* and *P. titteya* had shown a considerable variation for the depth. In contrast, the relative depth of *Tor khudree* was markedly different: it is cruised close to the bottom along the stream, but occasionally utilized upper-column zones, especially in fast, shallow water at run and riffle interfaces, while *Pethia cumingii* was usually found in relatively shallow waters near the substratum (Wickramanayake and Moyle, 1989; Batuwita, 2016). Benthic species like *Garra ceylonensis, Schistura notostigma* and *Lepidocephalichthys jonklassi* were recorded from a lower depth (except for juveniles of *Garra ceylonensis*) (Batuwita, 2016).

10.2.2.3 STREAM SUBSTRATE

Some species are mainly associated with fine substrates like detritus, mud, silt, and sand (e.g., *Awaous melanocephala, Lepidocephalichthyus jonckaassi*). In contrast, a few other fishes (e.g., *Garra ceylonensis, Sicyopterus lagocephalus*) are primarily associated with boulders and bedrock. According to Wickramanayake and Moyle (1989), only a few species of habitat specialists were observed in the substratum: bottom-associated *Pethia cumingii, Garra ceylonensis* and *Lepidocephalichthyus thermalis,* while most other species utilized a more comprehensive range of substrate type and also responded to depth and velocity, e.g., *Rasboroides pallidus, Devario malabaricus, Rasbora dandia* and *Belontia signata*. Batuwita (2016) reported that *Rasboroides* spp. was always in sandy substrates with extremely low silt and leaf debris and was sometimes found among submerged vegetations, where slightly fast velocity might have washed the silt particles in those microhabitats, while most of the time these fishes preferred to live closer to a waterfall, but away from the fast-flowing areas (i.e., close to river/stream banks) and among the selected sites of his study, preferable substrates were observed in the Kanneliya and Kottawa forests. Batuwita (2016) further confirmed that Kosmulla though situated in the Sinharaja World Heritage Site (WHS), did not possess the substrate types preferred by *Rasboroides*. Also, at the Madola site, *Rasboroides pallidus* was not found, maybe due to the high flow rate during rainy seasons that would have disturbed the fish and the spill (at the Hiyare reservoir), which might have facilitated larger predatory fishes to enter. According to the same source, Homadola site also did not show the necessary substrates for *Rasboroides,* and the stream was too swift, where the only observed common species in this stream was *Devario micronema*, and occasionally *Schistura notostigma* and *Sicyopterus lagocephalus*.

10.2.2.4 WATER TEMPERATURE

Temperature showed an inversely significant ($P<0.0001$) relationship with species richness and Shannon index (Perera et al., 2015). According to Batuwita (2016), the sampling sites' monthly water temperature pattern indicated bi-modal variation with two peaks in February and October (Figure 10.4). There had been a gradual decrease in water temperature from May to August because the southwestern wet zone received monsoon rains during this period, except for the disturbed sites (i.e., Madola, and Homadola), the other sites showed relatively lower water temperature (Figure 10.4). This parameter has significantly varied with the rainfall, especially in the streams within the wet zones of Sri Lanka (Figure 10.4). This rapid temperature drop in these stream waters may trigger most of these fish (Batuwita, 2016).

The existence of modest canopy cover affected water temperature, especially in the lotic habitats. A recent study of Batuwita (2016) showed that a high percentage of canopy cover in all but Madola sites. However, according to the results, canopy cover alone did not contribute to the existence of *Rasboroides pallidus* in the Gin River. Simultaneously, Homadola, and Kosmulla areas possessed a high canopy cover, but the flow rate and substrate compositions were not suitable compared to the sites in which this species was found. This may be the reason for the absence of *Rasboroides* in the latter two locations.

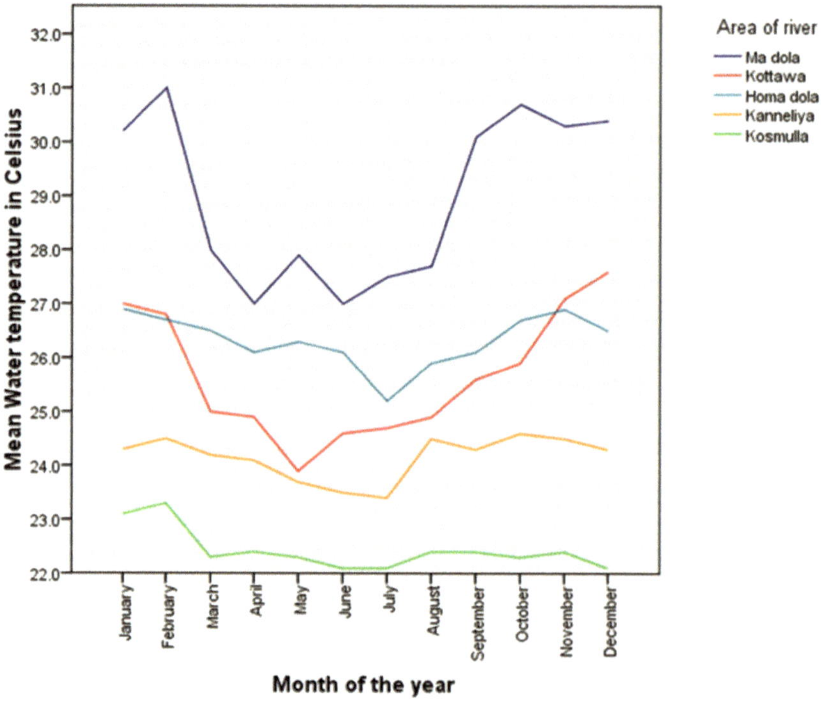

FIGURE 10.4 Variation of mean monthly water temperature of the sampling sites in Gin and Polathumodera rivers throughout the study period (from January to December).
Source: Reprinted from Batuwita (2016).

A recent study on the temperate fishes also revealed that the fishes' increased temperature to shift towards up-stream (Carosi et al., 2019).

10.2.2.5 WATER PH

Like temperature, pH also has a significant influence on the breeding of fish. The onset of monsoon in the south-western region of the island had lower pH values (Figure 10.5), whereas, during the dry season, pH values were markedly increasing (except for the site, Kosmulla; Figure 10.5) most probably because it received a continuous supply of water from Duvilli Ella, which falls through Sinharaja WHS. During the study (Batuwita, 2016), it was found that the electrical conductivity of the stream water in the forested sites (e.g., Kanneliya, Kosmulla) was consistent throughout the year, while as at the other three sites, it fluctuated mostly due to the intensive agricultural practices in the three sites. Although Kottawa was located close to a rainforest, it was also very close to Tea Research Institute, Kottawa (TRI), in a tea estate. Electrical conductivity fluctuations in the Kottawa sites may be due to agrochemical discharges from the surrounding catchment.

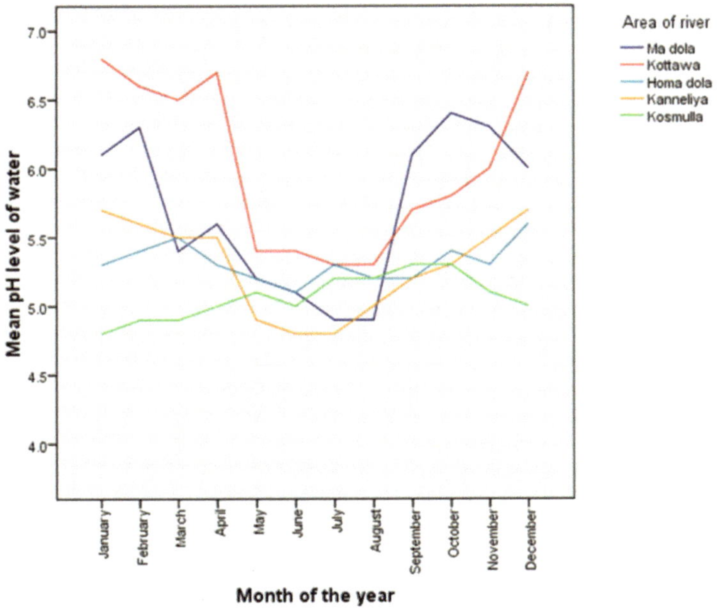

FIGURE 10.5 Fluctuation of mean monthly pH in the selected sites in Gin and Polathumodera rivers (from January to December).
Source: Reprinted from Batuwita (2016).

10.2.2.6 TURBIDITY

Turbidity changed rapidly due to rains, especially in the streams with great human influence (Figure 10.6). The Homadola site was the most disturbed site with high turbidity, significantly lower species richness (Batuwita, 2016). However, the other four sites had clear water. Still, with the onset of heavy rains, the Kottawa and Kanneliya sites' turbidity changed from good to average, and this was the period of spawning of *Rasboroides pallidus* (Batuwita, 2016). According to the researchers, during the rainy season, *Rasboroides pallidus* was neither observed in the turbid nor clear water, but was found in the up-streams and appeared again in clear water observed during the dry season. Thus, *Rasboroides pallidus* was showing Spatio-temporal distribution within this range due to its breeding behavior.

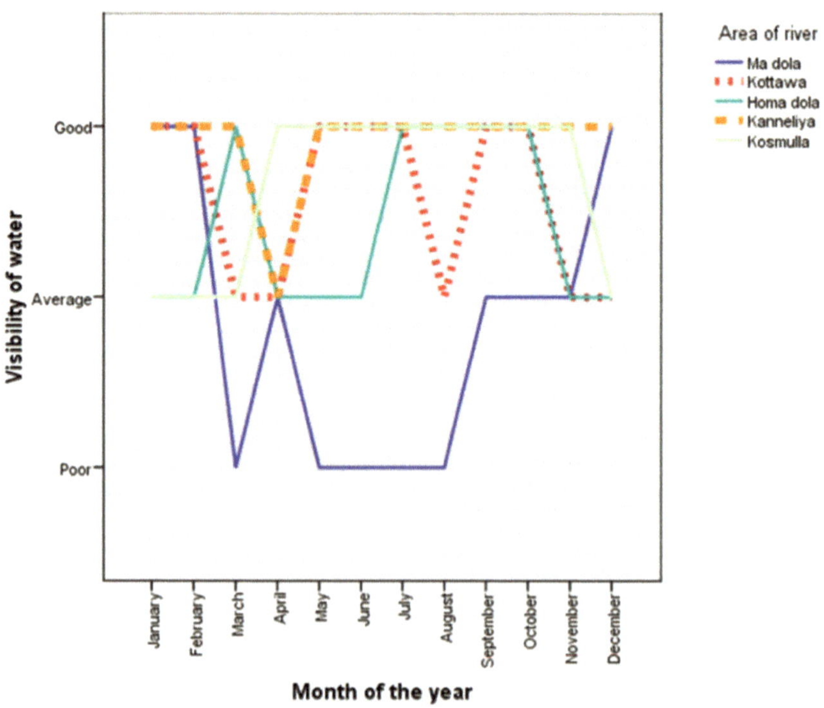

FIGURE 10.6 Variation of turbidity throughout the study period in Gin and Polathumodera rivers (from January to December).
Source: Reprinted from Batuwita (2016).

According to Piet (1996), in the lentic water bodies, the distribution of species and their size-classes differed considerably based on dry and wet conditions (in the wet season-high turbidity and the dry season low turbidity). The decreased turbidity during the drought period (period 2) affected the entire fish's daytime distribution, causing a shift towards the bottom (the deepest regions). During the wet season (period 1), fish with small size-classes and almost every species preferred the vegetated inshore zone (Piet, 1996). In this season (wet/period 1), all herbivorous/ detritivorous species occupied similar positions along the reservoir's vertical and horizontal gradients. The preference for either offshore or the inshore zone was not observed during this period. However, *Oreochromis* spp. had shown a distinct shift between these two periods, demonstrating a significant preference for the bottom half of the water column during the dry season. Interestingly, during both periods, the benthivores like *Puntius thermalis* and *P. dorsalis* were found close to the bottom and preferred to live in the inshore zone, while *Systomus sarana* preferred the column zone of the offshore, where a significant size-specific effect along the vertical gradient was observed during the wet season for all the benthivorous species. Piet (1996) showed that during both wet and dry seasons, the piscivorous species inhabited mainly the bottom layers where *Glossogobius giuris* and *Mystus* spp. Inhabited the inshore zones, daytime segregation was observed for them during the wet period along the horizontal gradient.

Piet (1996) observed significant spatial differences in the proportion of food items in the gut of major consumers along the vertical and horizontal gradients, where the general trend found along the vertical gradient for both periods respectively was increased availability of gastropods, micro-benthos (wet season) and detritus (dry season) towards the bottom, while a decreased availability of zooplankton (in both periods) and along the horizontal gradient the availability of insects, chironomids, and epiphytic algae were observed to increase towards the inshore zone (Kalk et al., 1979; Dvorak and Best, 1982; Rabe and Gibson, 1984; Piet, 1996). The two periods' trend only contradicted the availability of the food items: phytoplankton and detritus concentrations along the vertical gradient, which were inversely related (Piet, 1996).

10.2.3 NICHE DIMENSIONS

10.2.3.1 SPATIAL AND TEMPORAL SEGREGATIONS IN THE LOTIC ECOSYSTEMS (ON THE WATER COLUMN DEPTH, VELOCITY, SUBSTRATE, DIET, AND MORPHOLOGY)

Of the 190 species pairs, 167 (88%) had lower overlap values for either diet or relative depth or both (Moyle and Senanayake, 1984). Of the remaining 23 species pairs, seven were found in distinct habitats, thus less co-occurrence. The same researchers revealed that *Schistura notostigma* and *Sicyopterus lagocephalus* segregated based on the substrate, while *Mastacembelus armatus* and *Channa orientalis* are based on both substrate and water column depth. Also, 14 pairs observed by them did not segregate based on diet, microhabitat, or habitat. However, a few of these species' combinations showed overlap values of 0.67 or more for either diet or relative depth.

However, *Belontia signata* represented five of these pairs and thus classified as a habitat and food generalist. *Garra ceylonensis* and *Devario micronema,* were more specialized in their habitat requirements than the other fish (Moyle and Senanayake, 1984; Batuwita, 2016). Also, *Puntius dorsalis,* and *Garra ceylonensis* occurred in fast water in association with coarse substrates. Both of them fed on diatoms and detritus, but *Puntius dorsalis* was a suction feeder. *Garra ceylonensis* scraped material from rocks. *Rasbora dandia* and *Devario micronema* were fusiform in shape, found in the column zone, and fed on insects (Moyle and Senanayake, 1984). Still, *Devario micronema* preferred to hold the position in fast water, feeding on the drift, whereas *Rasbora dandia* tended to cruise about in all types of water. Indeed, the iridescent colors and barring pattern made *D. micronema* extremely difficult to observe in turbulent water, where it often held position.

In contrast, *Rasbora dandia* found to cruise through fast water but rarely held in it for extended periods. A significant correlation between relative depth and diet seems to have been caused by the large number of species pairs with low overlaps in both dimensions. Thus, the segregation of species by depth is related to feeding habits and suitable prey availability.

Wickramanayake and Moyle (1989) also categorized species based on feeding habits: column insectivores, column omnivores, benthic omnivores, benthic carnivores, and diatom feeders, whereas *Rasbora pallidus* and *Devario malabaricus* were grouped in the column insectivore category, which fed mostly on Ephemeroptera larvae and terrestrial invertebrates, while Ephemeroptera larvae somewhat frequently observed in the 'drift' component and were preyed on by both species mentioned above. Also, ants comprised a large proportion of the terrestrial invertebrate food component of *Rasboroides pallidus, R. vaterifloris* (Shirantha et al., 2005; Batuwita, 2016) and *D. malabaricus,* and a small proportion of diet also included seeds and algae for the latter species. However, most filamentous algae were not considered digested food for *Devario malabaricus* (Wickramanayake and Moyle, 1989).

Belontia signata, a column omnivore mostly fed on aquatic insects, crustaceans, and lesser amounts of algae and plant material, while *Tor khudree* and *Rasbora dandia* were mid to low-column zone omnivores, which devoured on a wide range of food types, but the former fed more on aquatic insects, especially chironomid larvae and the latter's diet included mostly terrestrial insects (Wickramanayake and Moyle, 1989). The benthic omnivore group comprised of *Pethia nigrofasciata, Puntius titteya,* and *Lepidocephalichthyus thermalis*. The diet of *Pethia nigrofasciata* had a comparatively broad variation, which mostly consisted of plant material, aquatic insects, diatoms, and algae. However, *Puntius titteya* had a much diverse diet than *P. nigrofasciata,* which fed on plant material, algae, insects (mostly Diptera, Ephemeroptera, and Trichoptera larvae) and crustaceans, though *Lepidocephalichthyus thermalis*' diet consisted of mostly chironomid larvae and filamentous algae (Wickramanayake and Moyle, 1989).

Channa orientalis, C. kelaartii, and *Schistura notostigma* were reported as benthic carnivores, which fed exclusively on aquatic insects such as larvae of Coleopterans, Dipterans, Ephemeropterans, Trichopterans, Odonate nymphs, and crustaceans (Wickramanayake and Moyle, 1989). *Channa kelaartii* and *C. orientalis* are secretive species (maybe nocturnal), often found to occur amongst submerged vegetation along stream margins and among boulders, respectively. Especially the latter species preferred to live in clear waters and nearby fast-flowing streams (Wickramanayake and Moyle, 1989).

As described by Wickramanayake and Moyle (1989) *Puntius bimaculatus, Pethia cumingii* and *Garra ceylonensis* can be grouped into the diatom feeding fishes. However, *Puntius bimaculatus* and *Pethia cumingii* also fed on some benthic insects. Based on their morphology, the species mentioned above are bottom feeders, and the *Garra ceylonensis* is well adapted to the stream bottom with gravels or boulders, a specialized benthic scraper (Moyle and Senanayake, 1984).

The most important feature of the internal morphology, which gives evidence on their dietary habit, is the gut length: herbivores and diatom feeders have relatively a very long gut, whereas carnivores or omnivores with a relatively short gut. Herbivores and diatom feeders having longer guts, with *the* gut length of *Garra ceylonensis* as much as six times the body length (Wickramanayake and Moyle, 1989). According to them, amongst most of the *Pethia* and *Puntius* species, the relative gut length was between two to three times the body length. They also reported that the gut length of a few species of insectivores was less than the body length, while the gut lengths of omnivores were as long as their body lengths.

The niche breadth values for the six niche dimensions (relative depth, water column depth, the velocity at fish, substrate, diet, and morphology) indicated that most species used a relatively narrow spectrum of available resources. However, there appeared to be at least some generalists and some extreme specialists in each dimension. Still, the division between them seemed to be the water column's position (relative depth) (Moyle and Senanayake, 1984). Wickramanayake and Moyle (1989) revealed that high species overlaps were associated with species pairs involving at least one introduced species (including *Pethia nigrofasciata*, i.e., certain fishes introduced to Mahaweli River). They further explained that the species mainly overlapped due to substrates, whereas the separation was reported mostly because of the focal point depth and the relative depth. Introduced species like *Pethia cumingii* and *P. nigrofasciata* had accounted for more microhabitat overlaps than those of the indigenous species (Wickramanayake and Moyle, 1989). Thus, the data suggested that the introduced species tend to interact more strongly with each other than some indigenous species.

Wickramanayake and Moyle (1989) mentioned that the three diatom feeding guild members, *Puntius bimaculatus, Pethia cumingii* and *Garra ceylonensis* overlapped highly due to the food. Interestingly, they had observed a dietary shift of *Pethia nigrofasciata* in Ceypotha Ela (from diatoms to plant material). Omnivorous *Rasbora dandia* and *Tor khudree* were grouped (from mid-to-low-column zone). However, they represented two different feeding habits: *Tor khudree* fed more on aquatic insects (chironomid larvae commonly associated with the benthos). *Rasbora dandia* had more terrestrial insects (Schut et al., 1984; Kumar and John, 1987; Wickramanayake and Moyle, 1989).

It was reported that out of 66 pairs, only 15 pairs overlapped moderately or highly for the relative depth, velocity, and food. Moreover, 11 of these 15 pairs overlapped for one of the above parameters. The following two pairs had the highest overlaps for all the three parameters: *Pethia cumingii* with *Puntius bimuculatus* and *Pethia nigrofusciata* with *Puntius bimaculatus*.

Batuwita (2016) showed that almost all the species were either omnivorous or herbivorous. None of them was a potential predator of *Rasboroides pallidus,* where sampling had been performed during the daytime; thus, predators of *Rasboroides pallidus* were not reported. However, the presence of *Channa orientalis* and *Clarias brachysoma,* which may be potential predators of *Rasboroides pallidus,* had been recorded in those sites (De Silva et al., 2015). Wickramanayake and Moyle (1989) recorded a high overlap between *Lepidocephalichthyus thermalis* and *Rasborides pallidus* (introduced populations) at their study site (Ginigathhena), though former species together with *Rasboroides pallidus* or with *R. vaterifloris* were not observed in the wet zone streams.

10.2.3.2 SPATIAL AND TEMPORAL SEGREGATIONS IN THE LENTIC ECOSYSTEMS (DIET, MORPHOLOGY, SEASONS, DIEL VARIATION OF DIET)

Based on feeding, species overlaps were reported in the reservoirs, as well. Piet (1996) discussed the species which inhabited the Tissawewa reservoir in Sri Lanka and divided the fish assemblage into four trophic groups: (i) herbivorous/detritivorous trophic group (detritus and to lesser extent phytoplankton and macrophytes), which included *Amblypharyngodon melettinus* and *Oreochromis mossambicus*; (ii) benthivorous trophic group (feed on chironomids, gastropods, and microbenthos) consisted of barbs, *Puntius thermalis, P. dorsalis,* and *Systomus sarana*; (iii) zooplanktivorous group, *viz., Hyporhamphus limbatus* and *R. dandia*; and (iv) piscivorous group which included *Glossogobius giuris* and *Mystus* spp.

Interestingly, the introduced species, *Oreochromis mossambicus* showed a dietary shift from zooplankton to detritus (Piet, 1996). On the other hand, according to the study, species with benthivorous diet showed differences in their diet quality and quantity, e.g., *Puntius thermalis* fed mainly on chironomids, microbenthos, and detritus. In contrast, *P. dorsalis* fed on chironomids, gastropods, and microbenthos, whereas gastropods and macrophytes were the principal food items of *Systomus sarana*. It has been documented that for most fish species, zooplankton was the main diet during their initial stage of ontogenetic development (Gilliam, 1982).

With the increase in body size, the proportion of zooplankton content in all barb species diet decreased. In contrast, the proportion of chironomids, gastropods, and fish increased respectively for *Puntius thermalis, P. dorsalis,* and *Systomus sarana* (Piet, 1996). The ontogenetic changes cause the shifts of diet from zooplankton to invertebrates in the piscivores species, which inhabit the temperate region, e.g., the Large-mouth bass (Gilliam, 1982). Piet (1996) also reported a decrease in plant matter with the increase in body size of *Puntius thermalis*. In contrast, in *Systomus sarana* the plant matter increased. With the increase in size, zooplanktivorous *Rasbora dandia* showed an increased proportion of insects in its diet, and the piscivores exhibited some distinct size-related changes in its diet.

Welianga and Amarasinghe (2003) identified seven main feeding guilds based on data from three different reservoirs in Sri Lanka, *viz.,* Minneriya, Udawalawe, and Victoria. They are: (i) the detritivorous and phytoplanktivorous guild—*Amblypharyngodon melettinus*, *Esomus danrica*, *Oreochromis mossambicus,* and *O. niloticus*; (ii) the zooplanktivorous guild—*Hyporhamphus limbatus*; (iii) the insectivorous guild—*Laubuka lankensis* and *Rasbora dandia*; (iv) the macrophytophagous guild—*Etroplus suratensis* and *Tilapia rendalli*; (v) the omnivorous guild—*Dawkinsia singhala*; (vi) the benthic invertebrate predators—*Puntius thermalis* and *P. dorsalis*; and (vii) the invertebrate and vertebrate predatory guild—*Glossogobius giuris*. However, according to the authors, a considerable variation had been observed between these groups: primarily based on the size groups and the seasons (for rising/receding water levels). *Dawkinsia singhala* exhibited such variations during both seasons; *Puntius thermalis* showed a size-dependent dietary shift mainly during the rising water level (Welianga and Amarasinghe, 2003). On the other hand, the diet of *Amblypharyngodon melettinus*, *Oreochromis mossambicus,* and *O. niloticus* were mainly detritus and a few phytoplankton. Welianga and Amarasinghe (2003) mentioned that body size did not affect the dietary habits of *Amblypharyngodon melettinus.* Still, a pelagic surface feeder,

Hyporhamphus limbatus is considered a species with a narrow nutritional pattern. The benthic invertebrate predators like *Puntius thermalis* and *P. dorsalis* exhibited similar diet items for both seasons but fed on a wide variety of feeds (e.g., ostracods, insects, insect larvae, detritus) (Welianga and Amarasinghe, 2003). They had further stated that *Puntius thermalis* and *P. dorsalis* consumed qualitatively different food items for their diet, and those food items had more significant variation according to their body size.

Kumara et al. (2009) also discussed the species distribution in the same three reservoirs, which were studied previously by Welianga and Amarasinghe (2003). Based on catch data, they grouped the species into two categories, inshore, and offshore example for the horizontal distribution of fish according to their habitat utilization. According to the latter authors *Amblypharyngodon melettinus, Dawkinsia singhala, Puntius thermalis,* and *Rasbora dandia* have shown the same distribution in both zones. Kumara et al. (2009) stated that phytoplankton is the most abundant food type in the above three reservoirs. *Amblypharyngodon melettinus, Puntius thermalis,* and *Dawkinsia singhala* feed mainly on them (Hofer and Schiemer, 1983; Weliange and Amarasinghe, 2003).

10.2.3.3 DIEL VARIATION OF DIET

Most interestingly, Piet (1996) demonstrated diel variation of the diet of different fish species in the Tissawewa reservoir and identified five other trophic groups: (i) the detritivorous *Amblypharyngodon melettinus*; (ii) the herbivorous *Oreochromis mossambicus* and *O. niloticus*; (iii) the benthivorous barbs *Puntius thermalis, P. dorsalis* and *Sytomous sarana*; (iv) the small pelagics *Hyporhamphus limbatus* and *Rasbora dandia,* which feed on insects; and (v) the piscivores species *Glossogobius giuris* and *Mystus* spp. They reported significant ($p<0.01$) diel changes in the diet for four species (*Puntius thermalis, Sytomous sarana, Hyporhamphus limbatus,* and *Oreochromis mossambicus*), which belonged to three trophic groups.

The two main food items (macrophytes and phytoplankton) that caused this variation showed inverse diel variations in their diet proportions. In contrast, the proportion of epiphytic algae in the gut of Oreochromis niloticus was significantly higher at midnight than during the rest of the day (but was not significant). The diel feeding behavior varied considerably for the two benthivorous species, *Puntius thermalis* and *Systomus sarana*. The proportion of zooplankton in the gut of *Puntius thermalis* was significantly higher at night than the daytime and *vise versa* for the proportion of fish eggs in its gut.

In the gut of *Systomus sarana* only insects and gastropods (Menon and Chacko, 1958) were observed and showed significant diel changes. Proportions of insects in the guts of *Systomus sarana* was highest at noon. The proportion of gastropods was highest at dusk, while some other species like *Hyporhamphus limbatus* had no macrophytes in their guts at night. Still, at noon the proportion of macrophytes was significantly higher, while the proportion of insects was lower than the proportion of insects found during the rest of the day (Piet, 1996).

Rasbora danconius had a higher proportion of insects at dusk, while some species did not show such variations for the diel changes, e.g., *Amblypharyngodon melettinus Glossogobius giuris* (Piet, 1996). According to the authors, active feeding time is also essential. The following variables independently determined it, i.e., gut fullness, the proportion of digested matter, and relative activity, except for Systomus sarana *Hyporhamphus limbatus, Mystus* spp. (Piet, 1996)and *Rasbora dandia,* in which gut fullness did not differ significantly between the time of day, where gut fullness was the best indicator for most of the species. Also, of those four above mentioned species, the relative activity differed significantly only between the times of day for *Hyporhamphus limbatus* and *Mystus* spp.

Six species, *viz., Amblypharyngodon melettinus, Puntius thermalis, Systomus sarana, Hyporhamphus limbatus, Mystus* spp. (Piet, 1996) and *Oreochromis mossambicus* showed a significant diel variation in their feeding activity. Analysis of the same researcher also showed that the largest diurnal species comprised of *Hyporhamphus limbatus, Amblypharyngodon melettinus, Puntius dorsalis, P. thermalis,* and the *Oreochromis* spp. Also, *Mystus* spp. was an exclusively nocturnal fish. At the same time, the third group consisted of species that showed a feeding activity that was more evenly distributed over the 24 hours with mainly diurnal *Systomus sarana,* nocturnal *Mystus* spp. and crepuscular *Rasbora dandia.*

The typical diurnal feeders are active from dawn until dusk. In contrast, most of the diurnal feeders, feeding was based on their gut fullness and inversely to the proportion of digested matter, and only the detritivorous Amblypharyngodon melettinus had a proportion of unrecognizable matter. It was not correlated with the gut fullness, which may be due to the unrecognizable ingested food (Piet, 1996). The relative activity of *Puntius dorsalis* suggested that it is more active at dawn, though all variables observed by Piet (1996) were attributed to diurnal species. *Oreochromis mossambicus* and *O. niloticus* displayed a similar period of active feeding during the daytime. Simultaneously, *Systomus sarana, Glossogobius giuris,* and *Rasbora dandia* did not show a distinct diel pattern, thereby suggesting that these species are evenly active throughout the 24 hours (Piet, 1996). He also had concluded that *Systomus sarana* is a diurnal species showing a relatively longer feeding activity because it was active from dawn until dusk. On the other hand, *Glossogobius giuris* is a nocturnal species, and *Rasbora dandia* is mainly crepuscular, while contrary to the above, *Mystus* spp the only exclusively nocturnal species.

Piet (1996) demonstrated diel spatial variation on the vertical distribution for species assemblage at Tissawewa tank in Southern Sri Lanka. He reported that except for *Mystus* spp., vertical distribution of *Hyporhamphus limbatus* and *Rasbora dandia* varied significantly through the 24 hours, and size-related diel differences were not observed in the vertical distribution for all the observed species. Additionally, except for insectivorous *Hyporhamphus limbatus* and *Rasbora dandia,* which occupied upper layers of the column zone throughout the 24 hours, all other species showed a significant ($p<0.01$) feeding status effect on the vertical distribution and especially at dawn when *Hyporhamphus limbatus* and *Rasbora dandia* were actively feeding, while *R. dandia* was found just below the upper layer, which was occupied by *Hyporhamphus limbatus* (Piet, 1996).

During feeding, some benthivores, e.g., *Puntius thermalis* and *P. dorsalis* inhabited the bottom layer similar to the introduced herbivorous grazers like *Oreochromis mossambicus* and *O. niloticus* though *Systomus sarana* showed diel variation and were found in the bottom layers at dawn and moved strongly towards the surface when it became dark and interestingly when actively feeding, piscivores like *Glossogobius giuris* and *Mystus* spp. (Piet, 1996).

10.2.3.4 SPECIES ABUNDANCE AND SPAWNING IN THE LOTIC ECOSYSTEMS

Ediriweera et al. (2003) reported that *Garra ceylonensis, Pethia nigrofasciata* and *Lepidocephalichthyus thermalis* had the highest species abundance upper, middle, and lower stream habitats. Moreover, it was mentioned that *Puntius titteya* was frequently observed in the tributaries of this stream, and *Pethia nigrofasciata* was the most common species throughout the stream. Jayaratne and Surasinghe (2010) stated that runs maybe the most suitable habitats for most of the fish species present in their study area. They also reported a higher abundance of *Dawkinsia singhala.* Two invasive species most preferred to runs were *Poecilia reticulata* and *Oreochromis mossambicus* (Jayaratne and Surasinghe, 2010), which is an important observation because these two invasive species live together most other indigenous and endemic species. Thus, species competition may occur with the increase of invasive species populations.

Batuwita (2016) sampled and described species assemblages from four different streams at Homadola, Kanneliya, and Kosmulla (Gin River), and Kottawa (Polathumodera River) in the two river systems of Sri Lanka. Based on their data, the total comprised about seven to eight species for each location. The lowest number of species was recorded from Homadola (four species) due to rapid water flow and low habitat complexity in the Homadola site. In contrast, Kosmulla, Kanneliya, and Kottawa showed a diverse fish fauna. Moreover, Kanneliya had the most incredible abundance during the study period. One of the sites where *Rasboroides pallidus,* an endemic freshwater fish, was observed had the highest abundance for the assemblages compared to the other sites (Figure 10.7).

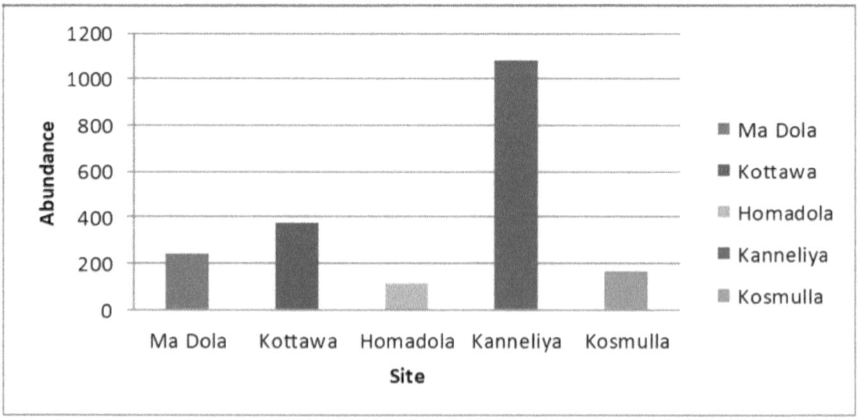

FIGURE 10.7 Cumulative number of individuals observed during the study period at each site.
Source: Reprinted from Batuwita (2016).

The total number of Kanneliya and Kosmulla species was comparatively higher, probably because it had a relatively undisturbed and more extensive forest cover than other sites. Six species (including *Rasboroides pallidus*) showed an abundance of ~40 (adult) individuals at Kanneliya: *Devario micronema, Dawkinsia singhala, Pethia nigrofasciata, Rasbora dandia,* and *Systomus pleurataenia.* The abundant distribution of the species across sampled sites showed a typical right skew (Figure 10.8), like studies of birds, butterflies, and other communities (Williams, 1964; Magurran, 2004). This means that most fish species are relatively rare, while a few species dominate an area in terms of their abundance. Fortunately, exotic invasive species like *Oreochromis* spp. were not collected from these sites, but there is a single record of *O. mossambicus* in the Hiyare reservoir (De Silva, 1997) near Madola.

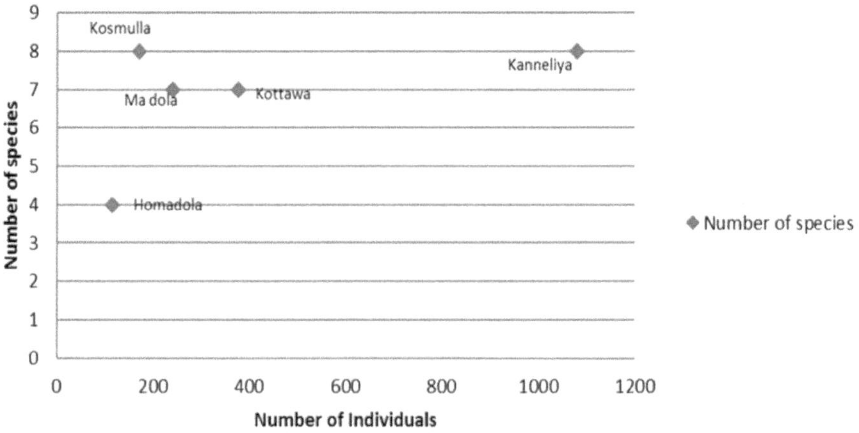

FIGURE 10.8 Distribution of species richness and abundance in all the sites.
Source: Batuwita (2016).

The mean abundance of *Rasboroides pallidus* was very high in the Kanneliya forest (Figure 10.9), which may be due to a greater forest cover, natural habitats, abundant feeds with minimal pollution. Human encroachment was very high in the Kottawa forest reserve, having only a few natural streams (often polluted). This species is now restricted to a few locations within the Kottawa forest. However, Rasboroides pallidus had been observed since the late '90s in the Kottawa forest (Batuwita, 2016) from various locations (from both the arboretum side and the other side, where a human-made pool had been built). Due to the human-made pool, the stream was diverted, which resulted in not having enough water to the downstream, where *Rasboroides pallidus* thrived around the late '90s. As a result, within ~10 years, all the fish, including *Rasboroides pallidus* in this stream, had disappeared (Batuwita, 2016).

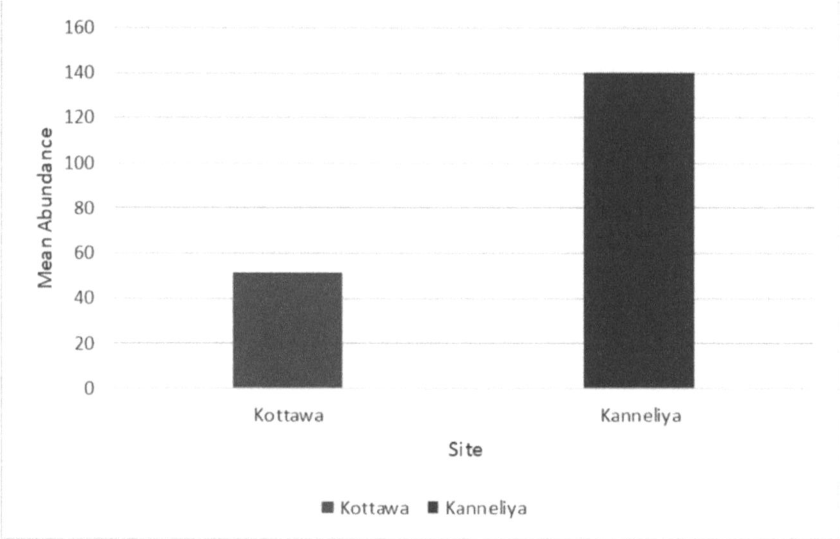

FIGURE 10.9 Mean abundance of *Rasboroides pallidus* in the two sites during the study period.
Source: Batuwita (2016).

Most fishes synchronize breeding activity with monsoon or flood seasons (De Silva et al., 1985). Batuwita (2016) showed that *Rasboroides pallidus* could be considered a seasonal spawner (Figure 10.10) because there was a peak in individuals between July to November. Between these months, juvenile numbers were very high. From March to July, all the sites received heavy rains, and this season was the breeding season of *Rasboroides pallidus*.

Rasboroides pallidus was challenging to collect or observe during the rainy season. Deraniyagala (1958) also reported that *Rasboroides vaterifloris* had extirpated at this type of locality. However, this species persists in small, shaded streams and rivulets in and around Gilimale (Batuwita et al., 2013; De Silva et al., 2015). Moreover, Pethiyagoda (1991) noted that the survey referred to by Deraniyagala (1958) might have been conducted during the rainy season, when streams were often turbid and swollen. *Rasboroides vaterifloris* is not easy to collect under such conditions. Many freshwater fishes in rivers and streams of Sri Lanka migrate within river systems, and that might frequently be linked with breeding (Silva and Davies, 1986). During Batuwitas (2016) study, several individuals of *Rasborides pallidus* were observed in the Kanneliya forest, where fishes swam up the smaller creeks to spawn in shallow waters during periods of heavy rain.

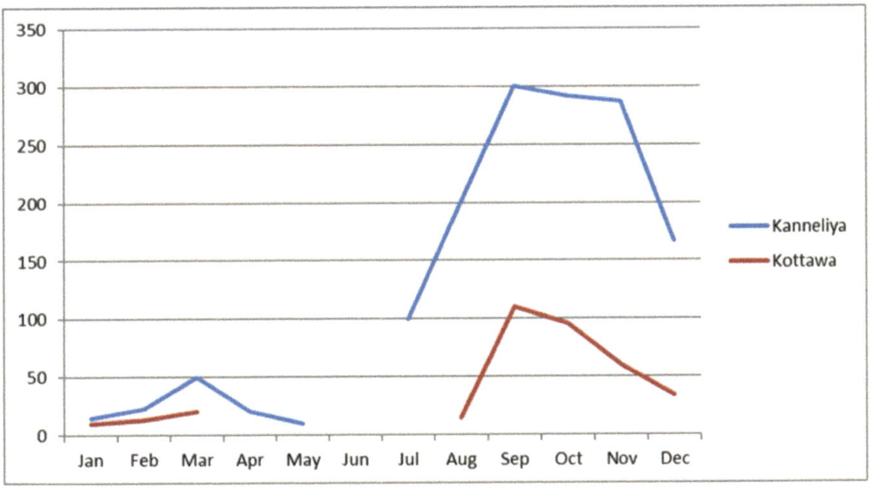

FIGURE 10.10 Number of individuals (including juveniles) of *Rasboroides pallidus* observed in the Kottawa and Kanneliya.
Source: Batuwita (2016).

10.2.3.5 SPECIES ABUNDANCE AND SPAWNING IN THE LENTIC ECOSYSTEMS

Species abundance in the reservoirs evaluated in certain studies has shown that some species prefer to live in offshore areas, whereas others prefer the inshore regions. Kumara et al. (2009) had identified *Amblypharyngodon melettinus*, *Puntius thermalis* and *Dawkinsia singhala* as the most abundant species in Minneriya, Udawalawe, and Victoria reservoirs, demonstrating that in the offshore areas, *Dawkinsia singhala* and *Puntius thermalis* were abundant. They concluded that the availability of preferred feeds in pools facilitated their colonization success as well.

Piet (1996) showed that the highest relative abundance (in terms of number and biomass) of the herbivorous guild was represented by *Amblypharyngodon melettinus* and its size-class 2 was the most abundant. This size-class was much smaller than *Oreochromis mossambicus* and *O. niloticus* (respectively for the size-classes of 4 and 6 [in terms of biomass]). The benthivorous guild, which accounted for the second most crucial guild, had *Puntius dorsalis* with the highest biomass. In the same guild, *Puntius thermalis* was the smallest species, which had 62% of the species biomass in size-class 3. In comparison, *Systomus sarana* was the largest species with over 80% of its biomass, representing size-classes 5 and 6. The piscivorous feeding guild was the least essential guild with *Mystus* spp.; however, slightly more abundant than *Glossogobius giuris.* Later species was comparatively more considerable than the former, with most of its biomass in size-class 6, whereas the main size-class of *Mystus* spp. was 4 and 5. Within the zooplanktivorous guild, *Rasbora dandia* was the most critical species with about twice the biomass of the slightly larger size class of *Hyporhamphus limbatus* (Table 10.1).

TABLE 10.1 Total Length Ranges of Each Size-Class of Fish

Size-Class	1	2	3	4	5	6
Total length (cm)	3.0–4.4	4.5–6.9	7.0–9.9	10.0–13.9	14.0–18.9	>19.0

Source: Piet (1996).

Reservoir fishes spawn in the littoral zone (inshore areas) during spawning (De Silva and Chandrasoma, 1980; De Silva and Sirisena, 1988; Pet and Piet, 1993). Thus, vegetation is considered an essential factor determining the spawning sites in the dry zone water bodies of Sri Lanka.

10.3 DISCUSSION

10.3.1 MORPHOLOGY

Motta et al. (1995) concluded that an ecomorphological study's initial goal is to identify the essential characters that are correlated with the features of its ecology. Fishes have functional morphological features to survive in different habitats (Moyle and Cech, 1982; Webb, 1984; Matthews, 1998; Pflieger, 2004). Based on the functional analysis, the causal mechanism can be explained to provide predictions on the role of morphological variation in determining the potential niche (Piet, 1996).

The diversity of body shape and morphological specializations associated with feeding is generally considered to reflect co-evolution to reduce competition for limited resources (Gatz, 1979). However, other studies have found a great deal of dietary similarity among morphologically diverse fishes in tropical streams (Lowe-McConnell, 1975). In a most extreme case, Knoppel (1972) found little dietary specialization among the Amazonian fishes he studied, despite the presence of specialized trophic structures. In contrast, morphological fields of tropical stream fishes (e.g., in Sri Lankan fishes) were associated with dietary and microhabitat specialization, i.e., among the seven Cyprinid species, which, as a group showed the highest amount of morphological overlaps, e.g., algae-feeding *Pethia nigrofasciata, Puntius dorsalis* and *Systomus pleurotaenia* were often found together, where the first two species were bottom oriented. Still, *Pethia nigrofasciata* devoured filamentous algae off rocks with its terminal mouth, whereas *Puntius dorsalis* sucked up diatoms and detritus, with its fleshy lips and sub-terminal mouth (Moyle and Senanayake, 1984). Accordingly, *Systomus pleurotaenia* was a streamlined column zone fish and picked algae from boulders and logs with its terminal mouth, especially in fast water. Thus, the morphological diversity of the fishes reflected their ecological diversity (i.e., habitat utilization). According to Moyle and Senanayake (1984), several species are habitat generalists and nevertheless showed strong segregation from the habitat specialists.

Alternate fish morphologies have also evolved in Sri Lankan fishes that occupy other niches, including the bottom dwellers (*Garra* spp., Gobies, *Schistura notostigma*), where flattened or sloping heads minimize energy usage and allow the fish to remain close to the bottom while moving through the water or staying stationary in flowing waters (Matthews, 1985; Webb, 1989; Webb et al., 1996; Moyle and Cech, 1982). Streamline fish have an elongated caudal peduncle (e.g., *Devario, Puntius, Systomus*). Moreover, fishes' functional morphology has been shown to influence the fish location in lakes or rivers (Ehlinger and Wilson, 1988; Douglas and Matthews, 1992) or habitat selection (Hoagstrom and Berry, 2007). These findings collectively suggest that body morphology in fishes relates to the type of habitat where they exist and provides clues to the theory of niche partitioning in fish communities (Schlosser, 1982; Douglas, 1987).

In the fish communities, resource partitioning chiefly exists along the trophic dimensions (Ross, 1986). Hence, to determine the fishes' potential niches from their morphology, examining related morphological features to their feeding ecology is crucial. Since fishes grow continuously, it is essential to study their ontogenetic stages, including their size-classes. Piet (1996) showed the correlations between principal morphological components and food characters that indicated the crucial morphological attributes, which accounted for their feeding ecology. Of the morphological variables, a causal relationship has been established for food particle size and mouth gape (Wainwright and Richard, 1995), while the length of the gut had distinct correlations with the digestive demands as well (De Silva et al., 1980; Zihler, 1982; Vergina, 1991; Sturmbauer et al., 1992).

According to Moyle and Senanayake (1984), for the fishes habitat utilization and their morphology and digestive demands: (i) the form and position of the mouth (terminal mouth-omnivorous fish; superior mouth-surface feeders or carnivorous fish; subterminal or inferior mouth-bottom or benthic feeders); (ii) the structure of the gill rakers (filter feeders-more and compact gill rakers; omnivorous and carnivorous fish with few and thick gill rakers); (iii) the functional aspects of the swim bladder (benthic fish-no or rudimentary swim bladder; column or top column zone fish-well developed swim bladder); and (iv) the disposition and form of the fins relative to an active or passive lifestyle in conjunction with the sensory apparatus for locating the prey are involved.

According to Piet (1996), mouth gape (size and shape) was the most crucial feature of fish, and it explained about 50% of the total variance. Fishes with largemouth devour large prey, and the largemouth gape is also due to fishes' allometric nature of growth. Several previous studies have shown this character with a significant correlation for fishes (Gatz, 1979; Wikramanayake, 1990; Winemiller, 1991; Nagelkerke et al., 1994; Wainwright and Richard, 1995). However, Piet (1996) mentioned that this character had a considerable variation during the ontogenetic stages; thus, his conclusion may be biased. Also, an argument described that prey size is primarily constrained by the pharyngeal gape instead of the mouth gape, but this hypothesis was again proven false by subsequent studies (Werner, 1974; Kislalioglu and Gibson, 1976; Osenberg and Mittelbach, 1989).

The advantages in having a larger mouth opening, mostly attributed to the carnivorous species, because their prey covered the largest size range and all other species (except for the herbivorous *Amblypharyngodon melettinus* and omnivorous *Oreochromis mossambicus*) showed an increase in the average size of their prey throughout ontogeny (Piet, 1996). Studies have shown that the size of the mouth opening determined the maximum size of the prey (Randall, 1967; Starck and Schroeder, 1971; Gladfelter and Johnson, 1983; Keast, 1985; Wainwright, 1987; Osenberg and Mittelbach, 1989; Piet, 1996).

Piet (1996) concluded that gut length and body depth are the other most essential characters in fish and are useful to distinguish herbivores from carnivores. The relatively long gut of herbivorous species is possibly due to the relatively energy-poor food resources (Hofer, 1991). Overall relative gut length is considered a good indicator of adaptation to herbivore diet (Gatz, 1979; Hofer, 1991; Winemiller, 1991), which is essential to digest cellulose in the plant cell wall.

It has been proven that the possession of barbels indicates the importance of non-optic senses in feeding (Gilbert and Bailey, 1972) and the benthic habit of the fish (Gosline, 1973; Aleev, 1969; Gatz, 1979; Nagelkerke et al., 1994). Some species like *Systomus sarana* and *Mystus* spp., (which often feed on non-benthic food items) with well-developed barbels, allows them to forage nocturnally (Pet and Piet, 1993, Piet, 1996).

Gatz (1979) showed that the mouth's orientation is an indicator of prey's location relative to the predator. Moreover, species with a ventral orientation mouth are associated with benthic feeding and a dorsal orientation with foraging in the upper column/surface (Wikramanayake, 1990; Winemiller, 1991). However, there are some unique species like the piscivore *Glossogobius giuris,* which has a dorsal oriented/superior mouth and prefers the bottom layer (Pet and Piet, 1993) from which it feeds on its prime target species, *Amblypharyngodon melettinus,* which lives in the lower part of the column (Pet and Piet, 1993; Piet, 1996). The negative correlation between the eyes' absolute size and the vertical position is proven (Piet, 1996). The larger eyed species are better adapted to forage closer to the bottom, where the light condition is poor. Comparatively smaller eyes and dorsally oriented mouths of *Rasbora dandia* and *Hyporhamphus limbatus* are often found in the surface layer (Pet and Piet, 1993, Piet, 1996), which provides an example of the influence of their spatial distribution. The body depth has significantly correlated with the vertical distribution of food items: species with a higher relative body depth forage on benthic food items (e.g., *Puntius thermalis, P. dorsalis*, and *Systomus sarana, Oreochromis mossambicus*) (Piet, 1996). This was following the study of Sibbing et al. (1994), who found that typical benthic molluscivore possessed a deep body shape, thereby improving its maneuverability.

Inter-specific and intra-specific resource partition can be found due to their development stages (Werner and Gilliam, 1984; Werner, 1986). Thus, the importance of the ontogenetic changes in the study of ecomorphology has also been proven (Motta and Kotrschal, 1992). However, the reviews on this aspect are a handful (Galis, 1993; Nagelkerke et al., 1995) and often not ascertained in most studies. Based on Piets (1996) research, only about one-fourth of the species (he had studied) showed intra-specific differences for the ontogenetic stages. It concluded that several absolute measures of morphological characters that were previously considered species-specific were size-specific.

Moreover, within a fish population, different size-classes may be represented differently at certain times of the year simply because of demographic characters (Miller, 1984). However, certain characters like mouth gape and size of eye determined the maximum prey size and sight (when feeding), respectively, affecting the feeding position and microhabitat depth (Gatz, 1979; Barel et al., 1989; Wikramanayake, 1990; Wainwright and Richard, 1995). Moreover, the morphological changes during ontogeny: herbivory to carnivory and small prey to large prey of carnivorous species also provide examples for morphological changes for the individual feeding behaviors (Gilliam, 1982; Wainwright and Richard, 1995).

Piet (1996) concluded that for several species, intra-specific competition between specific ontogenetic stages was smaller than the inter-specific competition, e.g., juveniles of *Mystus* spp. vs. *Glossogobius giuris* and adults of *Hyporhamphus limbatus* vs *Rasbora dandia* and further stated that this could be applied to the size-classes of the omnivorous fish species as well.

10.3.2 ABIOTIC FACTORS

Numerous physicochemical aspects of aquatic habitats (i.e., abiotic factors) provide structural complexity such as substrate (Lipcius and Hines, 1986;

Smith and Coull, 1987), litter (Ware, 1972), emergent macrophytes (van Dolah, 1978), and submerged macrophytes (Crowder and Cooper, 1982; Stoner, 1982; Coull and Wells, 1983). Abiotic factors are considered the base for species' existence and ecological communities' structure and function (Dunson and Travis, 1991). Temporal heterogeneity of the environment is another factor affecting communities (Menge and Sutherland, 1976). Being ectothermic organisms, fish depend on the surrounding temperatures for metabolic activity, which allows for mobility and growth. Fishes are differently adapted to different light intensities, which affect their relative competitive abilities (Bergman, 1988; Diehl, 1988), which is particularly important for fish communities in lakes, deep pools in rivers or reservoirs, where the light regime depends on the time of the season, time of day, and water depth. Also, watercolor, and turbidity affect the light intensity in the water column (Ranåker, 2012), which may differently affect different fish species (Guthrie and Muntz, 1993; Ranåker, 2012).

Metabolism of fish are also affected by abiotic factors, e.g., different metabolic adaptions to different temperatures, salmonid fish (lives in temperate waters) have a higher active metabolic rate than cyprinid fish, mostly living in tropical waters (**Cavender, 1991**) at 15°C, which is approximately the optimum temperature of many salmonid species (Johnston and Dunn, 1987; Clarke and Johnston, 1999). Several studies have also shown that cyprinid fish can instead benefit from temperatures warmer than 15°C (Persson et al., 1991; Holmgren and Appelberg, 2000; Graham and Harrod, 2009). Metabolic traits and metabolic costs (depend on body size) of moving fish and foraging efficiency also determine their distributions (Clarke and Johnston, 1999).

Wickramanayake and Moyle (1989) showed that the depth utilization pattern for combined streams (n=5) was consistent with those in individual streams, which was sufficient proof to conclude that these species might be consistently choosing the particular depths, while habitat generalists prefer to live in variable depths, but even for them predictability of their choices is possible. Also, they had classified the studied species according to the water velocity gradient: (i) fast (e.g., *Garra ceylonensis, Tor khudree* and *Devario micronema*); (ii) slow (e.g., *Puntius titteya, Rasboroides vaterfloris, Lepidocephalichhtyus thermalis, Belontia signata* and *Pethia cumingii*); (iii) moderate (e.g., *Pethia nigrofasciata, Schistura notostigma, Puntius bimaculatus* and *Rasbora dandia*). Though the focal point velocities of most species were less than their water velocities, it showed that fishes prefer to live in relatively slower water microhabitats within their respective habitats, e.g., *Garra ceylonensis* and *Tor khudree* (fast water species usually found on or close to substratum) and contrary to the above, *Danio micronema,* a species found in mid to upper column, had no difference between focal point velocities and mean water velocities. *Rasboroides pallidus* usually found in slow water habitats associated with submerged vegetation, dead branches, whereas *Belontia signata* and *Malpulutta kretseri* prefer to live in pool-like habitats, and near submerged tree roots and similar forms of cover. Slow water species, such as *Lepidocephalicththyus thermalis, Puntius titteya* and *Pethia cumingii* were bottom associated. They were found close to the stream margins and associated marginal vegetation with comparatively high canopy cover.

Certain species, for instance, *G. ceylonensis, D. malabaricus, Pethia nigrojasciata, Puntius bimaculatus, Tor khudree* and *Rasbora dandia* had a considerable variability in the focal point velocities and mean water velocities, showing a general velocity preference. In contrast, *Pethia nigrofasciata* and *Puntius bimaculatus* usually live in pool-like habitats and runs with faster velocities. *Tor khudree* covered a wide range of velocities by cruising along the stream bottom, and *Rasbora dandia* would roll through and even feed in fast water but were usually found in slow water close to the stream margins (Wickramanayake and Moyle, 1989). Hence, they concluded that many species exhibited high variability for their velocities, usually preferred to live in the moderate velocities. Usually, the individual species' velocity preferences were similar in all the sampled streams ($n = 5$). Also, unique stream patterns were identical to the combined data, and these species were consistent in their relative distributions along the velocity gradients. In contrast, some species always occupied faster waters, and others preferred to live in slower waters.

Ebner et al. (2015) described the habitat selection of a moray eel, *Gymnothorax polyurandon*. They observed that the fish mostly stayed within pool habitats (~ 70%), and a few individuals in runs (23%) and rarely (3%) in riffles. It specifically and positively preferred slow flow (< 0.05 ms^{-1}) and deeper water (> 80.0 cm) than the higher flows (> 0.5 ms^{-1}) and shallow waters (< 40.0 cm depth). Also, boulders were the substrate they positively selected, whereas pebbles were avoided, and the other substratum categories were neutrally selected. Larger *G. polyurandon* (> 800.0 mm in total length) tended to refuge under relatively more massive boulders because of having a streamlined and elongate body. Slow-water tropical freshwater fish, *Maccullochella ikei,* most often found in association with cover (i.e., boulders, woody debris, macrophytes, and bedrock shelves). Though habitat shifts occurred regularly for some individuals, others changed rarely and often for only short periods before returning to their original site (Butler et al., 2014). According to the authors, during high-discharge events, *Maccullochella ikei* mainly occupied ephemeral backwaters and edge habitats. Inundated riparian vegetation was often used as cover and demonstrated a significant preference for pools. Still, fast water, riffles, and glides were usually avoided. This fish has a marked tendency for large boulders and woody debris and a significant avoidance of waters less than one meter. Still, all other depths were used in relative proportion and most frequently associated with water depths between 1–3 m (67%) and only 10% of locations deeper than five meters (Butler et al., 2014).

10.3.3 NICHE DIMENSIONS

10.3.3.1 DIET AND MORPHOLOGY

Kortmulder (1987) mentioned sophisticated interactions between fish species and their environment, i.e., interactions between species for food, space, and spawning. He also discussed the direct human influence (erosion, pollution) or climatic factors on the fishes. Tests of the Sales (1969a) model confirmed that the intensity of searching behavior varied greatly depending on the cover's water depth and presence. Searching behavior was lowest for *Acanthurus triostegus sandvicensis*, when it had access to shallow water, the preferred habitat (Sale, 1969b). Thus, emigration events are likely to occur when environments lack adequate resources, which trigger continued exploration until animals eventually emigrate from an area searching for suitable conditions elsewhere. Those individual emigrants readily become residents when needed resources are supplied, and those residents' become emigrants when resources are limited (Matter et al., 1989; Nelson et al., 2002).

Species morphology determines its potential niche from its constraints on foraging efficiency on the available food resources. In contrast, the realized niche of a species is changed due to its foraging efficiency concerning the other assemblage members. Piet (1996) showed that the foraging efficiency on each food resource is assumed to depend on three food characters: (i) size of the food; (ii) food type; and (iii) fish's vertical position in the water column and concluded that a species potential of avoiding competition is directly related to its possibilities of utilizing vegetable food resources, in which the carnivorous species have the most significant competition.

Piet (1996) divided the Tissawewa fish assemblage (they had observed) as follows: *Amblypharyngodon melettinus* and *Oreochromis mossambicus* (as herbivorous species); *Puntius thermalis, Puntius dorsalis* and *Systomus sarana* (as omnivorous species); and *Hyporhamphus limbatus, Rasbora dandia, Mystus* spp. and *Glossogobius giuris* (as carnivorous), where herbivorous species also have considerable morphological differences in their digestive tract; for instance, *Oreochromis mossambicus* has a stomach, while *Amblypharyngodon melettinus* does not have. The stomach substances help to utilize the plant, algal, or detrital matter of *Oreochromis mossambicus* (Moriarty, 1973; Moriarty and Moriarty, 1973; Bowen, 1976, 1979). Stomachless *Amblypharyngodon melettinus* is not capable of digesting phytoplankton Bitterlich (1985). Still, certain cyprinidae species like *Ctenopharyngodon idella* (Grass carp) are known to possess the capability of breaking-down plant materials mechanically by their pharyngeal teeth (Sibbing, 1991). Interestingly, *Amblypharyngodon melettinus* has well-developed pharyngeal teeth with grinding surfaces (pers. obs.). Moreover, the shape of the pharyngeal teeth helps to crush formidable prey such as gastropods by carps. A morphological study on the *Systomus sarana* in the Sri Lankan reservoirs revealed that it has the most powerful pharyngeal musculature making it the species best equipped to exploit gastropod diet (Adamicka, 1983). Furthermore, the distinct association between dentition and diet is proven (Lauder, 1983; Kotrschal and Goldschmid, 1983; Stoner and Livingston, 1984; Norton, 1988; Motta, 1989).

Carnivorous species have well-developed teeth (e.g., *Mystus* spp. and *Glossogobius giuris*) and are best adapted for piscivory (Gatz, 1979; Winemiller, 1991; Piet, 1996). Piet (1996) demonstrated high competition among carnivorous and omnivorous fishes and observed a more considerable variation in their prey-size in Tissawewa. Thus, suggesting that this competition may be avoided by the potentials of non-herbivorous species such as mouth gape size. Therefore, obligate carnivores like *Glossogobius giuris* can devour the largest prey, while *Mystus* spp.

feed on intermediate size prey followed by the smaller-sized prey by *Hyporhamphus limbatus* and *Rasbora dandia.* According to Piet (1996), the maximum mouth gape of the omnivores increases from *Puntius thermalis* to *P. dorsalis,* and to *Systomus sarana,* but remains below that of *Glossogobius giuris* and even relatively large-gaped species like *Puntius dorsalis* and *P. thermalis* remain below the gape size of two species of *Mystus.* Thus, the resource partition of these species in this reservoir based on their gape size is evident.

Moreover, resource partition can also be seen according to the orientation of the mouth. Tissawewa fish assemblage could further be divided into two typical benthivores, *Puntius thermalis* and *P. dorsalis,* and three species, *Glossogobius giuris, Hyporhamphus limbatus,* and *Rasbora dandia,* which were adapted to feed on prey from surface waters to at least the upper level of the column (Piet, 1996). Thereby, species having a dorsal orientation of mouth is unable to exploit benthic food sources. In contrast, species with a ventral orientation of the mouth are prevented from feeding on organisms occurring on the surface water. Although it lives in the bottom layer, *Glossogobius giuris,* with the dorsally oriented mouth, lacking a swim bladder, utilizes resources from the bottom (Piet, 1996). Interestingly, *Glossogobius giuris* mostly prefer shallow water. Hence the superior/dorsally oriented mouth helps to feed on its prey.

Some species of Tissawewa, as mentioned by Piet (1996), such as *Puntius thermalis, P. dorsalis, Systomus sarana,* and *Mystus* spp. with barbels, provide a significant advantage when foraging in poor turbidity conditions (during the rainy season and dark conditions at night). He concluded that the realized niches of the species of the Tissawewa fish were not unremarkable because this ecological principle have also been proven both in lentic (Schiemer and Hofer, 1983; Trewavas, 1983; Kumar, 1987) and lotic ecosystems (Schut et al., 1984; De Silva et al., 1984; Kortmulder, 1987).

It has been proven that the endemic species represent well-established niche segregations and ecological adaptations compared to the indigenous species (De Silva et al., 1977, 1980; De Silva and Kortmulder, 1977, 1982; Kortmulder et al., 1978). Thus, habitat destruction may damage the endemic fauna in Sri Lanka (Wickramanayake and Moyle, 1989).

10.3.3.2 DIETS AND ABIOTIC FACTORS

There were higher species overlaps (≥0.67) for the variables like relative depth, mean velocity, and feed (Wickramanayake and Moyle, 1989). Most of the above-discussed species pairs exhibited relatively well-defined overlaps for some factors other than relative depth, mean velocity, and feed. However, based on the complementarity state of the variables, sometimes those species were separated. *Puntius bimaculatus, Pethia cumingii* and *P. nigrofasciata* were consistently overlapped for the microhabitat variables, although they demonstrated some complementarity for the other resources (Wickramanayake and Moyle, 1989). Shirantha et al. (2005) also showed that the latter two species moderately overlapped based on the diet in the Kalu and Kelani Rivers of Sri Lanka. *Puntius bimaculatus* and *Pethia cumingii* overlapped highly on food and all microhabitat variables. They were still separated based on microhabitat (distribution patterns): *P. cumingii* inhabited shallow, downstream reaches with a sandy substrate and close to the river confluence. *Puntius bimaculatus* had a wide range and lived in up-streams. At the same time, *P. nigrofasciata* is a wide-ranging taxon, when foraging it often observed in schools together with *P. cumingii* (Wickramanayake and Moyle, 1989). This schooling behavior has also been observed by Senanayake (1980) and Sudesh Batuwita (Pers. obs., 1999; at Nawalapitiya). Wickramanayake and Moyle (1989) further stated that even though microhabitat availability differed among streams, species distribution relative to each other in resource space was consistent across streams based on the velocity, mainly fast water species *vs.* slow water species.

Also, similar to lentic ecosystems, those species, which live in streams and rivers showed a pattern of vertical distribution as well (i.e., surface, column, and bottom zones) though when considering all the species, only the column zone/layer species had a consistent pattern for the vertical distribution (Wickramanayake and Moyle, 1989). The overall microhabitat data correctly described the assemblage structure in all the streams (n=5) studied. It was also concluded that fish assemblage structure and habitat and microhabitat utilization are predictable based on fishes' prevailing environmental conditions and the zoogeographic distributions. According to Wickramanayake and Moyle (1989), the high degree of complementarity along the microhabitat and food axes and the consistent pattern in the streams they sampled supported the competition based resource partitioning consistency as described by Schoener (1974).

Wickramanayake and Moyle (1989) speculated that even though the intensity of inter-specific competitive interactions is reduced through co-evolution, the larger size of adult *Pethia nigrofasciata* and *Rasboroides pallidus* in the Ceypotha Ela is most likely due to competitive release within the spacious stream assemblages. In the Ceypotha Ela, *Pethia nigrofasciata* showed a dietary shift in the presence of *Garra ceylonensis* (a morphologically specialized diatom/algae feeder and the most abundant species in this stream) (Wickramanayake and Moyle, 1989).

Moyle and Senanayake (1984) have shown that although several species were habitat generalists, most species occurred in distinct habitats and mentioned that within these fish communities' habitats, microhabitat overlap among co-occurring species was low, particularly concerning their position in the water column. They have also observed feeding specialization among the fish species co-occurring in the rainforest streams in Sri Lanka.

Wickramanayake and Moyle (1989) have suggested that fish assemblages in wet zone streams of Sri Lanka are co-evolved systems with competition serving as a fundamental structuring force to reinforce species segregation. Based on a study on the macro-and micro-habitats and feeding habits of several Cyprinid fishes (then as *Barbus*) in Sri Lanka, Kortmulder et al. (1990) stated that habitat characteristics are common among most endemic and indigenous species. Kortmulder (1987) mentioned that these Cyprinids (then as *Puntius* species) are distributed over a range of stream types. In the marshy brooks with relatively still water, *Puntius titteya* is found, while *P. bimaculatus* occurs in areas with strong currents. Also, in rocky creeks with generally strong currents, only adult individuals of *Puntius bimaculatus* inhabits. In the forested areas' irregular hill streams, *Systomus pleurotaenia* inhabits the deep parts with the rapid flow.

In contrast, *Pethia nigrofasciata* inhabits the shallow areas of the stream with the abated current. It was also observed that *Pethia cumingii* is found in deep pits of the lower courses of hill streams and torrents, with a relatively high current pattern. The latter species prefers the deep pits, where there is a balance between the verge of current and non-current. Kortmulder (1987) has provided the altitudinal distribution of 11 cyprinid species in Sri Lanka, beginning with headwaters and ending with brackish waters near the coast. The order of sequence was, *Puntius titteya, P.*

bimaculatus, Systomus pleurotaenia, Pethia nigrofasciata, P. cumingii, Puntius dorsalis, Dawkinsia singhala, Puntius vittatus, Systomus sarana, Puntius thermalis and *P. kamalika.*

The study of fish assemblages in the Tissawewa of Southern Sri Lanka by Piet (1996) is unique because most of the research on the freshwater fish assemblages of Sri Lanka were limited to wet zone streams. This study provides remarkable contributions to the ecology of freshwater fishes of Sri Lanka. It was also stated that the pronounced resource partitioning and complementarities along axes of macro-habitat and micro-habitat utilization and trophic status have resulted in assemblages that co-evolved to reduce interspecific competition. It was also pointed out that except for introduced species—which had more significant overlaps for both, amongst themselves and their associations with some indigenous species—low overlaps and resource complementarity are pronounced amongst the indigenous species.

Based on the species and size-classes' potential niches, a trophic structure can be created for the Tissawewa fish assemblages, describing how the different size-structured populations are expected to be segregated trophically based on their morphological characters (Piet, 1996). Also, variations in relatively few morphological characters account for most inter-specific and intra-specific ecological variations.

10.3.3.3 DIET AND DIEL VARIATION

Most diurnal fishes are either suspended in the water column in pools or hide underneath logs and rocks, often in shallow water. Thus, the time of activity is yet another niche dimension (e.g., nocturnal fishes) (Lowe-McConnell, 1964; Smith and Tyler, 1972). Most importantly, Piets (1996) study included a diel variation of Sri Lankan fishes. Abiotic or biotic factors or both may either cause the vertical distribution of the fish. However, the dissolved oxygen and temperature in Tissawewa did not show variations during a 24-hour period that could have pushed for the observed patterns of vertical distribution (Piet, 1996). As zooplanktons are abundant in the pelagic zones (Werner and Hall, 1988; Piet, 1996), zooplanktivorous fish species usually show a diel vertical distribution pattern (Begg, 1976; Gliwicz, 1986; Janssen and Brandt, 1980). However, biotic interactions such as avoidance of potential competitors or predators might help the observed diel patterns in the Tissawewa and survival among the potential competitors was due to the resource partition on the resource dimensions (i.e., temporal, and trophic) (Piet, 1996): *Puntius thermalis,* and *P. dorsalis* feed exclusively on benthic food items, whereas *Systomus sarana* found in the column utilized higher proportion of fish and insects. In addition, when they were not feeding; *Puntius thermalis, P. dorsalis,* and *Oreochromis* spp. Occupied the entire column zone, but *Systomus sarana* was only observed in the column zone's upper half.

A similar pattern and the daytime feeding activity of *Oreochromis mossambicus* and *O. niloticus* were reported previously by Trewavas (1983). Also, these two species had higher gut fullness around dusk (Hofer and Schiemer, 1983; Moriarty and Moriarty, 1973; Piet, 1996). Schiemer and Hofer (1983) concluded that *Oreochromis mossambicus* did not show any diel feeding pattern. Piet (1996) found that these two *Oreochromis* species were not spatially separated in the Tissawewa reservoir.

Generally, carnivore, and omnivore species cannot be restricted to a single zone (not spatially separated). Carnivore species such as *Glossogobius giuris* and *Mystus* spp. live in the bottom during the day-time. When feeding at dusk occurs around the surface and column zones, *Glossogobius giuris* was not strictly nocturnal; the latter is an example of a temporally segregated carnivore (Piet, 1996). This segregation also provides a model for carnivore species segregation, which is crucial for the resource partitioning of freshwater fishes of Sri Lanka.

When considering the surface and column zones species, *Hyporhamphus limbatus* and *Rasbora dandia* were both temporally and spatially segregated. Concerning temporal segregation, *Rasbora dandia* is active during crepuscular periods (Kumar, 1985), and *Hyporhamphus limbatus* is mainly active during daytime with a peak around noon (Piet, 1996). In contrast, spatial segregation occurred at dawn when *Rasbora dandia* was restricted to the column zone and *Hyporhamphus limbatus* to the surface. However, both fishes remained in the surface layer when they were not feeding (Piet, 1996).

Amblypharyngodon melettinus was the main prey of most piscivorous fishes (Piet, 1996). Due to its filter-feeding habit *Amblypharyngodon melettinus* (Schiemer and Hofer, 1983; Pethiyagoda, 1991) is segregated from all other species along the trophic spatial dimensions (Piet, 1996). The stomachless *Amblypharyngodon melettinus* was unable to digest a significant part of the algae (Bitterlich, 1985). In aquariums, *Amblypharyngodon melettinus* feeds in the column zone (Pethiyagoda, 1991), and when it is not feeding, it is usually observed in the surface layers. Even though there was a high concentration of suspended detritus near the bottom, *Amblypharyngodon melettinus* was not found in the bottom zone (Piet, 1996), which shows the avoidance of predators, a phenomenon known as the tradeoff between food intake and predation risk (Clark and Levy, 1988). The vertical distribution of fish is interrelated with the food source that they utilize. Resource partitioning avoids exploitative competition, which involves a species' position along both the spatial and the trophic resource dimensions (Piet, 1996).

It is concluded that the temporal dimension is due to the three independently determined variables: (i) gut fullness; (ii) proportion of digested matter; and (iii) relative activity of the first two. About half to two-thirds of the assemblage species are diurnal, one quarter to one third are nocturnal, and the remaining 10% or so are primarily crepuscular (Helfman, 1978; Piet, 1996).

By summarizing the data on diets of about 600 fish species, Love (1980) revealed that only 3% of fishes were detritivorous. However, this lesser percentage might be due to the overlooked studies of differentiating ingested detritus and digested matter (Maitipe and de Silva, 1985; Khoo and Tay, 1992). According to Fänge et al. (1979), herbivores have a higher digestion rate than non-herbivorous species.

10.3.3.4 HABITAT PROFITABILITY

Piet (1996) pointed out that by employing fish into a habitat, its profitability can be established. Further, the gut contents of different species from different habitats show spatial differences. However, all the fish had displayed a similar spatial trend of the leading food items observed in their digestive tract (Piet, 1996).

The fitness of a species is due to its foraging profitability and predation risk. Thus, concerning different habitats, these strategies may vary because each species tends to balance these strategies' trade-off. Many species show ontogenetic shifts in their diet and habitat, affecting foraging efficiency and predation risk (Werner and Gilliam, 1984). Even among predators (*Glossogobius giuris* or *Mystus* spp.) that co-exist in the Tissawewa, there were preferred prey size classes 1 and 2. Species belonging to the former two types were *Amblypharyngodon melettinus, Hyporhamphus limbatus*, and *Rasbora dandia.* Species predation usually increases due to vertical distribution (from surface to bottom) because of the distribution of major predators (e.g., *Glossogobius giuris* and *Mystus* spp.), but predation risk is reported to be low in habitats associated with abundant vegetation (Werner et al., 1983; Anderson, 1984; Hershey, 1985; Werner and Hall, 1988). Thus, the smallest size-classes of the fish assemblage preferred the vegetated areas in the wet or rainy season when the associated vegetation flourished in the inshore regions.

According to the optimal foraging theory, it is projected that species prefer the habitat during feeding, which maximizes foraging profitability. Spatial differences in gut fullness and gut contents are appropriate indicators of habitat profitability; thus, a positive correlation between the trends of habitat profitability and distribution indicated that the species could assess foraging profitability within the different habitats and adjust its distribution accordingly (Werner et al., 1983; Werner and Hall, 1988; Piet, 1996). About foraging activity, the fish assemblage was divided into two major groups: the nocturnal piscivores and diurnal non-piscivores (Pet and Piet, 1993; Piet, 1996).

Gorman and Karr (1978) found that species diversity in Central American forest streams increased with habitat diversity (i.e., depth, velocity, and substrate). Knoppel (1972) inferred some spatial segregation from feeding habits. Most studies of the feeding habits of tropical stream fishes have found a large amount of dietary overlap and a heavy reliance on allochthonous food, mainly terrestrial arthropods, and detritus of terrestrial origin (Costa and Fernando, 1967; Knoppel, 1970, 1972; Bishop, 1973; Saul, 1975; Lowe-McConnell, 1975). Conversely, Moyle and Senanayake (1984) observed that dietary overlap was low, and there was a heavy reliance on autochthonous materials, mainly based on algae and aquatic invertebrates.

The relative constancy of the environment (in the water bodies) is essential because autochthonous food is essential to the fishes (Lowe-McConnell, 1975). Presumably, functional morphology, habitat structure, and the availability of algae and aquatic invertebrates may influence fish species' existence and distribution. Hence, a combination of functional morphology, habitat structure, behavior (e.g., foraging, breeding) with the availability of feeds allows the exploitation of varying compartments of the resources and exhibits heterogeneous behavioral and distributional fish patterns.

However, stream fishes of temperate zone exhibit different morphologies (Winemiller, 1992; Matthews, 1998; Moyle and Cech, 1982; Pflieger, 2004), which appear to be directly or indirectly related to changing water velocities (Ross, 1986). According to Gaston et al. (2012), water flow (velocity) has influenced fish habitat selection in temperate rivers like Mississinewa, Salamonie, Wabash, and White in Indiana. Gaston et al. (2012) demonstrated that pool habitat was characterized by more significant numbers of fish (with a low body depth ratio); whereas, riffle habitats are characterized by more significant numbers of fishes (with a high body depth ratio), which are typically found in riffle areas.

10.4 CONSERVATION ISSUES AND RECOMMENDATIONS

Wickramanayake and Moyle (1989) pointed out that the assemblages of ichthyofauna of Sri Lanka have co-evolved to reduce interspecific competition: low overlaps and resource complementarity are pronounced amongst the indigenous species. However, some indigenous species have more significant overlap with the introduced species.

Although *Eutroplus suratensis* and *Tilapia rendalli* are not dominant species in Sri Lankan reservoirs (De Silva, 1988; Amarasinghe, 1998), they are known to be substrate brooders (Fryer and Illes, 1971; Ward and Samarakoon, 1981), so that competition for nesting sites of these two is inevitable. Moreover, the other two introduced cichlids (*Oreochromis mossambicus* and *O. niloticus*) known as mouthbrooders again show the efficiency and maximum brood survival within many of the ecosystems indigenous species live. Also, Maitipe and De Silva (1985) have shown that introduced cichlid species like *Oreochromis mossambicus* in Sri Lankan reservoirs exhibits switches between phytophagous, zoophagous, and detritivorous dietary habits depending on the availability of feeds. This may be due to the differences in prey availability in the habitats, as shown by Garcia-Berthou (2002) for *Micropterus salmoides* (Large-mouth bass).

For most species' zooplankton is the main diet during their initial stage of ontogenic development. Piet (1996) stated that the introduced species, *Oreochromis mossambicus* had zooplankton proportion that decreased with its size-class. In contrast, the large size-classes depended on more proportion of detritus feeds. Thus, it is evident that this species has a dietary shift, thus feed on two different feed sources throughout their life. *Oreochromis mossambicus* not only shows the dietary change but also showed a diel variation in feeding. These strategies inevitably cause the utilization of more resources by an introduced species. Hence, such introductions cause severe threats to the indigenous and endemic fish fauna of Sri Lanka.

Also, contrary to the most indigenous species, *Oreochromis* spp. have displayed a distinct shift between periods of dry and wet seasons (rainy and drought), but a significant preference for the bottom half of the column zone during the drought and in the wet seasons. However, indigenous species such as benthivorous *Puntius thermalis* and *P. dorsalis* were found close to the bottom during both seasons. They displayed a distinct preference for the inshore zone, whereas *Systomus sarana* mainly lived in the offshore region column. Analyzing the above strategies of the species assemblage, it can be concluded that the introduced species tend to shift from shallow waters to deep waters during drought periods, while some indigenous species (*Puntius thermalis* and *P. dorsalis*) retained in the shallow waters throughout the year. This also provides an example of the comparatively low fitness of certain riverine indigenous fishes in the lentic ecosystems of Sri Lanka.

Also, certain invasive species, e.g., Knifefish (*Chitala ornata*), *Alligator gar (Atractosteus spatula), and* Tank cleaner fish (*Pterygoplichthys* spp.) have also drastically influenced the indigenous species, especially those inhabiting the wet zones of Sri Lanka.

Welianga and Amarasinghe (2003) stated that almost all lowland reservoirs in the dry zone of Sri Lanka are irrigation reservoirs; thereby, the hydrological regimes are significantly influenced by irrigation of agricultural lands (De Silva, 1988). On the other hand, in the wet zone, perennial rivers and streams are also subjected to drastic changes due to mini-hydropower plants' construction to uplift local electricity consumption. This is one of the significant threats for the survival of endemic and indigenous fauna, including freshwater fishes (anadromous, catadromous, and amphidromous fishes and torrent inhabiting fishes).

Wet zone stream forest fishes of Sri Lanka are not harvested either in large-scale or small-scale fisheries. However, fishes from the dry zone and the wet zone's down-river systems are regularly harvested throughout the year. Thus, threats to most freshwater fishes within these regions are evident. In Sri Lanka, except for the wet zone areas, fishers have certain ways of protecting the fishery resources through their Fisheries Co-operative Societies (e.g., fishers do not fish during full-moon days even in non-Buddhist fisher communities (Sudesh Batuwita, pers. obs., 2012; Bathalagoda reservoir, Sri Lanka: where there was only a single Buddhist fisher at that time). Moreover, there is an old norm that the anglers should not catch larger brood-stock fish and small-sized juveniles, and it was not allowed to fish in some unique areas in a reservoir, where the brood-stock and juveniles live. Thus, the local knowledge of the freshwater fish fauna helped protect this invaluable resource even during ancient times. Therefore, present government authorities/stakeholders should manage the fishery resources by providing urgent measures to utilize this resource sustainably.

For anglers who live in the wet zone and the dry zone, the government can introduce a standard mesh size for fishing. It is primarily well documented that certain introduced species (e.g., *Oreochromis* spp. and *Tilapia* spp.) have utilized more resources and can withstand the harsh environmental conditions easily than that of the indigenous and endemic fishes of Sri Lanka. Thus, the stakeholders can provide methods to harvest more introduced fishes than the local endangered fish.

10.5 CONCLUSIONS

Inter-specific and intra-specific interactions are always needed for the mutual survival of species. The environmental factors influence the species morphology and thereby their adaptations to the environment.

Species body/parts changes through their ontogenic development, and species whole dimensions of their morphological characters increases with increasing size.

The most correlated variables for species segregation are velocity at fish, mean water column velocity, depth at fish, water column depth, and the dominant substrate. The turbidity of the water column had a significant contribution to the species spatial and temporal segregation.

Based on diet, relative depth, or both, most stream fish species show more significant overlaps. Dietary shifts of fishes occur due to their ontogenetic changes.

Forest stream fish distribution across several sampling sites shows typical right skew distribution similar to birds, butterflies, and other communities, implying that most of the fish species are relatively rare. In contrast, a few species dominate an area in terms of their abundance.

Survival among the potential competitors is due to the resource partitioning on the resource dimensions (i.e., temporal, and trophic). Carnivorous and omnivorous species do not remain restricted to a single zone (spatially not segregate). The vertical distribution of fish is inter-related with the food source that they utilize. Resource partitioning avoids exploitative competition and positions a species along both the spatial and the trophic resource dimensions.

The fitness of a species is due to its foraging profitability and predation risk. Thus, concerning different habitats, these strategies may vary because each species tend to balance the trade-off between these strategies.

KEYWORDS

- **digestive system**
- **freshwater ecosystems**
- **geodiversity**
- **morphology**
- ***Schistura notostigma***
- **world heritage site**

REFERENCES

Adamicka, P., (1983). Contributions to the functional anatomy of the feeding apparatus of five cyprinids of Parakrama Samudra (Sri Lanka). In: Schiemer, F., (ed.), *Limnology of Parakrama Samudra, Sri Lanka* (pp. 171–175). W. Junk Publ., Den Haag.

Aleev, Y. G., (1969). *Function and Gross Morphology in Fish*. Israel program for scientific translations, Jerusalem.

Anderson, O., (1984). Optimal foraging by largemouth bass in structured environments. *Ecology, 65*, 851–861.

Anonymous, (1991). *Natural Resources of Sri Lanka: Conditions and Trends*. Natural Resources, Energy and Science Authority of Sri Lanka, Colombo, Sri Lanka.

Barel, C. D. N., Anker, G. C., Witte, F., Hoogerhoud, R. J. C., & Goldschmidt, T., (1989). Constructional constraint and its ecomorphological implications. *Acta Morphologica Neerlandica-Scandinavica, 27*, 83–109.

Batuwita, A. S. G., (2016). *A Taxonomic Study on the South Asian Cyprinid Genera Horadandia and Rasboroides (Ostariophysi: Cyprinidae: Danioninae) and Evaluation of the Present Distribution Pattern of Rasboroides vaterifloris Pallidus in Gin River* (p. 153). Unpublished MPhil thesis. The Postgraduate Institute of Agriculture, University of Peradeniya.

Batuwita, S., De Silva, M., & Udugampala, S., (2017). A review of the genus *Devario* in Sri Lanka (Teleostei: Cyprinidae), with description of two new species. *FishTaxa, 2*(3), 156–179.

Begg, G. W., (1976). The relationship between the diurnal movements of some of the zooplankton and the sardine *Limnothrissa miodon* in Lake Kariba, Rhodesia. *Limnology and Oceanography, 21*, 529–539.

Beier, U. Habitat Use in Fish Communities: Size- and Environment-dependent Mechanisms Affecting Biotic Interactions and Population Regulation. 2016. https://www.semanticscholar.org/paper/Habitat-Use-in-Fish-Communities-.-Size-and-Biotic-Beier/cef09bbb1a26769befde967eddafe758b041d2a5

Bencala, K. E., (2000). Hyporheic zone hydrological processes. *Hydrological Processes, 14*(15), 2797–2798.

Bergman, E., (1988). Foraging abilities and niche breadths of two percids, *Perca fluviatilis* and *Gymnocephalus cernua*, under different environmental conditions. *Journal of Animal Ecology, 57*, 443–453.

Berumen, M. L., & Pratchett, M. S., (2006). Effects of resource availability on the competitive behavior of butterflyfishes (*Chaetodontidae*). *Proceedings of the 10th International Coral Reef Symposium* (pp. 644–650). Okinawa, Japan.

Berumen, M. L., & Pratchett, M. S., (2008). Trade-offs associated with dietary specialization in corallivorous butterflyfishes (*Chaetodontidae*: *Chaetodon*). *Behavioral Ecology and Sociobiology, 62*, 989–994.

Bishop, J. E., (1973). Limnology of a small Malayan River, Sungae Gombak. *Monographiae Biologicae, 22*, 1–435.

Bitterlich, G., (1985). The nutrition of stomach less phytoplankton fish in comparison with *Tilapia*. *Hydrobiologia, 121*, 173–179.

Bossuyt, F., Meegaskumbura, M., Beenaerts, N., Gower, D. J., Pethiyagoda, R., Roelants, K., Mannaert, A., et al., (2004). Local endemism within the Western Ghats-Sri Lanka biodiversity hotspot. *Science, 306* (5695), 479–481.

Bowen, S. H., (1976). Benthic diatom distribution and grazing by *Sarotherodon mossambicus* in Lake Sibaya, South Africa. *Freshwater Biology, 8*, 449–453.

Bowen, S. H., (1979). A nutritional constraint in detrivory by fishes: The stunted population of *Sarotherodon mossambicus* in Lake Sibaya, South Africa. *Ecological Monographs, 49*, 17–31.

Brooker, R. M., Munday, P. L., Mcleod, I. M., & Jones, G. P., (2013a). Habitat preferences of a corallivorous reef fish: predation risk versus food quality. *Coral Reefs, 32*, 613–622.

Brown, J. S., (1988). Patch use as an indicator of habitat preference, predation risk, and competition. *Behavioral Ecology and Sociobiology, 22*, 37–47.

Butler, G. L., Rowland, S. J., & Baverstock, P. R., (2014). Movement patterns and habitat selection of the endangered eastern freshwater cod Maccullochella ikeiin the Mann River, Australia. Endangered Species Research, 23, 35–49.

Carosi, A., Padula, R., Ghetti, L., & Lorenzon, M., (2019). Endemic freshwater fish fange shifts related to global climate changes: A long-term study provides some observational evidence for the Mediterranean area. *Water, 11*, 2349. doi: 10.3390/w11112349w.

*Cavender, T. M., (*1991*).* The fossil record of the *Cyprinidae*. In: Winfield, I. J., & Nelson, J. S., (eds.). *Cyprinid Fishes: Systematics, Biology and Exploitation* (pp. 35–53). Chapman and Hall, New York.

Cerri, R. D., & Fraser, D. F., (1983). Predation and risk in foraging minnows: balancing conflicting demands. *The American Naturalist, 121*, 552–561.

Chandler, J. F., Burn, D., Berggren, P., & Sweet, M. J., (2016). Influence of resource availability on the foraging strategies of the Triangle butterflyfish *Chaetodon triangulum* in the Maldives. *PLOS One,* 11.

Charnov, E. L., (1976). Optimal foraging: the marginal value theorem. *Theoretical Population Biology, 9*(2), 129–136..

Clark, C. W., & Levy, D. A., (1988). Diel vertical migrations by juvenile sockeye salmon and the antipredation window. *American Naturalist, 131*, 271–290.

Clarke, A., & Johnston, N. M., (1999). Scaling of metabolic rate with body mass and temperature in teleost fish. *Journal of Animal Ecology, 68*, 893–905.

Cole, A. J., Pratchett, M. S., & Jones, G. P., (2008). Diversity and functional importance of coral-feeding fishes in tropical coral reefs. *Fish Fish., 9*, 286–307.

Costa, H. H., & Fernando, E. C. M., (1967). The food and feeding relationships of the common meso and microfauna in the Maha Oya, a small mountainous stream at Peradeniya, Ceylon. *Ceylon Journal of Science, 7*, 75–90.

Coull, B. C., & Wells, J. B. J., (1983). Refuges from fish predation: experiments with phytal meiofauna from the New Zealand rocky intertidal. *Ecology, 64*, 1599–1609.

Crowder, L. B., & Cooper, W. E., (1982). Habitat structural complexity and the interaction between bluegills and their prey. *Ecology, 63*, 1802–1813.

Daily, G. C., (1997). *Nature's Services.* Island Press. *Washington DC.*

De Silva, M. A., (1997). The study report of fresh water fish in Ma-Dola Galle District Southern Sri Lanka. *Loris, 21*(3), 101–103.

De Silva, M. A., Hapuarachchi, N., & Jayaratne, T., (2015). *Sri Lankan Freshwater Fishes*. Wildlife Conservation Society, Galle.

De Silva, S. S., & Chandrasoma, J., (1980). Reproductive biology of *Sarotherodon mossambicus*, an introduced species in an ancient man-made lake in Sri Lanka. *Environmental Biology of Fishes, 5*, 253–259.

De Silva, S. S., & Kortmulder, K., (1977). Some aspects of the biology of three species of *Puntius* (=*Barbus*) (Pisces, Cyprinidae), endemic to Sri Lanka. *Netherland Journal of Zoology, 27*, 182–194.

De Silva, S. S., & Sirisena, H. K. G., (1988). Observations on the nesting habits of *Oreochromis mossambicus* (Peters) (Pisces: Cichlidae) in Sri Lankan reservoirs. *Journal Fish Biology, 33*, 689–696.

De Silva, S. S., (1983). Reproductive strategies of some major fish species in Parakrama Samudra reservoir and their possible impact on the ecosystem-a theoretical consideration.

In: Schiemer, F., (ed.), *Limnology of Parakrama Samudra, Sri Lanka* (pp. 185–192). The Hague: Dr. W. Junk Publishers.

De Silva, S. S., (1988). *Reservoirs in Sri Lanka and Their Fisheries* (Vol. 298, p. 128). FAO Fisheries technical paper.

De Silva, S. S., Cumaranatunga, P. R. T., & De Silva, C. D., (1980). Food, feeding ecology and morphological features associated with feeding of four co-occurring cyprinids (Pisces: *Cyprinidae*). *Netherlands Journal of Zoology, 30*, 54–73.

De Silva, S. S., Kortmulder, K., & Wijeyaratne, M. J. S., (1977). A comparative study of the food and feeding habits of *Puntius bimaculatus* and *P. titteya* (Pisces: Cyprinidae). *Netherland Journal of Zoology, 27*, 253–263.

De Silva, S. S., Schut, J., & Kortmulder, K., (1985). Reproductive biology of six *Barbus* species indigenous to Sri Lanka. *Environmental Biology of Fishes, 12*(3), 201–218.

Diehl, S., (1988). Foraging efficiency of three freshwater fishes: Effects of structural complexity and light. *Oikos, 53*, 207–214.

Domrös, M., (1998). Variabilities of the temporal and spatial organization of rainfall in Sri Lanka. In: Domrös, M., & Roth, H., (eds.), *Sri Lanka: Past and Present Arachaeology, Geography, Economics – Selected Papers on German Research* (pp. 82–102). Margraf Verlag, Weikersheim.

Douglas, M. E., & Matthews, W. J., (1992). Does morphology predict ecology? Hypothesis testing within a freshwater stream fish assemblage. *Oikos, 65*, 213–224.

Douglas, M. E., (1987). An ecomorphological analysis of niche packing and niche dispersion in stream fish clades. In: Matthews, W. J., & Heins, D. C., (eds.), *Community and Evolutionary Ecology of North American Stream Fishes* (pp. 144–149). University of Oklahoma Press, Norman, OK.

Dunson, W. A., & Travis, J., (1991). The role of abiotic factors in community organization. *The American Naturalist, 138*, 1067–1091.

Dvorak, J., & Best, E. P. H., (1982). Macro-invertebrate communities associated with the macrophytes of Lake Vechten: structural and functional relationships. *Hydrobiologia, 95*, 115–126.

Ebenman, B., & Persson, L., (1988). *Size-structured Populations, Ecology and Evolution.* Berlin: Springer-Verlag.

Ebner, B. C., Fulton, C. J., Donaldson, J. A., & Schaffer, J., (2015). Distinct habitat selection by freshwater morays in tropical rainforest streams. *Ecology of Freshwater Fish, 2015*, 1–7.

Ehlinger, T. J., & Wilson, D. S., (1988). Complex foraging polymorphism in bluegill sunfish. *Proceedings of the National Academy of Science, 85*, 1878–1882.

Fänge, R., & Grove, D., (1979). Digestion. In: Hoar, W. S., Randall, D. J., & Brett, J. R., (eds.), *Fish Physiology* (Vol. VIII, pp. 161–260). Academic Press. New York. FAO-1972.

Fausch, K. D., & White, R. J., (1981). Competition between brook trout (*Salvelinus fontinalis*) and brown trout (*Salmo trutta*) for positions in a Michigan stream. *Canadian Journal of Fisheries.*

FishBase, (2020). http://www.fishbase.org/(accessed on 14 December 2021).

Fretwell, S. D., & Lucas, H. L. J., (1970). On territorial behavior and other factors influencing habitat distribution in birds. *Acta Biotheoretica, 19*,16–36.

Fryer, G., & Iles, T. D., (1972). *The Cichlid Fishes of the Great Lakes of Africa: Their Biology and evolution*. Oliver and Boyd, Edinburgh, U.K.

Galis, F., (1993). Interactions between the pharyngeal jaw apparatus, feeding behaviors, and ontogeny in the cichlid fish, *Haplochromis piceatus*: a study of morphological constraints in evolutionary ecology. *Journal of Experimental Zoology, 267*, 137–154.

Gassner, P., Sprenger, D., & Anthes, N., (2012). Red Sea butterflyfish responds to changing coral cover. *Biodiversity Science,* (6).

Gatz, A. J., (1979a). Community organization in fishes as indicated by morphological features. *Ecology, 60*, 711–718.

Gatz, A. J., (1979b). Ecological morphology of freshwater stream fishes. *Tulane Studies in Zoology and Botany, 21*, 91–124.

Gause, G. F., (1934). *The Struggle for Existence*. Hafner, New York.

Geistdoerfer, P., (1973). Diet of Atlantic and Mediterranean macrouridae (Teleosteens, Gadiformes), in relation to the morphology of the digestive tract. *Bulletin of the National Museum of Natural History. Paris., Series 3 (161) d'Ecologie Generalae, 17,* 285–295.

Geistdoerfer, P., (1978). Feeding ecology of Macrouridae. *Review of the work of the Institut des Peches Maritimes, 42,* 177–261.

Gilbert, C. R., & Bailey, R. M., (1972). Systematics and zoogeography of the American cyprinid fish *Notropis (Opsopoeodus) emiliae. Occasional Papers of the Museums of Zoology, 664*, 1–35. University of Michigan.

Gilliam, J. F., & Fraser, D. F., (1987). Habitat selection under predation hazard: test of a model with foraging minnows. *Ecology, 68*, 1856–1862.

Gilliam, J. F., (1982). *Habitat use and Competitive Bottlenecks in Size-Structured fish Populations*. Thesis Michigan State University, East Lansing.

Gladfelter, W. B., & Johnson, W. S., (1983). Feeding niche separation in a guild of tropical reef fishes (Holocentridae). *Ecology, 64*, 552–563.

Gliwicz, Z. M., (1986). A lunar cycle in zooplankton. *Ecology, 67*, 883–897.

Gorman, D. T., & Karr, J. R., (1978). Habitat structure and stream fish communities. *Ecology, 59*, 507–515.

Gosline, W. A., (1973). Considerations regarding the phylogeny of cypriniform fishes with special reference to structures associated with feeding. *Copeia, 1955*, 267–290.

Graham, C. T., & Harrod, C., (2009). Implications of climate change for the fishes of the British Isles. *Journal of Fish Biology, 74*, 1143–1205.

Grossman, G. D., Moyle, P. B., & Whitaker, J. O., (1982). Stochasticity in structural and functional characteristics of an Indiana stream fish assemblage: a test of community theory. *The American Naturalist, 120*, 423–454.

Guthrie, D. M., & Muntz, W. R. A., (1993). Role of vision in fish behavior. *Fish and Fisheries Series, 7*, 89–128. London: Chapman and Hall Ltd.

Helfman, G. S., (1978). Patterns of community structure in fishes: Summary and overview. *Environmental Biology of Fishes, 3*, 129–148.

Hershey, A. E., (1985). Effects of predatory sculpin on the chironomid communities in an arctic lake. *Ecology, 66*, 1131–1138.

Hoagstrom, C., & Berry, C., (2007). Morphological diversity among fishes in a Great Plains River drainage. *Hydrobiologia, 596*, 367–386.

Hofer, R., & Schiemer, F., (1983). Feeding ecology, assimilation efficiencies and energetics of two herbivorous fish; *Sarotherodon (Tilapia) mossambicus* (Peters) and *Puntius filamentosus* (Cuv. et Val.). In: Schiemer, F., (ed.), *Limnology of Parakrama Samudra, Sri Lanka*. W. Junk Publ., The Hague.

Hofer, R., (1991). Digestion. In: Winfield, I. J., & Nelson, J. S., (eds.), *Cyprinid Fishes: Systematics, Biology and Exploitation* (pp. 55–79). Chapman & Hall. London.

Holmgren, K., & Appelberg, M., (2000). Size structure of benthic freshwater fish communities in relation to environmental gradients. *Journal of Fish Biology, 57*, 1312–1330.

Janssen, J., & Brandt, S. B., (1980). Feeding ecology and vertical migration of adult alewives (*Alosa pseudoharengus*) in lake Michigan. *Canadian Journal of Fisheries and Aquatic Sciences, 37*, 177–184.

Jayaratne, R., & Surasinghe, T., (2010). General ecology and selectivity of fresh water fishes of the Rawan Oya, Kandy, Sri Lanka. *Sabaragamuwa University Journal, 9*(1), 11–43.

Johnston, I. A., & Dunn, J. F., (1987). Temperature acclimation and metabolism in ectotherms with particular reference to teleost fish. *In:* Bowler, K., & Fuller, B. J., (eds.), *Temperature and Animal Cells* (pp. 67–93). Society for Experimental Biology Symposium 41.

Kalk, M., McLachlan, A. J., & Howard-Williams, C., (1979). Lake Chilwa, studies of change in a tropical ecosystem. *Monographiae Biologicae* (Vol. 35, p. 462). Dr W. Junk Publishers, The Hague.

Keast, A., (1985). The piscivore feeding guild of fishes in small freshwater ecosystems. *Environmental Biology of Fishes, 12*, 119–129.

Khoo, H. W., & Tay, S., (1992). Detritus production and its utilization by *Oreochromis mossambicus* in Selatar Reservoir. In: De Silva, S. S., (ed.), *Reservoir Fisheries of Asia* (pp. 62–72). IDRC. Ottawa.

Kislalioglu, M., & Gibson, R. N., (1976). Prey "handling time" and its importance in food selection by the 15 spined stickleback, *Spinachia spinachia* (L.). *Journal of Experimental Marine Biology and Ecology, 25*, 115–158.

Knoppel, H. A., (1970). Food of central Amazonian fishes. *Amazonia, 2*, 257–352.

Knoppel, H. A., (1972). For food of tropical freshwater fish from South America. *Amazonia, 3*(23), 1–246.

Kortmulder, K., (1972). A comparative study in color patterns and behavior in seven Asiatic *Barbus* species (Cyprinidae, Ostariophysi, Osteichthyes): A progress report. *Behavior, Supplement,* (19), 1–331.

Kortmulder, K., (1982). Etho-ecology of seventeen *Barbus* species (Pisces, *Cyprinidae*). *Netherlands Journal of Zoology, 32*(2), 144–168.

Kortmulder, K., (1987). Ecology and behavior of tropical freshwater fish communities. *Archiv. für Hydrobiologie, 28*, 503–513.

Kortmulder, K., Feldbrugge, E. J., & De Silva, S. S., (1978). A combined field study of *Barbus* (= *Puntius*) *nigrofasciatus* Gunther (Pisces, *Cyprinidae*) and water chemistry of its habitats in Sri Lanka. *Netherlands Journal of Zoology, 28*(1), 111–131.

Kortmulder, K., Padmanabhan, K. G., & De Silva, S. S., (1990). Patterns of distribution and endemism in some cyprinid fishes as determined by the geomorphology of South-West Sri Lanka and South Kerala (India). *Ichthyological Explorations in Freshwaters, 1*(2), 97–112.

Kotrschal, K., & Goldschmid, A., (1983). Food preferences, morphology and arrangement of teeth of 14 species of Adriatic blennies (Pisces, Teleostei). *Thalassia, 19*, 217–219.

Kumar, K. P., & John, P. A., (1985). Seasonal variations of the allochthonous fauna and flora and their influence on the diet of *Rasbora daniconius* (Hamilton). *Archives for Hydrobiology, 102*, 537–542.
Kumar, K. P., & John, P. A., (1987). Feeding ecology of the allochthonous feeder *Rasbora daniconius* (Hamilton) (Cyprinidae -Teleostei). *International Review of the Entire Hydrobiology, 72*, 325–337.
Kumar, K. P., (1985). Diurnal variations in the feeding of *Rasbora daniconius* (Hamilton). *Aquatic Biology, 5*, 51–54.
Kumara, P. A. D. A., Amarasinghe, U. S., Schiemer, F., Winkler, G., & Schabuss, M., (2009). Distribution and abundance of unexploited fish species in three Sri Lankan reservoirs, *Asian Fisheries Science, 22*, 867–884.
Lauder, G. V., (1983). Neuromuscular patterns and the origin of trophic specialization in fishes. *Science, 219*, 1235–1237.
Lipcius, R. N., & Hines, A. H., (1986). Variable functional responses of a marine predator in dissimilar homogeneous microhabitats. *Ecology, 67*, 1361–1371.
Love, R. M., (1980). *The Chemical Biology of Fishes* (Vol. 2). Advances 1968–1977. Academic Press, London.
Lowe-McConnell, R. H., (1964). The fishes of the Rupununi savanna district of British Guiana, South America. Part 1. Ecological groupings of fish species and effects of the seasonal cycle on the fish. *Journal of Linnaean Society London (Zoology), 45*, 103–144.
Lowe-McConnell, R. H., (1975). *Fish communities in Tropical Freshwaters.* London: Longman.
Lowe-McConnell, R. H., (1987). *Ecological Studies of Tropical Fish Communities.* Cambridge University Press, Cambridge, U.K.
MacArthur, R. H., & Pianka, E. R., (1966). On optimal use of a patchy environment. *American Naturalist, 4*, 100.
Maitipe, P., & De Silva, S. S., (1985). Switches between zoophagy, phytophagy and detritivory of *Sarotherodon mossambicus* (Peters) populations in twelve man-made Sri Lankan lakes. *Journal of Fish Biology, 26*, 49–61.
Matter, W. J., Mannan, R. W., Bianchi, E. W., McMahon, T. E., Menke, J. H., & Tash, J. C., (1989). A laboratory approach for studying emigration. *Ecology, 70*, 1543–1546.
Matthews, W. J., (1985). Critical current speed and microhabitats of the benthic fishes *Percina roanoka* and *Etheostoma flabellare*. *Environmental Biology of Fishes, 12*, 303–308.
Matthews, W. J., (1998). *Patterns in Freshwater Fish Ecology*. New York: Chapman and Hall.
McLellan, T., (1977). Feeding strategies of the macrourids. *Deep-Sea Research, 24*, 1019–1036.
Menge, B. A., & Sutherland, J. P., (1976). Species diversity gradients: synthesis of the role of predation, competition, and temporal heterogeneity. *American Naturalist, 110*, 351–369.
Menon, M. D., & Chacko, P. T., (1958). Food and feeding of freshwater fishes of the Madras state. *Journal of the Bombay Natural History Society, 55*, 117–124.
Miller, G. L., (1984). Seasonal changes in morphological structuring in a guild of benthic stream fishes. *Oecologia, 63*, 106–109.

Mittermeier, R. A., Myers, N., Thomsen, J. B., Da Fonseca, G. A. B., & Olivieri, S., (1998). Biodiversity hotspots and major tropical wilderness areas: approaches to setting conservation priorities. *Conservation Biology, 12*(3), 516–520.

Moriarty, C. M., & Moriarty, D. J. W., (1973). Quantitative estimation of the daily ingestion of phytoplankton by *Tilapia nilotica* and *Haplochromis nigripinnis* in Lake George, Uganda. *Journal of Zoology, 171*, 15–23. London.

Moriarty, D. J. W., (1973). The physiology of digestion of blue-green algae in the cichlid fish, *Tilapia nilotica. Journal of Zoology, 171*, 25–39. London.

Morris, D. W., (1988). Habitat-dependent population regulation and community structure. *Evolutionary Ecology, 2*, 253–269.

Motta, P. J., & Kotrschal, K. M., (1992). Correlative, experimental, and comparative evolutionary approaches in ecomorphology. *Netherland Journal of Zoology, 42*, 400–415.

Motta, P. J., (1989). Dentition patterns among Pacific and Western Atlantic butterflyfishes (Perciformes, Chaetodontidae): relationship to feeding ecology and evolutionary history. *Environmental Biology of Fishes, 25*, 159–170.

Motta, P. J., Clifton, K. B., Hernandez, P., & Eggold, B. T., (1995). Ecomorphological correlates in ten species of subtropical seagrass species: diet and microhabitat utilization. *Environmental Biology of Fishes, 44*, 37–60.

Moyle, P. B., & Cech, J. J. Jr., (1982). *Fishes: An Introduction to Ichthyology.* Englewood Cliffs, N.J.: Prentice Hall Inc.

Moyle, P. B., & Senanayake, F. R., (1984). Resource partitioning among the fishes of rainforest streams in Sri Lanka. *Journal of Zoology, 202*, 195–223. London.

Myers, N., Mittermeier, R. A., Mittermeier, C. G., Da Fonseca, G. A. B., & Kent, J., (2000). Biodiversity hotspots for conservation priorities. *Nature, 403*, 853–858.

Nagelkerke, L. A. J., Sibbing, F. A., & Osse, J. W. M., (1995). Morphological divergence during growth in the large barbs (*Barbus* spp.) of Lake Tana, Ethiopia. *Netherland Journal of Zoology, 45*, 431–454.

Nagelkerke, L. A. J., Sibbing, F. A., Boggart, J. G. M., Lammens, E. H. R. R., & Osse, J. W. M., (1994. The barbs (*Barbus* spp.) of Lake Tana: a forgotten species flock? *Environmental Biology of Fishes, 39*, 1–22.

Nelson, A. R., Johnson, C. L., Matter, W. J., & Mannan, R. W., (2002). Tests of emigration in small mammals under experimental conditions. *Canadian Journal of Zoology, 80*, 2056–2060.

Norton, S. F., (1988). The role of the gastropod shell and operculum in inhibiting prédation by fishes. *Science, 241*, 92–94.

Okamura, O., (1970). Studies of the macrourld fishes of Japan-morphology, ecology and phylogeny. *Report of the USA Marine Biological Station, 17*, 1–179. Kochi University.

Oksanen, T., (1990). Exploitation ecosystems in heterogenous habitat complexes. *Evolutionary Ecology, 4*, 220–234.

Osenberg, C. W., & Mittelbach, G. G., (1989). Effects of body size on the predator-prey interaction between pumpkinseed sunfish and gastropods. *Ecological Monographs, 59*, 405–432.

Paine, R. T., (1966). Food web complexity and species diversity. *American Naturalist, 100*, 65–75.

Perera, P. A. C. T., Sundarabarathy, T. V., Sivananthawerl, T., & Edirisinghe, U., (2015). Potential impact of water quality and hydrology on fish community in upper Malwathu Oya, Sri Lanka. *Journal Of Global Ecology and Environment, 3*(4), 218–229.

Persson, L., (1988). Asymmetries in competitive and predatory interactions in fish populations. In: Ebenman, B., & Persson, L., (eds.), *Size-Structured Populations, Ecology and Evolution* (pp. 203–218). Berlin: Springer-Verlag.

Persson, L., Diehl, S., Johansson, L., Andersson, G., & Hamrin, S. F., (1991). Shifts in fish communities along the productivity gradient of temperate lakes-patterns and the importance of size structured interactions. *Journal of Fish Biology, 38*, 281–293.

Pet, J. S., & Piet, G. J., (1993). The consequences of habitat occupation and habitat overlap of the introduced tilapia *Oreochromis mossambicus* (Peters) and indigenous fish species for fishery management in a Sri Lankan reservoir. *Journal of Fish Biology,43*, 193–208.

Pethiyagoda, R., (1991). *Freshwater Fishes of Sri Lanka* (p. 362). Wildlife Heritage Trust, Colombo, Sri Lanka.

Pethiyagoda, R., Silva, A., Maduwage, K., & Meegaskumbura, M., (2008). A review of the genus *Labuca* in Sri Lanka, with description of three new species (Teleostei: Cyprinidae), *Ichthyological Exploration of Freshwaters, 19*, 7–26.

Pflieger, W. L., (2004). *The Fishes of Missouri*. Jefferson City: Missouri Department of Conservation.

Piet, G. J., (1996). On the ecology of a tropical fish community (p. 189). Doctoral Thesis, Agriculture University, Wageningen.

Poff, N. L., Allan, J. D., Bain, M. B., Karr, J. R., Prestegaard, K. L., Richter, B. D., Sparks, R. E., & Stromberg, J. C., (1997). The natural flow regime. *BioScience, 47*, 769–784.

Polis, G. A., (1988). Exploitation competition and the evolution of interference, cannibalism, and intraguild predation in age/size-structured populations. In: Ebenman, B., & Persson, L., (eds.), *Size-Structured Populations, Ecology and Evolution* (pp. 185–202). Berlin: Springer-Verlag.

Polis, G. E., (1991). Complex interactions in deserts: an empirical critique of food-web theory. *The American Naturalist, 138*, 123–155.

Pulliam, H. R., & Danielson, B. J., (1991). Sources, sinks, and habitat selection: a landscape perspective on population dynamics. *The American Naturalist, 137*, S50–S66.

Rabe, F. W., & Gibson, F., (1984). The effect of macrophyte removal on the distribution of selected invertebrates in an inshore environment. *Journal of Freshwater Ecology*, 359–371.

Ranåker, L., (2012). *Piscivore-Prey Fish Interactions - Consequences of Changing Optical environments*. Doctoral Thesis, Department of Biology. Lund: Lund University.

Randall, J. E., (1967). Food habits of reef fishes of the West Indies. *Studies on Tropical Oceanography, 5*, 665–847.

Ross, S. T., (1986). Resource partitioning in fish assemblages: a review of field studies. *Copeia, 2*, 352–388.

Sale, P. F., (1969a). A suggested mechanism for habitat selection by the juvenile manini, *Acanthurus triostegus sandvicensis* streets. *Behavior, 35*, 27–44.

Sale, P. F., (1969b). Pertinent stimuli for habitat selection by the juvenile manini, *Acanthurus triostegus sandvicensis*. *Ecology, 54*, 616–623.

Saul, W. G., (1975). An ecological study of fishes at a site in upper Amazonian Ecuador. *Proceedings of the Academy of Natural Sciences of Philadelphia, 127*, 93–134.

Schiemer, F., & Hofer, R., (1983). A contribution to the ecology of the Parakrama Samudra reservoir. In: Schiemer, F., (ed.), *Limnology of Parakrama Samudra, Sri Lanka* (pp. 135–148). W. Junk Publ., The Hague.

Schlosser, I. J., (1982). Fish community structure and function along two habitat gradients in a headwater stream. *Ecological Monographs, 52*, 395–414.

Schoener, T. W., (1983). Field experiments on interspecific competition. *American Naturalist, 122*, 240–285.

Schut, J., De Silva, S. S., & Kortmulder, K., (1984). Habitat, associations and competition of eight Barbus (=Puntius) species (Pisces, Cyprinidae) indigenous to Sri Lanka. *Netherland Journal of Zoology, 34*, 159–181.

Shirantha, R. R. A. R., Wijeyaratne, M. J. S., & Amarasinghe, U. S., (2005). Food niche segregation among co-occurring endemic fish species in Sri Lanka. *Sri Lanka Aquatic Science, 10*, 19–34.

Sibbing, F. A., (1991). Food capture and oral processing. In: Winfield, I. J., & Nelson, J. S., (eds.), *Cyprinid Fishes: Systematics, Biology and Exploitation* (pp. 377–412). Chapman & Hall, London.

Sibbing, F. A., Nagelkerke, L. A. J., & Osse, J. W. M., (1994). Ecomorphology as a tool in fisheries: identification and ecotyping of Lake Tana barbs (*Barbus intermediuscomplex*), Ethiopia. *Netherlands Journal of Agricultural Science. 42*, 77–85.

Silva, A., Maduwage, K., & Pethiyagoda, R., (2010). A review of the *Rasbora* in Sri Lanka, with description of two new species (Teleostei: Cyprinidae). *Ichthyological Exploration of Freshwaters, 21*(1), 27–50.

Silva, E. I. L., & Davis, R. W., (1986). Movement of some indigenous riverine fish in Sri Lanka. *Hydrobiologia, 137*, 263–270.

Smith, C. C., & Tyler, J. C., (1972). Space resource sharing in a coral reef fish community. *Bulletin of the Los Angeles County Museum of Natural History Science, 14*, 125–170.

Smith, L. D., & Coull, B. C., (1987). Juvenile spot (Pisces) and grass shrimp predation on meiobenthos in muddy and sandy substrata. *Journal of Experimental Marine Biology and Ecology, 105*, 123–136.

Southwood, T. R. E., (1977). Habitat, the template for ecological strategies? *Journal of Animal Ecology, 46*, 337–365.

Starck, W. A., & Schroeder, R. E., (1971). Investigations on the grey snapper, *Lutjanus griseus. Studies on Tropical Oceanography, 10*, 1–150.

Stoner, A. W., & Livingston, R. J., (1984). Ontogenetic patterns in diet and feeding morphology in sympatric sparid fishes from seagrass meadows. *Copeia, 174*–187.

Stoner, A. W., (1982). The influence of benthic macrophytes on the foraging behavior of pinfish, *Lagodon rhomboides* (Linnaeus). *Journal of Experimental Marine Biology and Ecology, 58*, 271–284.

Strauss, S. Y., (1991). Indirect effects in community ecology: their definition, study and importance. *Trends in Ecology and Evolution, 6*, 206–210.

Sturmbauer, C., Mark, W., & Dallinger, R., (1992). Ecophysiology of aufwuchs-eating cichlids in Lake Tanganyika: niche separation by trophic specialization. *Environmental Biology of Fishes, 35*, 283–290.

Sudasinghe, H., Pethiyagoda, R., & Meegaskumbura, M., (2020). Evolution of Sri Lanka's giant danios (Teleostei: Cyprinidae: *Devario*): teasing apart species in a recent diversification. *Molecular Phylogenetics and Evolution, 149*, 1–15.

Svärdson, G., (1976). *Interspecific Population Dominance in Fish Communities of Scandinavian Lakes, 55*, 144–171. Institute of Freshwater Research. Drottningholm Report.

Trewavas, E., (1983). *Tilapiine Fishes of the Genera Sarotherodon, Oreochromis and Danakilia*. British Museum (Natural History), London.

UNESCO (United Nations Educational, Scientific and Cultural Organization), (2010). UNESCO Headquarters, 7, Place de Fontenoy, 75352, Paris, 07 SP, France. http://whc.unesco.org/en/list/1203/documents (accessed on 14 December 2021).

Van, D. R. F., (1978). Factors regulating the distribution and population dynamics of the amphipod *Garnmarus palustris* in an intertidal salt marsh community. *Ecological Monographs, 48*, 191–217.

Vergina, I. A., (1991). Basic adaptations of the digestive system in bony fishes as a function of diet. *Journal of Ichthyology, 31*, 8–20.

Wainwright, P. C., & Richard, B. A., (1995). Predicting patterns of prey use from morphology of fishes. *Environmental Biology of Fishes, 44*, 97–113.

Wainwright, P. C., (1987). Biomechanical limits to ecological performance: Molluscs crushing by the Caribbean hogfish, *Lachnolaimus maximus* (Labridae). *Journal of Zoology, 213*, 283–297. London.

Ware, D. M., (1972). Predation by rainbow trout (*Salmo gairdneri*): the influence of hunger, prey density, and prey size. *Journal of the Fisheries Research Board of Canada, 29*, 1193–1201.

Webb, P. W., (1984). Form and function in fish swimming. *Scientific American, 251*, 72–82.

Webb, P. W., (1989). Station-holding by three species of benthic fishes. *Journal of Experimental Biology, 145*, 303–320.

Webb, P. W., Gerstner, C. L., & Minton, S. T., (1996). Station-holding by the mottled sculpin, *Cottus bairdi* (Teleostei: Cottidae), and other fishes. *Copeia, 1996*, 488–493.

Welcomme, R. L., & Halls, A., (2001). Some considerations of the effects of differences in flood patterns on fish populations. *Ecohydrology and Hydrobiology, 1*, 313–321.

Weliange, W. S., & Amarasinghe, U. S., (2003). Seasonality in dietary shifts in size-structured freshwater fish assemblages in three reservoirs of Sri Lanka. *Environmental Biology of Fishes, 68*, 269–282.

Werner, E. E., & Gilliam, J. F., (1984). The ontogenetic niche and species interactions in size-structured populations. *Annual Review of Ecology and Systematics, 15*, 393–425.

Werner, E. E., & Hall, D. J., (1988). Ontogenetic habitat shifts in bluegill: the foraging rate-predation risk trade-off. *Ecology, 69*, 1352–1366.

Werner, E. E., Gilliam, D. J., Hall, D. J., & Mittelbach, G. G., (1983b). An experimental test of the effects of prédation risk on habitat use in fish. *Ecology, 64*, 1540–1548.

Werner, E. E., Mittelbach, G. G., Hall, D. J., & Gilliam, J. F., (1983a). Experimental tests of optimal habitat use in fish: the role of relative habitat profitability. *Ecology, 64*, 1525–1539.

Wikramanayake, E. D., & Moyle, P. B., (1989). Ecological structure of tropical fish assemblages in wet-zone streams of Sri Lanka. *Journal of Zoology, 218*, 503–526. London.

Wikramanayake, E. D., (1990). Ecomorphology and biogeography of a tropical stream fish assemblage: evolution of assemblage structure. *Ecology, 71*(5), 1756–1764.

Winemiller, K. O., (1991). Ecomorphological diversification in lowland freshwater fish assemblages from five biotic regions. *ecological monographs, 61*, 343–365.

Winemiller, K. O., (1992). Ecomorphology of freshwater fishes. *National Geographic Research & Exploration, 8*, 308–327.

Zaret, T. M., & Rand, A. S., (1971). Competition in tropical stream fishes: support for the competitive exclusion principle. *Ecology, 52*, 336–342.

Zihler, F., (1982). Gross morphology and configuration of the digestive tracts of *Cichlidae* (*Teleostei*, *Perciformes*): phylogenetic and functional significance. *Netherland Journal Zoology, 32*, 544–571.

CHAPTER 11

Freshwater Biodiversity: Indicators of Eutrophication and Pollution

MAHROOFA JAN,[1] SUMAYA HASSAN,[2] and ASMAT ARA[3]

[1]*Department of Botany, Government Degree College, Ganderbal – 191201, Jammu and Kashmir, India*

[2]*Department of Biochemistry, University of Kashmir, Hazratbal, Srinagar – 190006, Jammu and Kashmir, India*

[3]*Division of Genetics and Plant Breeding, SKUAST-K, Jammu and Kashmir, India*

ABSTRACT

Freshwater ecosystems offer crucial resources for life and represent the lone territory for diverse, widespread, and complex biota. Environmental stressors from natural and human sources affect these freshwater ecosystems. Every kind and level of the aquatic food web is affected, which affects various ecological processes, efficiency, and nutrient cycling of the aquatic life. Eutrophication and water pollution also add to the agony of aquatic biodiversity. The damaging consequences of eutrophication on biodiversity have been well studied in larger aquatic ecosystems. However, this influence may vary in water bodies maintaining naturally high nutrient levels in lowlands like ponds and lakes. Eutrophication causes higher plant and animal biomass, augmented algal blooms, propagation of rooted plants and species diversity decline. It also results in an intensified water turbidity causing an anoxic state. The high density of aquatic biomass because of eutrophication leads

Biodiversity of Freshwater Ecosystems: Threats, Protection, and Management. Suhaib A. Bandh, PhD, Basharat Mushtaq, PhD, Mohammad Farooq Mir, PhD, & Shahid A. Wani (Eds.)

to competition for resources among the living organisms, hence, struggle for existence that is further complicated by resulting high chemical and physical stresses. The cumulative effect has resulted in significant population decline and extensive decline of freshwater biodiversity throughout the world. Such a reduction in biodiversity is much more effective in freshwater ecosystems than in terrestrial ecosystems.

11.1 INTRODUCTION

We inhabit a world where it is essential to find ideal means to balance the rising human need for fresh water and the proper functioning of freshwater ecosystems to support the aquatic biodiversity and keep sustaining the services they provide. Every organism living on the globe require water to stay alive. Still, approximately 100,000 and more of the living species need a superior kind of water found in a few places and is rarely available: freshwater. We as humans also require this kind of water, a treasured resource of this planet (Carpenter et al., 2011). Freshwater ecosystems are extremely active and complex with low salt content. Freshwater comprises just 0.01% of the total Earth's water and approximately 0.8% of its surface. Yet, such a minute portion of Earth's water supports the life of more than 1.8 million species – nearly 6% of the reported species (Schmidt et al., 2019). Freshwater bodies support the life of two important ecological groups, namely planktons and benthos. Planktons dwell in the open water, and benthos occupies the bottom sediments of the water. Despite being small and unremarkable, most of these species are considered to have enormous importance as they provide essential services to the ecosystem (Carpenter et al., 2011). Freshwater biodiversity delivers an extensive range of unique benefits to humans (Covich et al., 2004a). The wealth of this biodiversity has numerous gains: through input in the economy; 'insurance' significance in the context of unpredicted occasions; repository of genetic information and assist the establishment of ecosystem services (Pearce, 1998; Heal, 2000; Covich et al., 2004b).

11.2 FRESHWATER BIODIVERSITY

Biodiversity is described as a wide variety of life present on Earth. It can be, more specifically, referring to all of the species, including plants, animals, bacteria, and humans found in a particular region or ecosystem. Biodiversity is regarded as the foundation of ecosystem services that are closely linked to human well-being. Living organisms occupy the most complex, dynamic, and diverse layer of the Earth that is exposed to the more intense alteration at humans' hands. This layer, called the biosphere, unites the atmosphere, hydrosphere, and geosphere through the cumulative metabolic activities of its countless plants, animals, and microorganisms, into one such environmental system that supports the life of millions of species, including humans. The breathable air, clean water, productive soils, fertile lands, magnanimous seas, and the most suitable climate are a few manifestations of life's workings on Earth. It is believed that extensive human influences over this biota have remarkable impacts on our well-being. It also surveys that the nature of these influences, either favorable or unfavorable, is within humans' control (Figure 11.1).

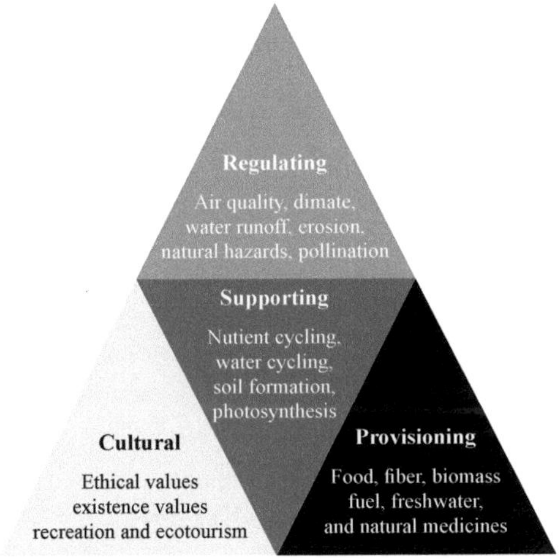

FIGURE 11.1 Biodiversity-foundation of ecosystem services.

Freshwater ecosystems, including wetlands, rivers, and lakes, are regarded as dynamic resources for humans and are the solitary habitat for diverse and sensitive biota. They are the rich source of drinking water, food, and recreation.

Freshwater ecosystems are on their way to joining the most endangered ones of the globe. Declines in freshwater biodiversity are remarkably higher than in the most known influenced land ecosystems (Sala et al.,

2000). Freshwater habitats and their respective biodiversity supported by them are more exposed to environmental variation and human activities. Almost 10,000 species of fish inhabit freshwater ecosystems (Lundberg et al., 2000) that make up nearly 40% of the worldwide fish diversity and one-fourth of worldwide vertebrate diversity. Just one-third of all vertebrates are restricted to these freshwaters (Gleick, 1996).

11.3 FRESHWATER ECOSYSTEMS TYPES

Freshwater ecosystems have been classified into two categories based on the aqueous phase mobility: Lotic ecosystems and lentic ecosystems (Figure 11.2).

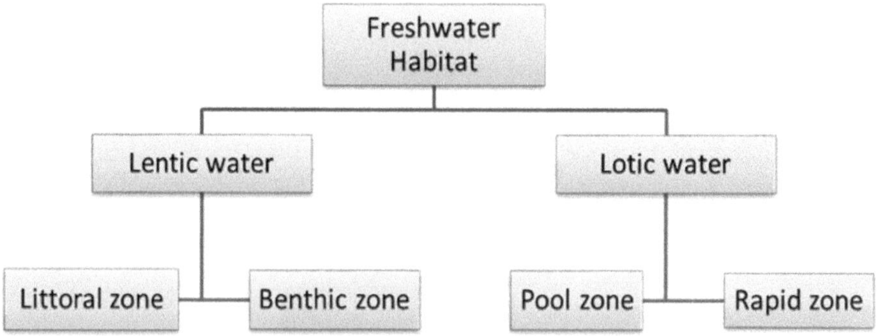

FIGURE 11.2 Types of freshwater habitats.

11.3.1 LOTIC FRESHWATER ECOSYSTEM

Freshwater ecosystems having continual and unidirectional water flow are called lotic systems (Thorp et al., 2008). This constant mixing of the water column in lotic freshwater ecosystems makes them less stratified. These ecosystems have two major water zones, rapids, and pools. Rapids are the zones where the water currents are fast enough to keep the bottom clear of any substances. Simultaneously, pools are comparatively deeper water areas where the winds are slower, resulting in a buildup of silt. Brooks, creeks, channels, streams, and rivers are a few examples of lotic freshwater ecosystems (Figure 11.3).

FIGURE 11.3 Lotic habitats.

11.3.2 LENTIC FRESHWATER ECOSYSTEM

A standing water body (for instance, lakes, ponds, and wetlands) characterized by convectional *in situ* water flow represents a lentic ecosystem (Winter, 2004). These ecosystems are highly stratified because of their moderately immobile nature. Every zone in lentic water systems has unique chemistry because of the variation in dissolved oxygen concentration, organic matter, and content of light penetration. Consequently, the diversity of aquatic organisms in these zones of lentic systems varies considerably (Figure 11.4).

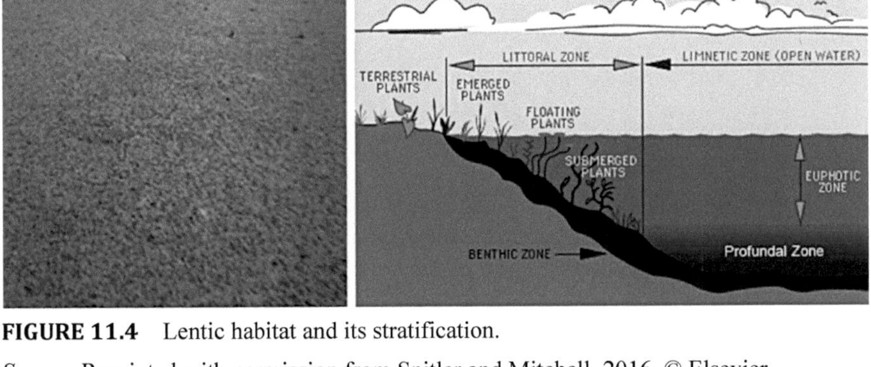

FIGURE 11.4 Lentic habitat and its stratification.

Source: Reprinted with permission from Spitler and Mitchell, 2016. © Elsevier.

11.4 THREATS TO THE BIODIVERSITY OF FRESHWATER

Globally, freshwater ecosystems are considered among the most endangered ones. Biodiversity decline is more remarkable in freshwater than in terrestrial and marine ecosystems. Apart from global changes, anthropogenic activities also pose a potent threat to freshwater habitats. The freshwater ecosystem consisting of vulnerable, endangered, and critically endangered species from all the taxa is the most threatening world ecosystem (Revenga et al., 2005). Freshwater ecosystems are susceptible because of the new species invasion and alteration of physical, chemical, and ecological settings. There has been a rapid increase in biodiversity loss since 2006. Freshwater biodiversity is gradually risked at the hands of multiple factors worldwide (Thieme et al., 2010; Vorosmarty et al., 2010). Dudgeon et al. (2006) recognized five chief grounds of freshwater biodiversity decline viz water pollution, loss of habitat, modification in water flow, overexploitation, and invasion of exotic aquatic species. Reid et al. (2019) identified 12 fresh and intensified previous threats to freshwater ecosystems, including climate change, developing contaminants, e-commerce, pollution, and invasion, harmful algal blooms, salinization of freshwater, diminishing calcium and communicable ailments. The freshwater biodiversity is susceptible to loss because freshwater provides a resource available for extraction and diversion. It could also be contained or contaminated by humans, thereby compromising its value, and posing a threat to its biodiversity (Figure 11.5).

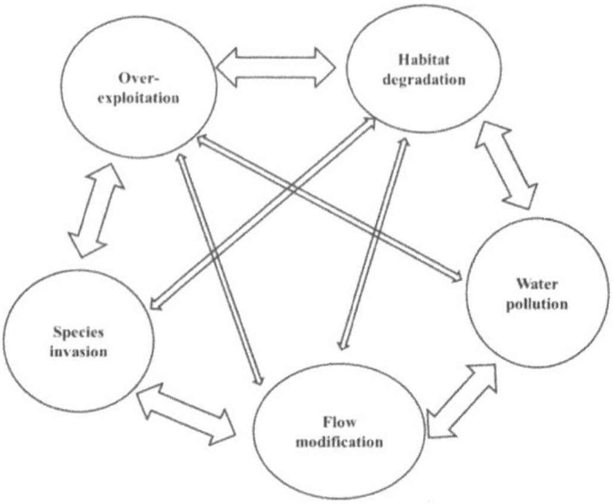

FIGURE 11.5 Major threats to freshwater biodiversity.

Source: Redrawn from Dudgeon et al., 2006.

11.5 EUTROPHICATION

Eutrophication, seen as the enrichment of water bodies by nutrients, causes physical alterations to its ecosystem, for instance: amplified algae and aquatic flora, reduction of fish species and an overall decline of water value. The excessive aquatic growth could be due to the increased accessibility of one or more limiting growth factors required for photosynthesis (Schindler, 2006), such as carbon dioxide, sunlight, and nutrient fertilizers. Eutrophication occurs naturally over times as lakes age and is stuffed with residues. It is an environmental problem caused by the augmentation of nutrients, such as N and P, in aquatic ecosystems. These nutrients are obtained from various sources, extending from urban and industrial sewage to agricultural wastes (Johnson et al., 2007; Donald et al., 2011).

Eutrophication has been classified into natural and artificial (also called anthropogenic) (Bartoszek and Czech, 2014). The natural form of eutrophication takes place slowly and gradually, starting next to water body formation. Artificial type of eutrophication is initiated by intensive land exploitation, mismanagement of fertilizers and products used for crop protection, deforestation, progressively increasing rural and urban areas and release of crude sewage directly into the surface waters (Figure 11.6) (Pajewski, 2016; Huang et al., 2017; Sender and Jaruga, 2017; Serrano et al., 2017; Bhagowati and Ahamad, 2018).

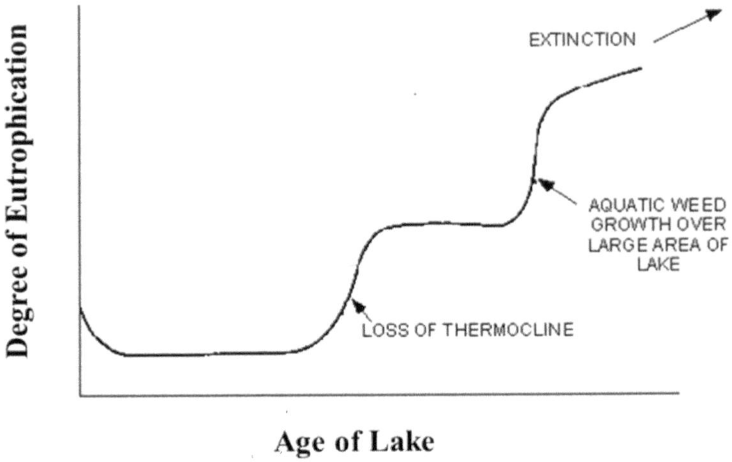

FIGURE 11.6 Natural eutrophication.

Enhanced nutrient concentration causes a sharp boost in phytoplanktons' primary production (Drupp et al., 2011) that subsequently changes its physical, chemical, and biological characteristics extensively (Carneiro et al., 2014; Carpenter, 2005). Phytoplanktons are sensitive to any variation in nutrient content, making them an essential indicator of eutrophication if present. The environmental consequences of eutrophication include biodiversity loss, water pollution, increased discharge of greenhouse gases (GHGs), and water treatment plans' financial overburden (Tilman et al., 2001).

The waters of eutrophic bodies are nutrient-rich and hence very productive. Lakes are regularly classified following their extent of nutrients and organic matter enrichment. Classification based on the trophic state includes oligotrophic, mesotrophic, eutrophic, and dystrophic lakes. Several water bodies are naturally eutrophic because they arise in inherently fertile watersheds (Carpenter, 2005). Eutrophication can be caused in inland and marine waters, even though the problem is seen more in freshwaters.

The main consequence of eutrophication is the disproportion caused in the food web, leading to enhanced phytoplankton microalgae, mainly of diatom biomass in stratified water bodies, finally causing algal blooms (Drupp et al., 2011). Shallow water bodies may also show a strong growth of aquatic plants called macrophytes. The amplified productivity of algae and macrophytes can influence the higher trophic levels and be more productive. The aquatic invertebrates, fish, and waterfowl, can also be found in abundance in such water bodies. However, hypertrophic (extremely eutrophic) waters may develop lethal blooms of cyanobacteria and algae during the summer season, leading to an off-flavor in water, making it unsuitable for drinking. Additionally, decomposition of the algal biomass demands a large quantity of oxygen that causes anoxic, extraordinarily stressful, and fatal circumstances for aquatic animals. Severe eutrophication may render the water unsuitable for drinking, fishery, or recreation, thereby degrading the ecological qualities of waters (Figure 11.7).

11.5.1 EUTROPHICATION AND FRESHWATER BIODIVERSITY

Eutrophication modifies the structure and diversity of aquatic flora, affecting the food web and ecosystem organization. The structure, purpose, and diversity of freshwater populations are often predisposed by the biological behaviors of the species and their interactions in the ecosystem. These ecosystems' fluctuations are influenced by the extent of the stress-causing agents (like the degree of chemical pollution, exploitation of resources, hydrological instabilities) and their ecological pliability (Sanganyado, 2020). The diversity of the system is affected by the nutrient loading, and (Deegan et al., 2002) it has been reported that nitrogen (in the form of ammonia and nitrate) loading has a strong association with the dynamics of macrophytes. The droughts cause a decrease in nutrient loading in the lakes, hence triggering variations in species diversity (Tracy et al., 2003). Eutrophication is established to have decreased the total of rare species and augmented the richness of meso- to hyper-eutrophic ones. The contaminated ponds with biological matter and sewage cover the aquatic surface with weeds, and hence the preying of mosquito larvae by larvivorous fishes is dodged. Therefore, the larval population in such ponds is increased (Lee and Lee, 2002; Nehring and Welling, 2003). Biodiversity loss is also caused by oxygen depletion, fluctuating phytoplankton content, and diminishing transparency in the water (Smith, 2003). Eutrophication because of nitrogen and phosphorus contamination is believed to enhance toxic algal blooms that inescapably lead to biodiversity loss (Whitehead and Crossman, 2013).

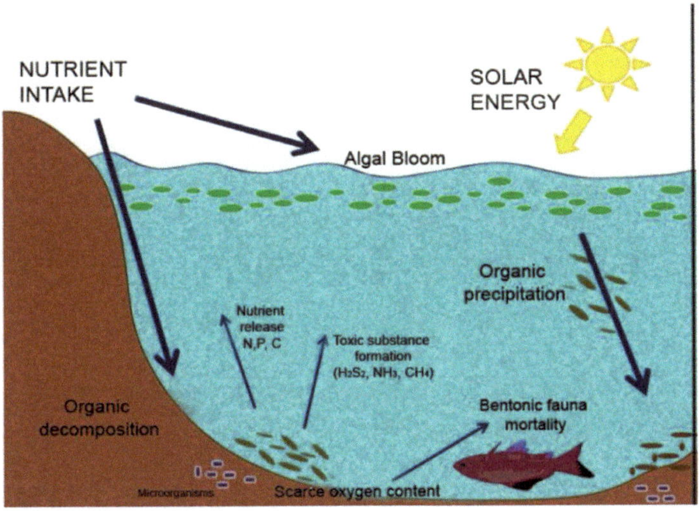

FIGURE 11.7 Eutrophication process representation.
Source: Feem re-elaboration from Arpa Umbria (2009).

There is a rapid increase in eutrophication of lakes, making them among the most potent drivers of the overall deterioration of aquatic biodiversity (Dudgeon, 2014; Dudgeon et al., 2006; Vorosmarty et al., 2010). Dissolved inorganic nitrogen and dissolved reactive phosphorus represent the leading culprits of eutrophication, restricting the growth of autotrophs and heterotrophs (Camargo and Alonso, 2006; Conley et al., 2009; Ferreira et al., 2015; Smith and Schindler, 2009). These elements can be directly lethal to freshwater creatures at much higher concentrations (Camargo and Alonso, 2006; Camargo et al., 2005; Hickey and Martin, 2009). If their decomposition increases disproportionately, it can lead to hypoxic conditions. Nutrients encourage the growing microbes that decompose dead biological matter. This process utilizes a lot of oxygen, thereby dramatically dropping the dissolved oxygen content and altering pH that subsequently affects the inhabiting invertebrates and fishes (Dodds and Smith, 2016; Ferreira et al., 2015; Hilton et al., 2006; Smith et al., 2006). Additionally, the microbes can mine their surroundings for nutrients and create nutrient-poor debris, consecutively increasing the taste for detritus-processing invertebrates and dismissing nutrient restricted growing pattern (Figure 11.8) (Evans-White and Halvorson, 2017; Ferreira et al., 2015; Guo et al., 2016; Hessen et al., 2013).

FIGURE 11.8 Algal blooms in Dal Lake.

11.6 INFLUENCE OF ENHANCED NITROGEN CONTENT ON FRESHWATER BIODIVERSITY

Nitrate concentration in water is an essential element for water quality assessment. Nitrate is released into freshwater from waste disposed at dumpsites or agroindustry products (Ogbonna, 2014). Excess quantities of nitrogen in water enhance eutrophication and algal bloom production (Smitha et al., 2013). The average concentration of nitrate comes from rocks, plant, and animal matters and land drainage. Extremely high levels of nitrate in freshwater are lethal (Ogbonna, 2014). It is rarely abundant in natural shallow water because it is assimilated into the cells and is chemically modified by microbes into atmospheric free nitrogen (Nighat et al., 2020; Ogbonna, 2014). It is a highly oxidized form of nitrogen and occupies surface and groundwater zones (Uqab et al., 2017). Naturally, nitrates are found in the water bodies as end products of the aerobic decomposition of organic nitrogenous matter (Adesuyi et al., 2015). The concentration of nitrate is usually low in surface and ground waters but can increase due to leaching or runoff from the agricultural fields (Bhat et al., 2017). Its levels are also enhanced by the oxidation of ammonia from human or animal wastes (Uqab et al., 2017). Oxidation of nitrites and the river biomass's mineralization also add to water nitrates (Tog et al., 2017). Chief sources of nitrate pollution include sewage leaks, agrochemicals, landfills, industrial wastes, use of wastewater for irrigating fields, human, and animal wastes, etc. (Adesuyi et al., 2015). Such contaminated drinking water has also been associated with major health complications (Krishnan and Indu, 2006; WHO, 2007; Jagessar and Sooknundun, 2011).

11.7 INFLUENCE OF ENHANCED PHOSPHORUS LEVELS ON FRESHWATER BIODIVERSITY

Phosphorus is regarded as an indispensable nutrient and exists in water in particulate and dissolved forms (Ogbonna, 2014). It is released into the water body through sources like leached or weathered igneous rocks, detergent phosphates from industrial emissions, runoff from fertilized farmlands and domestic sewage containing human excreta (Abowei et al., 2010; Bhat et al., 2017; Eunice et al., 2017). The higher level of phosphates in the surface water is regarded as a significant environmental issue because of its involvement in eutrophication (Adesuyi et al., 2015). Such elevated phosphate levels have been associated with swelling rates of plant growth, variations in species composition and propagation of planktonic and epiphytic and epibenthic algae (Nighat et al., 2020). A credible reason behind the differential concentration is its exclusive behavior in shallow waters (Adesuyi et al., 2015). Phosphorus as phosphate rapidly gets adsorbed at the surface of mud and re-enters the groundwater (Onwugbuta-Enyi et al., 2008). The high concentration of phosphates in water is attributed to outflow from the dumpsites, domestic wastes and agricultural flow-off (Ogbonna, 2014). Increased phosphate concentration is one of the major causes of eutrophication because phosphorus is a limiting nutrient for the growth of algae (Eunice et al., 2017). All polyphosphates are ultimately hydrolyzed into their ortho-forms, and the rate of hydrolysis is enhanced by increased temperature, low pH, and bacterial enzymes (WHO, 2004). Phosphate stimulates microbial growth, molds, aquatic weeds, algae, and is returned to the environment via industrial, agricultural, and human sources (Figure 11.9) (Eunice et al., 2017; Tog et al., 2017).

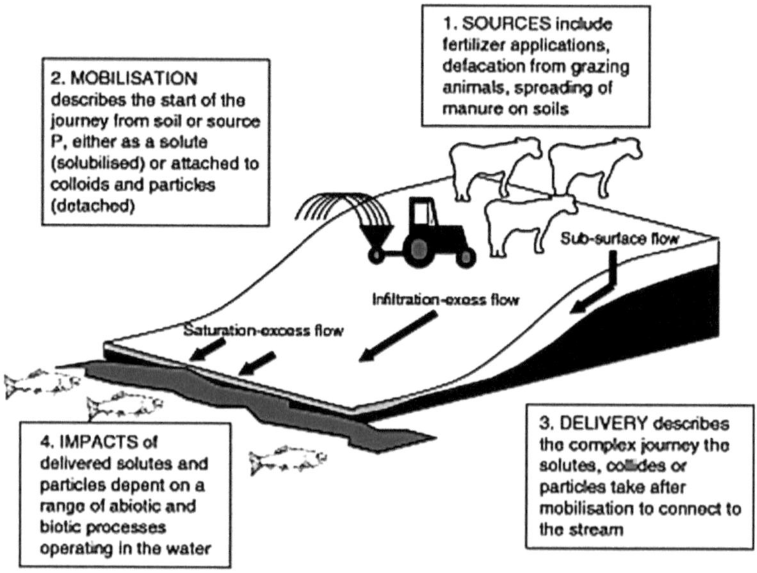

FIGURE 11.9 The components of the source-mobilization-delivery-impact continuum for phosphorus transfer from agricultural land to water.
Source: Reprinted from Withers and Haygarth (2007). © John Wiley. Open access.

11.8 IMPACT OF HYPOXIA ON FRESHWATER BIODIVERSITY

Oxygen is an essential requirement of all living forms on Earth. Plants release it during the process of photosynthesis in the presence of sunlight.

During the night, the aquatic animals and aerobic microorganisms consume the dissolved oxygen causing depletion of dissolved oxygen levels in the water. Algal blooms are responsible for significant variations in dissolved oxygen content in the water body (Chislock et al., 2013). The fast-growing algal blooms shield the water body from sunlight and prevent oxygen synthesis, causing decreased concentration of dissolved oxygen. The decomposition of algae also consumes oxygen and further decreases the oxygen levels. Hypoxia or oxygen depletion co-exists with other stressors, including hydrogen sulfide and raised temperature. It is an overall outcome of eutrophication, especially in freshwaters. It mainly affects the benthic creatures and bottom-dwelling fishes, which are thus more severely crushed than the species dwelling in the upper water zones. Benthic sessile organisms are most affected by hypoxia, and the effect could be lethal very often. Fish requiring high volumes of dissolved oxygen are equally affected by hypoxia (Vaquer-Sunyer and Duarte, 2008). Effects are usually sub-lethal for the rest of the organisms and comprise reduced growth and reproduction, forced relocation to some appropriate habitats, and physiological stress (Nighat et al., 2020).

11.9 POLLUTION AND FRESHWATER BIODIVERSITY

Pollution is a global challenge of the 21st century, as pure water is becoming less available with increasing water pollution. Currently, water pollution is considered a universal challenge for developed and developing countries. It affects the environmental health of people throughout the world. Much attention is being paid to water quality globally, but water conservation and its distribution should also be equally monitored because inappropriate dealing of wastewater worsens the water catastrophe globally (FAO, 2017). The most exclusive benefit of the aquatic environment is that life exists in it, and the unique characteristic of this life is its 'biodiversity. The significance of water quality was acknowledged in the 2030 Agenda for Sustainable Development. The main focus was laid on future policies to ensure that water pollution control strategies be raised in national and international urgencies.

Aquatic biodiversity encounters various harms and, therefore, is exposed to severe declines in several countries. According to UNEP-International water quality Guidelines for ecosystems 2016, the awful matter is that nearly primitive ecosystems are amongst the most vulnerable ones. Though aquatic ecosystems are the most diverse, the millennium ecosystem assessment (MEA 5) proclaimed in 2005 that biodiversity deprivation in freshwater ecosystems happens twice than in other systems. Hence, their competence to make ecosystem services available declines causing negative impacts on human health. The present loss of biodiversity has a prime reason linked to ecological function's loss. Regardless of the intensive research about associations between functions of ecosystems and biodiversity, significant missing links were detected while trying to recognize the effects of biodiversity loss on our ecosystem (Hooper et al., 2005), especially in freshwaters (Gessner et al., 2004; Dudgeon, 2006). Most aquatic organisms are sensitive to slight variation in their surroundings and respond to these variations differently. The most extreme responses are shown as death or relocation to some other place. Very few responses include decreasing reproductive ability and quashing a few enzyme systems required for metabolism (Chapman, 1992). The importance of zooplankton and macrobenthic constituents in the trophic dynamics of freshwater ecosystems is well recognized. Such organisms modify productivity by inhabiting intermediate levels of the food chain. They also designate environmental status in a definite period (Xie et al., 2008). An additional reason for gaining importance is their definite species capability for acting as indicators of any decline in water quality due to eutrophication and pollution (Khan, 2003; Hassan, 2008). Any disturbance in the food chain resulting from biodiversity loss results in a reduction in fish statistics at the top of food webs. Various water pollutants found in the surroundings affect biodiversity significantly, which is evident in different ways.

Aquatic life has been dramatically influenced by pollution with heavy metals. According to a report published in 1992, 50% of the industrial waste comes from the metallurgical industry, while weaving, spinning, and dyeing reason 30% of industrial waste. Though certain chemicals are crucial for supporting life in freshwater ecosystems at particular concentrations, few naturally occurring and artificial chemicals can present a potential risk to the environment. Anthropogenic activities like aquaculture, agriculture, mining, and other domestic activities often release chemicals into water (Sanganyado et al., 2018). Numerous studies indicate that chemical pollution stands as one of the principal reasons behind the loss of biodiversity in freshwater systems (Dudgeon et al., 2006). The release of endocrine-disrupting chemicals like polyfluorinated materials, halogenated flame retardants, other dynamic pharmaceutical elements into freshwater has been revealed to end up in reproductive and developmental damage in the dwelling species (Sanganyado et al., 2017; Teaf et al., 2017). A study carried in Bulawayo, Zimbabwe, reported extensive feminization of masculine fish in peri-urban rivers and lakes influenced by effluent from industrial and municipal wastewater treatment plant (Teta et al., 2018).

Microbes are the most hazardous contaminants of water streams, be it fresh or marine (FAO, 2006; WHO, 2012). Microbial freshwater populations play a vital role in biogeochemical recycling practice for phosphorus, nitrogen, and several other categories of metals. Spatial and temporal modifications in ecosystem services follow any chemical pollution (Zhuang et al., 2019). Primarily, freshwater ecosystems near urban zones are greatly threatened by releases of deadly chemicals. Tolerant microbes may sustain in these conditions, and therefore, studies are currently trying to develop approaches for using them in treating various effluents (Sathishkumar et al., 2020). The destiny and transport of chemical contaminants are essential for qualitative evaluation of the health risks (Fono et al., 2006; Sanganyado et al., 2016). The establishment of chemical pollution alleviation strategies remains a primary issue for guarding freshwater ecosystems, endorsing clean water and cleanliness (Chingono et al., 2018; Zhuang et al., 2020).

The cumulative impacts of the five major threats (Figure 11.5) have led to a decline in worldwide freshwater biodiversity. Data compiled from various studies indicate reductions in different wetland and water margin vertebrate species (birds, reptiles, fishes, and other mammals). In contrast, the population pattern shows regressions averaging 54% among freshwater vertebrates with an inclination toward greater values in tropical latitudes (Groombridge and Jenkins, 2000). Moreover, 32% of the biosphere's amphibian species are now threatened with extinction, a higher percentage than threatened birds (12%) or mammals (23%) (Amphibia Web, 2005).

The first sign of plastic pollution influencing freshwater fishes in the Amazon was reported by Andrade et al. (2019). Plastic bags, plastic bottles flow in the Amazonian waters and break into meso and microplastic particles integrated into the food chains via fishes. Examination of gastric fillings of 172 specimens of fish revealed the cause of their death to be the ingestion of these plastic particles.

11.10 CONCLUSIONS

Regardless of the tremendous significance of freshwater ecosystems for biodiversity and humans, these are considered the most vulnerable ecosystems. The biodiversity loss in rivers and streams is caused due to climate change, habitat destruction and pollution (Abell et al., 2011; Carpenter et al., 2011). Freshwater invertebrates play crucial roles in preserving water ecosystem facilities, primarily secondary production, and flow of energy. Aquatic insects have a tremendous significance to freshwater monitoring, conservation, and defense due to their sensitivity to disapproving of environmental alterations.

KEYWORDS

- **aquatic ecosystem**
- **biodiversity**
- **biomass**
- **competition**
- **eutrophication**
- **freshwater**
- **nutrients**
- **pollution**

REFERENCES

Abell, R., Thieme, M., Ricketts, T. H., Olwero, N., Ng, R., Petry, P., Dinerstein, E., et al., (2011). Concordance of freshwater and terrestrial biodiversity. *Conservation Letters, 4*, 127–136.

Abowei, J. F. N., Davies, O. A., & Eli, A., (2010). Physicochemistry, morphology and abundance of fin fish of Nkoro River, Niger Delta, Nigeria. *International Journal of Pharma and Bio Sciences, 6*, 1–11.

Adesuyi, A. A., Nnodu, V. C., Kelechi, L. N. K. N., & Anuoluwapo, J., (2015). Nitrate and phosphate pollution in surface water of Nwaja Creek, Port Harcourt, Niger Delta, Nigeria. *International Journal of Geology, Agriculture and Environmental Sciences, 3*, 14–20.

Amphibia Web, (2005). *AmphibiaWeb Species Numbers*. AmphibiaWeb: Information on Amphibian Biodiversity and Conservation. Berkeley, California, U.S.A.

Andrade, M. C., Winemiller, K. O., Barbosa, P. S., Fortunati, A., Chelazzi, D., Cincinelli, A., & Giarrizzo, T., (2019). First account of plastic pollution impacting freshwater fishes in the Amazon: Ingestion of plastic debris by piranhas and other serrasalmids with diverse feeding habits. *Environmental Pollution, 244*, 766–773.

Bartoszek, L., & Czech, D., (2014). The susceptibility of the Solina dam reservoir to degradation. *Journal of Civil Engineering, Environment and Architecture, 61*, 35–53.

Bhagowati, B., & Ahamad, K. U., (2019). A review on lake eutrophication dynamics and recent developments in lake modeling. *Ecohydrology & Hydrobiology, 19*, 155–166.

Camargo, J., & Alonso, Á., (2006). Ecological and toxicological effects of inorganic nitrogen pollution in aquatic ecosystems: A global assessment. *Environment International, 32*, 831–849.

Camargo, J., Alonso, A., & Salamanca, A., (2005). Nitrate toxicity to aquatic animals: A review with new data for freshwater invertebrates. *Chemosphere, 58*, 1255–1267.

Carneiro, F. M., Nabout, J. C., Vieira, L. C., Roland, F., & Bini, L. M., (2014). Determinants of chlorophyll-a concentration in tropical reservoirs. *Hydrobiologia, 740*, 89–99.

Carpenter, S. R., (2005). Eutrophication of aquatic ecosystems: bistability and soil phosphorus. *Proceedings of the National Academy of Sciences, 102*, 10002–10005.

Carpenter, S. R., Stanley, E. H., & Vander, Z. M. J., (2011). State of the world's freshwater ecosystems: physical, chemical, and biological changes. *Annual review of Environment and Resources, 36*, 75–99.

Chapman, D., (1992). Water quality assessments: A guide to the use of biota. *Sediments and Water in Environmental Monitoring.*

Chingono, K. E., Sanganyado, E., Bere, E., & Yalala, B., (2018). Adsorption of sugarcane vinasse effluent on bagasse fly ash: A parametric and kinetic study. *J, Environ, Manage., 224*, 182–190.

Chislock, M. F., Doster, E., Zitomer, R. A., & Wilson, A. E., (2013). Eutrophication: causes, consequences, and controls in aquatic ecosystems. *Nature Education Knowledge, 4*, 10.

Cottingham, K. L., Carpenter, S. R., & Amand, A. L. S., (1998). Responses of epilimnetic phytoplankton to experimental nutrient enrichment in three small seepage lakes. *Journal of Plankton Research, 20*, 1889–1914.

Covich, A. P., Austen, M. C., Bärlocher, F., Chauvet, E., Cardinale, B. J., Biles, C. L.. & Moss, B., (2004b). The role of biodiversity in the functioning of freshwater and marine benthic ecosystems. *BioScience, 54*, 767–775.

Covich, A. P., Ewel, K. C., Hall, R. O., Giller, P. E., Goedkoop, W., & Merritt, D. M., (2004a). Ecosystem services provided by freshwater benthos. *Sustaining Biodiversity and Ecosystem Services in Soils and Sediments, 64*, 45.

Deegan, L. A., Wright, A., Ayvazian, S. G., Finn, J. T., Golden, H., Merson, R. R., & Harrison, J., (2002). Nitrogen loading alters seagrass ecosystem structure and support of higher trophic levels. *Aquatic Conservation: Marine and Freshwater Ecosystems, 12*, 193–212.

Dodds, W., & Smith, V., (2016). Nitrogen, phosphorus, and eutrophication in streams. *Inland Waters, 6*, 155–164.

Donald, D. B., Bogard, M. J., Finlay, K., & Leavitt, P. R., (2011). Comparative effects of urea, ammonium, and nitrate on phytoplankton abundance, community composition, and toxicity in hypereutrophic freshwaters. *Limnology and Oceanography, 56*, 2161–2175.

Drupp, P., De Carlo, E. H., Mackenzie, F. T., Bienfang, P., & Sabine, C. L., (2011). Nutrient inputs, phytoplankton response, and CO_2 variations in a semi-enclosed subtropical embayment, Kaneohe Bay, Hawaii. *Aquatic Geochemistry, 17*, 473–498.

Dudgeon, D., (2014). Accept no substitute: biodiversity matters. *Aquatic Conservation: Marine and Freshwater Ecosystems, 24*, 435–440.

Dudgeon, D., Arthington, A. H., Gessner, M. O., Kawabata, Z. I., Knowler, D. J., Lévêque, C.. & Sullivan, C. A., (2006). Freshwater biodiversity: importance, threats, status and conservation challenges. *Biological Reviews, 81*, 163–182.

Eunice, O. E., Frank, O., Voke, U., & Godwin, A., (2017). Assessment of the impacts of refinery effluent on the physico-chemical properties of Ubeji creek, Delta State, Nigeria. *Journal of Environmental & Analytical Toxicology, 7*, 428.

Evans-White, M. A., & Halvorson, H. M., (2017). Comparing the ecological stoichiometry in green and brown food webs – A review and meta-analysis of freshwater food webs. *Frontiers in Microbiology, 8*, 1–14.

FAO, (2006). *Livestock's Long Shadow*. Rome, Food and Agriculture Organization of the United Nations.

FAO, I., (2017). *Water Pollution from Agriculture: A Global Review*. Published by the Food and Agriculture Organization of the United Nations, Rome and the International Water Management Institute on Behalf of the Water Land and Ecosystems Research Program, Colombo.

Ferreira, V., Castagneyrol, B., Koricheva, J., Gulis, V., Chauvet, E., & Graça, M. A. S., (2015). A meta-analysis of the effects of nutrient enrichment on litter decomposition in streams. *Biological Reviews, 90*, 669–688.

Ferrington, L. C., (2008). Global diversity of non-biting midges (Chironomidae; Insecta-Diptera) in freshwater. *Hydrobiologia, 595*, 447–455.

Fono, L. J., Kolodziej, E. P., & Sedlak, D. L., (2006). Attenuation of Wastewater-Derived Contaminants in an Effluent-Dominated River. *Environ Sci Technol., 40*, 7257–7262.

Gessner, M. O., Inchausti, P., & Persson (2004). Biodiversity effects on ecosystem functioning: Emerging issues and their experimental test in aquatic environments. *Oikos,104*, 423–436.

Gleick, P. H., (1996). Water resources. *Encyclopedia of Climate, Weather, 817*–823.

Groombridge, B., & Jenkins, M. D., (2000). *Global Biodiversity: Earth's Living Resources in the 21st Century*. World Conservation Press.

Guo, F., Kainz, M. J., Sheldon, F., & Bunn, S. E., (2016). Effects of light and nutrients on periphyton and the fatty acid composition and somatic growth of invertebrate grazers in subtropical streams. *Oecologia, 181*, 449–462.

Hassan, M. M., (2008). *Ecological Studies on Zooplankton and Macrobenthos of Lake Edku, Egypt*. Doctoral dissertation, Ph D. thesis Ain Shams University, Faculty of Science, Zoology Dep., Cairo, Egypt.

Heal, G., (2000). *Nature and the Marketplace: Capturing the Value of Ecosystem Services*. Island Press.

Heino, J., & Paasivirta, L., (2008). Unravelling the determinants of stream midge biodiversity in a boreal drainage basin. *Freshwater Biology, 53*, 884–896.

Hessen, D. O., Elser, J. J., Sterner, R. W., & Urabe, J., (2013). Ecological stoichiometry: An elementary approach using basic principles. *Limnology and Oceanography, 58*, 2219–2236.

Hickey, C. W., & Martin, M. L., (2009). *A review of Nitrate Toxicity to Freshwater Aquatic Species*. Environment Canterbury.

Hilton, J., O'Hare, M., Bowes, M. J., & Jones, J. I., (2006). How green is my river? A new paradigm of eutrophication in rivers. *Science of the Total Environment, 365*, 66–83.

Hooper, D. U., Chapin, III. F. S., Ewel, J. J., Hector, A., Inchausti, P., Lavorel, S., & Wardle, D. A., (2005). Effects of biodiversity on ecosystem functioning: a consensus of current knowledge. *Ecological Monographs, 75*, 3–35.

Huang, J., Xu, C. C., Ridoutt, B. G., Wang, X. C., & Ren, P. A., (2017). Nitrogen and phosphorus losses and eutrophication potential associated with fertilizer application to cropland in China. *Journal of Cleaner Production, 159, 17*1–179.

Jagessar, R. C., & Sooknundun, L., (2011). Determination of Nitrate anion in waste water from nine selected areas of coastal Guyana via a spectrophotometric method. *JJRRAS, 7*, 203–212.

Johnson, P. T., Chase, J. M., Dosch, K. L., Hartson, R. B., Gross, J. A., Larson, D. J., & Carpenter, S. R., (2007). Aquatic eutrophication promotes pathogenic infection in amphibians. *Proceedings of the National Academy of Sciences, 104,* 15781–15786.

Khan, R. A. (2003). Faunal Diversity of Zooplanton in freshwater wetlands of Southeastern West Bengal. Rec. zool. Surv. India, Occ. Paper No. 204, 1–107 (Published by the Director, Zool. Surv. India, Kolkata).

Krishnan, S., & Indu, R., (2006). *Groundwater Contamination in India: Discussing Physical Processes, Health and Socio Behavioral Dimensions*. IWMI-Tata, Water Policy Research Programs, Anand, India.

Lee, D. K., (2002). Biological control of *Culex pipiens* pattens (Diptera, Culicidae) by the release of fish muddy loach, *Misgurnus mizolepis* in natural ponds, Korea. *Entomological Research, 32,* 43–47.

Lencioni, V., Marziali, L., & Rossaro, B., (2012). Chironomids as bioindicators of environmental quality in mountain springs. *Freshwater Science, 31,* 525–541.

Lundberg, J. G., Kottelat, M., Smith, G. R., Stiassny, M. L., & Gill, A. C., (2000). So many fishes, so little time: an overview of recent ichthyological discovery in continental waters. *Annals of the Missouri Botanical Garden,* 26–62.

Lunde, K. B., & Resh, V. H., (2012). Development and validation of a macroinvertebrate index of biotic integrity (IBI) for assessing urban impacts to Northern California freshwater wetlands. *Environmental Monitoring and Assessment, 184,* 3653–3674.

Nehring, S., & Welling, M., (2003). Alien species in German waters-a risk for biodiversity. *Schriftenreihe des BMVEL Angewandte Wissenschaft, 498*, 40–52.

Nighat, M., Dig, V. S., Rouf, A. B., Moonisa, A. D., & Omar, B. H., (2020). Freshwater contamination: Sources and hazards to aquatic biota. *Fresh Water Pollution Dynamics and Remediation,* 27–50.

Ogbonna, D. N., (2014). The impact of untreated sewage wastes discharge on the physico-chemical properties of rivers in Port Harcourt metropolis. *World Journal of Scientific Research and Reviews, 2*(2), 1–19.

Ogbonna, J. O., Lawal, F. A., Owoeye, L. D., & Udeh, M. U., (2008). Chemical characteristics and fertilizing value of primary sludge from tannery effluent treatment plant. In: *6th Ann. Natl. Conf. Nigerian Inst. Sci. Technol. Univ. Ibadan,* (pp. 24–27).

Onwugbuta-Enyi, J., Zabbey, N., & Erondu, E. S., (2008). Water quality of Bodo Creek in the lower Niger Delta basin. *Advances in Environmental Biology, 2*, 132–136.

Pajewski, T., (2016). Water pollution as a negative effect of agricultural activities. *Annals scientific of Polish Association of Agricultural Economists and Agribusiness, 18*, 191–195.

Reid, A. J., Carlson, A. K., Creed, I. F., Eliason, E. J., Gell, P. A., Johnson, P. T., & Cooke, S. J., (2019). Emerging threats and persistent conservation challenges for freshwater biodiversity. *Biological Reviews, 94*, 849–873.

Revenga, C., Campbell, I., Abell, R., De Villiers, P., & Bryer, M., (2005). Prospects for monitoring freshwater ecosystems towards the 2010 targets. *Philosophical Transactions of the Royal Society B: Biological Sciences, 360*, 397–413.

Roque, F. O., Siqueira, T., Bini, L. M., Ribeiro, M. C., Tambosi, L. R., Ciocheti, G., & Trivinho-Strixino, S., (2010). Untangling associations between chironomid taxa in Neotropical streams using local and landscape filters. *Freshwater Biology, 55*, 847– 865.

Sala, O. E., Chapin, F. S., Armesto, J. J., Berlow, E., Bloomfield, J., Dirzo, R., & Wall, D. H., (2000). Global biodiversity scenarios for the year 2100. *Science, 287*, 1770–1774.

Sanganyado, E., Bi, R., Teta, C., Buruaem, M. L., Yu, X., Yajing, S., & Liu, W., (2020). Toward an integrated framework for assessing micropollutants in marine mammals: Challenges, progress, and opportunities. *Critical Reviews in Environmental Science and Technology*, 1–48.

Sanganyado, E., Fu, Q., & Gan, J., (2016). Enantiomeric selectivity in adsorption of chiral β-blockers on sludge. *Environ. Pollut., 214*, 787–794.

Sanganyado, E., Lu, Z., Fu, Q., et al., (2017). Chiral pharmaceuticals: A review on their environmental occurrence and fate processes. *Water Res., 124*, 527–542.

Sanganyado, E., Rajput, I. R., & Liu, W., (2018). Bioaccumulation of organic pollutants in Indo-Pacific humpback dolphin: A review on current knowledge and future prospects. *Environ. Pollut., 237*, 111–125.

Sathishkumar, K., Li, Y., & Sanganyado, E., (2020). Electrochemical behavior of biochar and its effects on microbial nitrate reduction: Role of extracellular polymeric substances in extracellular electron transfer. *Chem. Eng. J., 395*, 125077.

Schindler, D. W., (2006). Recent advances in the understanding and management of eutrophication. *Limnology and Oceanography, 51*, 356–363.

Schmidt-Kloiber, A., Bremerich, V., De Wever, A., Jähnig, S. C., Martens, K., Strackbein, J., & Hering, D., (2019). The Freshwater Information Platform: a global online network providing data, tools and resources for science and policy support. *Hydrobiologia, 838*, 1–11.

Sender, J. J., & Jaruga, C., (2017). Eutrophication of water reservoirs and role of macrophytes in this process. *Inżynieria Ekologiczna, 18*, 227–244.

Serrano, L., Reina, M., Quintana, X. D., Romo, S., Olmo, C., Soria, J. M., & Pätzig, M., (2017). A new tool for the assessment of severe anthropogenic eutrophication in small shallow water bodies. *Ecological Indicators, 76*, 324–334.

Smith, V. H., (2003). Eutrophication of freshwater and coastal marine ecosystems a global problem. *Environmental Science and Pollution Research, 10*, 126–139.

Smith, V. H., Joye, S. B., & Howarth, R. W., (2006). Eutrophication of freshwater and marine ecosystems. *Limnology and Oceanography, 51*, 351–355.

Smitha, A. D., & Shivashankar, P., (2013). Physico chemical analysis freshwater at river of the Kapila, Nanjangudu industrial area, Mysore. *International Research Journal of Environment Sciences, 2*, 59–65.

Spitler, J. D. Spitler and Mitchell, J. S. Chapter 8: Surface water heat pump systems. In Advances in Ground-Source Heat Pump Systems, ed. Rees, S.J. 2016. Pages 225-246, ISBN 9780081003114, https://doi.org/10.1016/B978-0-08-100311-4.00008-X.

Tang, H., Song, M. Y., Cho, W. S., Park, Y. S., & Chon, T. S., (2009). Species abundance distribution of benthic chironomids and other macroinvertebrates across different levels of pollution in streams. *Annales de Limnologie - International Journal of Limnology, 46*, 53– 66.

Teaf, C. M., Garber, M. M., Covert, D. J., & Tuovila, B. J., (2019). Perfluorooctanoic acid (PFOA): environmental sources, chemistry, toxicology, and potential risks. *Soil Sediment Contam., 28*, 258–273.

Teta, C., Holbech, B., Norrgren, L., & Naik, Y., (2018). Occurrence of oestrogenic pollutants and widespread feminization of male tilapia in peri-urban dams in Bulawayo, Zimbabwe. *African J. Aquat. Sci.*

Thieme, M. L., Turak, E., McIntyre, P., Darwall, W., Tockner, K., Cordeiro, J., & Butchart, S. H. M., (2010). Freshwater ecosystems under threat: the ultimate hotspot. *Fresh Water: The Essence of Life,* 123–151.

Thorp, J., Thoms, M., & Delong, M., (2008). Ecological implications of the riverine ecosystem synthesis: Some proposed biocomplexity tenets. *The Riverine Ecosystem Synthesis* (pp. 103–131). Elsevier, London, UK.

Tilman, D., Fargione, J., Wolff, B., D'antonio, C., Dobson, A., Howarth, R., & Swackhamer, D., (2001). Forecasting agriculturally driven global environmental change. *Science, 292*, 281–284.

Togue, F. K., Kuate, G. L. O., & Oben, L. M., (2017). Physico-chemical characterization of the surface water of Nkam River using the principal component analysis. *JMES, 8*, 1910–1920.

Tracy, M., Montante, J. M., Allenson, T. E., & Hough, R. A., (2003). Long-term responses of aquatic macrophyte diversity and community structure to variation in nitrogen loading. *Aquatic Botany, 77*, 43–52.

UNEP, (2016). International water quality guidelines for ecosystems (IWQGES). *How to Develop Guidelines for Healthy Freshwater Ecosystems: A Policy Oriented Approach.*

United Nations, (2016). *Report of the Inter-Agency and Expert Group on Sustainable Development Goal Indicators.* 47th Session of the United Nations Statistical Commission. New York, USA.

Uqab, B., Singh, A., & Mudasir, S., (2017). Impact of sewage on physico-chemical water quality of Tawi River in Jammu city. *Environmental Risk Assessment and Remediation, 1*, 56–61.

Vaquer-Sunyer, R., & Duarte, C. M., (2008). Thresholds of hypoxia for marine biodiversity. *Proceedings of the National Academy of Sciences, 105*, 15452–15457.

Verdonschot, R. C. M., Keizer-Vlek, H. E., & Verdonschot, P. F. M., (2012). Development of a multimetric index based on macroinvertebrates for drainage ditch networks in agricultural areas. *Ecological Indicators, 13*, 232–242.

Vorosmarty, C. J., McIntyre, P. B., Gessner, M. O., Dudgeon, D., Prusevich, A., Green, P., & Davies, P. M., (2010). Global threats to human water security and river biodiversity. *Nature, 467*, 555–561.

Whitehead, P. G., Crossman, J., Balana, B. B., Futter, M. N., Comber, S., Jin, L., & Read, D. S., (2013). A cost-effectiveness analysis of water security and water quality: impacts of climate and land-use change on the River Thames system. *Philosophical Transactions of the Royal Society A: Mathematical, Physical and Engineering Sciences, 371*, 20120413.

WHO, (2004). *Environmental Chemistry* (2nd edn.). New York: WH Freeman and Co.

WHO, (2012). *Animal Waste, Water Quality and Human Health.* Geneva, Switzerland, World Health Organization.

Winter, T. C., (2004). The hydrology of lakes. In: O'Sullivan, P. E., & Reynolds, C. S., (eds.), *The Lakes Handbook* (Vol. 1, pp. 61–78). Blackwell Science Ltd, Malden, MA, USA.

Withers, P. J. A., & Haygarth, P. M., (2007). Agriculture, phosphorus and eutrophication: a European perspective. *Soil Use and Management, 23*, 1–4.

Xie, Z., Xiao, H., Tang, X., Lu, K., & Cai, H., (2008). Interactions between red tide microalgae and herbivorous zooplankton: effects of two bloom-forming species on the rotifer *Brachionus plicatilis (*OF Muller). *Hydrobiologia, 600*, 237–245.

Zhuang, M., Sanganyado, E., Li, P., & Liu, W., (2019). Distribution of microbial communities in metal-contaminated nearshore sediment from Eastern Guangdong, China. *Environ Pollut., 250*, 482–492.

Zhuang, M., Sanganyado, E., Xu, L., et al., (2020a). High throughput sediment DNA sequencing reveals azo dye degrading bacteria inhabit nearshore sediments. *Microorganisms, 8*,1–15.

CHAPTER 12

Conservation of Biodiversity and Freshwater Ecosystems

GRESH CHANDER,[1] SONALI VERMA,[1] RUCHI SHAH,[2] and RAKESH KUMAR[1]

[1]Indian Council of Medical Research-Center for Advance Research, Shri Mata Vaishno Devi University, Katra, Jammu and Kashmir, India, E-mails: gresh.chander@smvdu.ac.in (G. Chander), Sonali.verma@smvdu.ac.in (S. Verma)

[2]Department of Biotechnology, Kashmir University, Jammu and Kashmir, India, E-mail: scientistdobt@gmail.com

ABSTRACT

Freshwater ecosystems provide services and goods of critical importance to human societies, yet they are among the most profoundly reformed ecosystems with over proportionate biodiversity loss. Freshwater biodiversity major threats include overutilization, fragmentation, water pollution, destruction or degradation of habitats, and assaults by non-native species. Amendments of natural flow ruled by land-use variations, river impoundments, human-made dams, and water abstraction often reflect the lotic systems. Comprehensive functional communications and developments in freshwater ecosystems provide a major challenge for scientists, but it is crucial for effective and sustainable restoration. The perception of 'integrative freshwater ecology and biodiversity conservation' (IFEBC) projected herein addresses the communications between biotic and abiotic factors on diverse stages of association quantitatively and qualitatively.

Biodiversity of Freshwater Ecosystems: Threats, Protection, and Management. Suhaib A. Bandh, PhD, Basharat Mushtaq, PhD, Mohammad Farooq Mir, PhD, & Shahid A. Wani (Eds.)

On the planet earth, it is important to get familiar with the perfect methods of balancing the human need for freshwater and its safety, which persists in supporting biodiversity. As per the recent climatical predictions within the next few eras, there will be a substantial impact of climate change in most of the freshwater bodies. Influences caused by a change in climate on freshwater ecosystems are noticeable chemically as well as physically. Core queries comprise modeling of the procedures in key aquatic habitats and their functionality based on the quantification and identification of aspects that control the Spatio-temporal dispersal of biodiversity and efficiency in aquatic biomes.

12.1 INTRODUCTION

This chapter analyzes the recent information and thoughts on the organization of biodiversity in aquatic ecosystems, which affect its Spatio-temporal distribution, its values for humankind, and major threats related to human activity. 'Integrative freshwater ecology and biodiversity conservation' (IFEBC) is proposed as a novel concept for addressing the need to link the applied conservation and ecological theory Geist (2011).

Global valuation of the freshwater biodiversity indicates that the perspective of biodiversity is greater than predicted. These systems' biodiversity potential is very much bigger than that would be expected from the area covered by inland waters (Balian et al., 2007). In their study, Balian, and his co-workers anticipated that 126,000 (approx. 10%) of the total 1.32 million species identified on Earth reside in freshwater. India's freshwater habitats support a remarkable ratio of the overall diversity of organisms, representing the highest taxonomic groups. Around 9.7% of the total fauna of India is related to freshwater ecosystems. However, it is predicted that faunal diversity may be overlooked due to the insufficient identification and exploration of several faunal groups, particularly invertebrates. As a consequence of the nearby freshwater diversity's authoritative status, these groups are yet to be appreciated (Chandra et al., 2017).

We live in the age of the phenomenal impact of human measures on global environments, natural developments, and the creatures that faith in them. In the stratigraphy of Earth, the range of social reach is universal and conspicuous (Waters et al., 2016). It is a warning to research scientists to propose the designation of a new geological period: the Anthropocene

(Cruzen, 2002). Biodiversity is at risk in this era of human dominance, including all life on Earth, with fresh water at the top level of this threat. A freshwater ecosystem occupies less than 1% of the Earth's surface.

Nevertheless, it contains 12% of all known species (Garcia-Moreno et al., 2014). Every creature on Earth for its survival needs water, but more than one lakh species, including us, require a distinct kind of water found in very rare supply in certain places, i.e., freshwater. The freshwater ecosystem parts are microbes, rocks, sunlight, animals, plants, water, and soil, which are found around this valuable resource.

To regulate the scale of existing changes within the freshwater ecosystems, the FLPI (freshwater living planet index) documents that the cumulative rates drop in the freshwater population, increasing from an average of 76% decline between 1970 and 2010 (World Wide Fund for Nature, 2014) to an 83% decline between 1970 and 2014 (World Wide Fund for Nature, 2018; Flitcroft et al., 2019). These declines are more than those reported for terrestrial and marine species. Similarly, in the 20th century, the inland wetlands have reduced by 64–71% (Davidson, 2014; Bastin et al., 2019; Reis et al., 2017). This fall in freshwater species and ecosystems are the outcome of the myriad pressures, such as greater water sewage from natural systems for the direct human consumption of irrigated farming; the change of physical traits such as the quantity, superiority, and timing of inflow, and related to this, water temperature and sediment loads; the indemnify of longitudinal and lateral connectivity; and the over-harvesting of species of commercial or subsistence importance (Dudgeon et al., 2006; Reid et al., 2019). Besides climate change, the seasonal pattern of temperature and river deliverance also increases the stress on species and habitats. The deterioration in wetlands, through agriculture drainages, is further degraded by shifting precipitation rules that change natural water availability (Kingsford, 2011).

12.2 THREATS AND PERCEPTIONS OF AQUATIC BIODIVERSITY

Among the non-specific communities, the biological diversity organization's level became popular since the Rio Summit of 1992. In a human remark, biodiversity is typically understood as the number and types of species in a given ecosystem or habitat, matching the concept of species richness (Krebs, 2009)—the estimations of species richness range from 3 to 30 million globally. MacArthur (1965) revealed that there is a spatial distribution in a given ecosystem based on the number of species, provincial species pool (γ-diversity), the number of species in each habitat type (α-diversity), and the profits of species between habitat types (β-diversity) (MacArthur, 1965). More elegant ideas of biodiversity increase the meaning of biodiversity beyond species' level and cover a much wider range of diversities, crossing the habitats, ecotones, and whole ecosystems to their species and biological communities to the discrepancy within species on genetic and phenotypic level (Figure 12.1). Worldwide, aquatic ecosystems are among the varied types of ecosystems. The freshwater ecosystems include the lotic and lentic surface waters, riparian systems in floodplains, phreatic, and hyporheic subsurface waters, and the ecotones (Ward and Tockner, 2001).

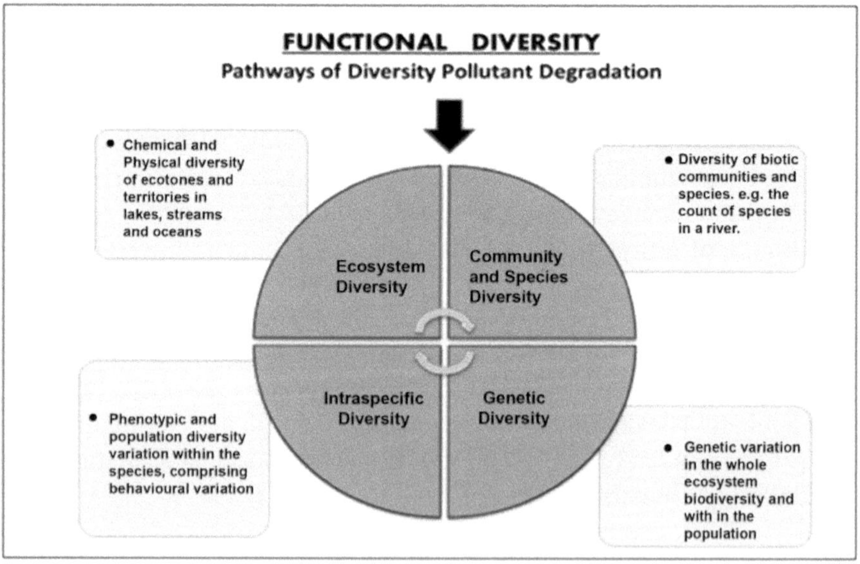

FIGURE 12.1 Biodiversity hierarchical levels in aquatic ecosystems.

Gradients in temperature, dissolved gasses, availability of light, salinity, and nutrients, along with biogeographic processes (e.g., isolation) have all contributed to the diversity of biological communities and species in aquatic ecosystems. About 10% of all animal species on Earth were projected to lie in freshwater systems (Balian et al., 2007).

The convention on biological diversity (CBD)-2019 distinguishes that the conservation wise usage of wetlands and freshwater ecosystems were gaps in the CBD 2011–2020 premeditated Plan. The International Union specifies for Conservation of Nature (IUCN) that extra stress needs to be given to conserving post-2020 freshwater biodiversity. In IUCN (2019a), it is mentioned that a sustainable future depends upon targeted activities (Flitcroft et al., 2019).

12.3 APPROACHES OF FRESHWATER CONSERVATION

The goals of adaptation of climate change and freshwater conservation unified water resource management provide an ideal platform. Ecological flow valuation and efficient conservation arrangements have evolved separately in respective terrestrial and aquatic kingdoms. Both are essential to the conservation of freshwater and can notify integrated water resources management. Environmental flow assessment contemplates dynamic flow provisions, measuring economic, ecological, and social costs of growth scenarios. The integration also proposes the chances to observe the freshwater conservation climate change impact (Nel et al., 2011). Within the larger context, the management of freshwater signifying the present geological age is regarded as the period through which human activity has had the maximum impact on the environment. Conservation of biodiversity is interlinked with the living of a limited number of people exposed in large provinces such as the Amazon (Reis et al., 2019), Lake Victoria (Van et al., 2019), and the Andes (Tognelli et al., 2019). Planning for water and ecosystem protection also seems to have upright influences among groundwater and surface water in aquifers and the regular lateral networks. The instantaneous impairment of freshwater biodiversity and enhancement of human dependencies on water resources demands that strict action must be taken to stop the worsening freshwater ecosystems and to sustain their valuable services. To be practical, these steps should be taken by a well-planned strategy, aimed at arranging freshwater for both people and biodiversity governed by sound recognized national procedures and policies. Preservation through systematic planning and proposals, structured, efficient, and scientifically defensible conservation framework is necessary. For achieving these objectives, positioning of significant geographic areas, and applying appropriate conservation activities is a must (Margules and Pressey, 2000).

Environmental planning's adaptive management tactic usually reflects the public and environmental needs and is generally effective in marine applications. However, a data-intensive management scheme permits the investigation of different managing inventions through composite modeling that supports decision-making (Langhans et al., 2019).

12.4 FACTORS INFLUENCING AQUATIC ECOSYSTEMS

In any aquatic ecosystem, inconsistency, and change is a natural process, and ecosystem groups and specific organisms have in numerous cases altered to different ecological circumstances. Human activities affect aquatic ecosystems by heavy pollution, variations in hydrological systems, and large-scale influences like global climate change. It is difficult to predict the disturbances and complications of aquatic ecosystems and the linkages within them. These linkages mean that impairment to one constituent of the environment can impact other components of the ecosystem. Our sympathetic action in an aquatic ecosystem can lead to good practices that minimize the influences of aquatic environments. The human factors which administer biodiversity are key to the progress of integrated aquatic ecosystem conservation. Impacts on biodiversity comprise of both evolutionary and ecological facets, including evolutionary speed, interspecific interactions, productivity, and disturbances (Krebs, 2009; Currie, 1991; Willig, Kaufman, and Stevens, 2003).

The hypothesis advises it of evolutionary speeds that biodiversity is the product of evolution. Evolutionary rates depend on temperature and, consequently, rely on the period through which the biota has established in a continuous method (Fischer, 1960; Turner, 2004). In Russia, Siberia, and Baikal lakes are the classic examples of the oldest lakes globally and comprise a diverse fauna compared to the newer lakes of Canada (Kozhov, 2013; Sanders, 1968).

Human insights for the quality of water and aquatic ecosystems have been reviewed to a smaller extent than precise services such as safe drinking water (Ioana-Toroimac et al., 2020). It is important to know how humans perceived the aquatic ecosystems because it can impact the communal support for management strategies (Genskow and Born, 2006; Larned et al., 2006; Kondolf and Yang, 2008; Ives and Kendal, 2014). The improved considerate of features that affect the perception, and role of esthetic preferences in observations, would support the human opinion of better aquatic ecosystems and enhance the management strategies. This could rationally be brought to increased community engagement with aquatic ecosystem facilities and, subsequently, improve social support and satisfaction of management activities (Moyle and Weiler, 2017). Some studies have revealed that in what way the human assessment of aquatic ecosystems can be motivated by the setting (Rooney et al., 2015), condition (Steinwender, Gundacker, and Wittmann, 2008), and type (Artell, Ahtiainen, and Pouta, 2013) of the resource itself.

Numerous demographic aspects were found to affect human perceptions of aquatic ecosystems (Figure 12.2). There is a mixed impact of age, but the younger population usually tends to be more critical and distrustful of water quality. As age increased, the alignment of human perceptions and scientific discernments of aquatic ecosystems traditionally improved. These findings somewhat emphasize education. Education was usually found to increase the position of social perceptions with scientifically derived assessments (Steinwender, Gundacker, and Wittmann, 2008).

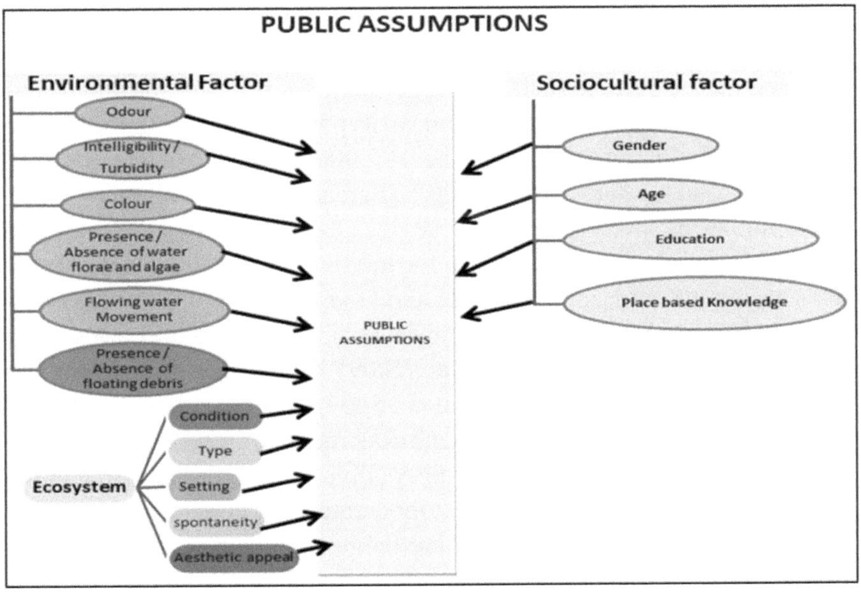

FIGURE 12.2 Factors inducing public notions of aquatic ecosystems.

Relations between ambient energy, biodiversity, and ecosystem productivity are also evident since fewer species can tolerate climatically unfavorable circumstances. In river ecosystems, the transitional disturbance hypothesis does not give the impression to apply under the circumstances with extremely mutable water flows and temperatures, which can significantly decline invertebrate diversity in the ecosystem (Death and Winterbourn, 1995). Some other instances exemplify that moderate disturbance by floods may be vital for the stream substratum's abundant variety (Geist and Auerswald, 2007). It has to be well-thought-out that high species productivity and diversity reason to non-specialized species can be undesirably connected with environment fitness and is thus not always required for managing conservation (Geist, Porkka, and Kuehn, 2006; Geist et al., 2007; Marchetti and Moyle, 2001).

12.5 BIODIVERSITY CRISIS AND EXTINCT

As shown by fossil records, many species have eventually become extinct (Townsend, Begon, and Harper, 2003). The recent observation of the rate of extinction in many taxonomic groups in this century is about 100–1,000 times and is comparable to the so-called mega extinction events caused by rapid climate changes, such as the Permian decline and the biodiversity loss in the Pleistocene (Rahel, 2002). Reports of species extinction are characteristically inferred from the 100,000 well-known species to the estimated total numbers (May, 2000). There is a lot of ambiguity about the total number of species, so that these rates can vary with a factor of about 100 (Pimm et al., 1995). Hotspots for biodiversity are freshwater ecosystems across the landscapes (Strayer and Dudgeon, 2010). They may be the most endangered ecosystems at the same time in the world (Dudgeon et al., 2006; Gleick, 2003; Gleick et al., 2009). Organisms of fresh water are the most threatened biota (Balian et al., 2007; Strayer and Dudgeon, 2010; Stein and Flack, 1997; Ricciardi and Rasmussen, 1999). Freshwater is a reserve for people that may be delimited, diverted, or contaminated in ways that settle down its value as a habitat for creatures. Approximately 0.01% of the world's water contains freshwater habitats and cover only about 0.8% of the Earth's surface (Schneider, 1996). Around 6% of all species and 1/3rd of all vertebrate species are limited to freshwater habitats (Dudgeon et al., 2006). North America's freshwater extinction rates are based on assorted data sets for unionid mussels, gastropods, crayfishes, fishes, and amphibians. That was estimated to average 0.5% per decade and projected to be as much as 4% per decade. It is five times higher than species losses calculated for any terrestrial fauna (Ricciardi and Rasmussen, 1999). The biodiversity crisis is responsible for streams and river ecosystem changes (Postel and Richter, 2012). Freshwater species of invertebrates have the highest extinction rates among all freshwater organisms. In aquatic biodiversity conservation concepts, they are often insufficiently considered despite their important functional roles in freshwater ecosystems (Strayer, 2006; Vaughn and Hakenkamp, 2001). Generating a conservative estimate of current extinction rates in freshwater mussels would be in a range of 2–7% per century. As a result, range among the highest current extinction rates among all taxonomic groups (Strayer et al., 2004; Dickman, Pimm, and Cardillo, 2007). A reduced number of conservative estimates have predicted future mussel extinction rates in continental North America in the order of more than 6% per decade (Ricciardi and Rasmussen, 1999).

12.6 BIOMARKERS FOR THE RECOGNITION OF STRESS AND TOXIC EFFECTS

Disturbances of aquatic ecosystems by pollutants are considered one of the five major threats to freshwater biodiversity (Dudgeon et al., 2006). Until now, studies of aquatic ecotoxicology characteristically focus on isolated populations (without an appreciation of diversity within species and genetic differentiation) and single terminal molecular responses.

IFEBC approaches take into account that the effects of the poisonous substance on aquatic species influence diverse levels of the biotic organization, from the molecular level (i.e., variations in the gene expression and transcription patterns) to effects on all communities. Based on the perception of IFEBC, one of the main challenges in aquatic ecotoxicology is to interlink the ecological relevance of a stressor on the one hand and the sensitivity of the response on the other hand.

Recently, environmental sciences and molecular biology have begun to fertilize mutually each other by recognizing the interactions between molecular events, the entire animal, and the populations' response (Steinberg, Stürzenbaum, and Menzel, 2008). Using the heavy metal (copper) and the pyrethroid insecticide esfenvalerate, links between tissue specific transcription of stress response genes and ecologically relevant endpoints such as mortality could be established, showing that integrative approaches are likely to be most successful (Geist and Auerswald, 2007). In a case study on striped bass (*Morone saxatilis),* cellular stress response markers for proteotoxicity (HSP70, HSP90), phase I detoxification mechanism (CYP1A1), metal-binding (metallothionein), as well as immune-function and pathogen-defense (TGF-þ, Mx-protein, nRAMP) were established. Quantitative real-time TaqMan PCR was used to examine tissue-specific changes in the transcriptome of liver, spleen, white muscle, anterior kidney, and gills after exposure to Cu and esfenvalerate and to link the observed responses to lethal (mortality) and sublethal (growth and swimming behavior) effects.

In standardized tests for regulatory ecotoxicology, the objective is to consider the adverse effect of a chemical stressor on a population rather than on the individual. Therefore, these tests are designed to evaluate endpoints in individuals that directly impact the population level, including sublethal effects. Specifically, effects on survival, growth, and reproduction are the most relevant endpoints. Data from these endpoints can be directly used in ecological risk assessment models to evaluate potential impacts. Whereas the ecological relevance of mortality of aquatic species is doubtless, the interpretation of links between sublethal effects, e.g., alterations of swimming behavior and growth, is more difficult. Experiments in larval fathead minnow (*Pimephales promelas*) can serve as examples that demonstrate that links between sublethal behavioral and development effects can be successfully validated in predation-prey exposures (Floyd, Geist, and Werner, 2008). This study illustrated that larval fish experiencing environmentally relevant, short term sublethal exposures to pyrethroids exhibited significant behavioral impairment, leading to reduced growth and increased susceptibility to predation, with potentially severe consequences for their ecological fitness. Consequently, the use of standardized tests that consider these sublethal effects should continue to be included in the toxicity testing of aquatic organisms.

The effects of pyrethroids have also been discussed in the context of the decline of delta smelt (*Hypomesus transpacificus),* an aquatic fish species endemic to the Sacramento-San Joaquin estuary in California, USA, which is listed under both the USA and California State Endangered Species Acts. A case study included developing a microarray to investigate the genome-wide effects of potentially causative stressors and applying this tool to assess the impacts of esfenvalerate on delta smelt. Selected genes were further analyzed using quantitative PCR analyzes. Pyrethroid exposure affected swimming behavior of larval delta smelt at concentrations as low as 0.0625 μg/l, which could be linked to significant differences in expression in genes involved in the neuromuscular activity. Alterations in the expression of genes associated with immune response, apoptosis, redox, osmotic stress, detoxification, growth, and development were also affected. Swimming impairment correlated significantly with the expression of aspartoa-cyclase (ASPA), an enzyme involved in brain cell function (Connon et al., 2009). This study shows that microarray technology is a useful approach in generating molecular biomarkers in non-model species by identifying specific genes that can be directly linked with sublethal toxicological endpoints, ultimately resulting in population-level impacts, e.g., variations in appearance levels of neuromuscular genes resulting in assessable swimming damages. Future research in the field of ecotoxicology and ecotoxicogenomics is likely to be successful if mechanistic links between population relevant effect response and the underlying molecular and cellular mechanisms can be established. The proposed integration of molecular and ecological responses may benefit from consideration of the approaches in the metabolic ecological theory, which try to link metabolic rate and resource allotment to survival, development, and reproduction with ecological procedures at all group levels from individuals to the biosphere (Brown et al., 2004). As demonstrated in the above examples, such research can greatly benefit from the inclusion of responses on multiple biological organization levels. Ecologically relevant interpretations require the inclusion of keystone and indicator species from natural aquatic ecosystems, typically non-model organisms.

12.7 CONCLUSIONS AND APPROACHES

The outcome of the available case studies advises that the conservation policies for the sustainable biodiversity of freshwater can be of utmost effectiveness if they integrate various biotic organization levels. An instantaneous contemplation of the dynamics of the availability of actual habitat availability, connectivity, and quality, along with plans for the conservation of the stressors of the environment, and the basis of genetic biodiversity, is likely to be most effective in the long run. IFEBC field subsequently needs a wide spectrum of ecological and molecular techniques to model, analyze, and interpret the functional methods in freshwater bio networks, which form the source of the ecosystem facilities resultant from the biodiversity of freshwater. At present, climate change is the major concern for the scientific society and peoples of all sections of the world. Industrialization, an increase in human population, agricultural, and unscientific practices have put an excessive burden on the climate associated entities. In the atmosphere, greenhouse gasses (GHGs) are increasing directly or indirectly due to the generation of municipal solid waste (MSW), the building of artificial dams, vehicular waste, human activities, and the burning of fossil fuels. Strategies of conservation and protecting entire fundamentals of freshwater biodiversity would assure that the use of water for peoples is sustainable while, in contrast, the scale of the threat to and loss of biodiversity is an indicator of the level to which present practices are not able to be maintained. Numerous current tactics have tried to transfer ideas to spatial freshwaters prioritization, preservation, and planning using different software and modeling techniques. Systematic tactics in the conservation of freshwater planning based on comprehensiveness, adequacy, representativeness, and efficiency (CARE) are considered a well-planned way onward. Yet, they are presently not consolidative, e.g., they do not understand genetics level of biodiversity. Developmental and population genetic studies can significantly contribute to the significances of conservation. Successful management is the part of the joint participation of managers and stakeholders to bridge a gap between research and implementation.

KEYWORDS

- **Anthropocene**
- **aquatic biodiversity**
- **biodiversity conservation**
- **convention on biological diversity**
- **greenhouse gases**
- **integrative freshwater ecology**

REFERENCES

Artell, J., Ahtiainen, H., & Pouta, E., (2013). Subjective vs. objective measures in the valuation of water quality. *Journal of Environmental Management, 130*, 288–296.

Balian, E. V., et al., (2007). The freshwater animal diversity assessment: an overview of the results. In: *Freshwater Animal Diversity Assessment* (pp. 627–637). Springer.

Bastin, L., et al., (2019). Inland surface waters in protected areas globally: Current coverage and 30-year trends. *PLoS One, 14*(1), e0210496.

Brown, J. H., et al., (2004). Toward a metabolic theory of ecology. *Ecology, 85*(7), 1771–1789.

Chandra, K., Gopi, K., Rao, D., Valarmathi, K., & Alfred, J. (2017). *Freshwater Faunal Diversity in India.*

Connon, R. E., et al., (2009). Linking mechanistic and behavioral responses to sublethal esfenvalerate exposure in the endangered delta smelt; *Hypomesus transpacificus* (Fam. *Osmeridae*). *BMC Genomics, 10*(1), 608.

Cruzen, P. J., (2002). Geology of mankind: the Anthropocene. *Nature, 415*, 23.

Currie, D. J., (1991). Energy and large-scale patterns of animal-and plant-species richness. *The American Naturalist, 137*(1), 27–49.

Davidson, N. C., (2014). How much wetland has the world lost? Long-term and recent trends in the global wetland area. *Marine and Freshwater Research, 65*(10), 934–941.

Death, R. G., & Winterbourn, M. J., (1995). Diversity patterns in stream benthic invertebrate communities: the influence of habitat stability. *Ecology, 76*(5), 1446–1460.

Dickman, C. R., Pimm, S. L., & Cardillo, M., (2007). The pathology of biodiversity loss: the practice of conservation. *Key Topics in Conservation Biology*, 1–16.

Dudgeon, D., et al., (2006). Freshwater biodiversity: importance, threats, status, and conservation challenges. *Biol. Rev. Camb. Philos. Soc., 81*(2), 163–182.

Fischer, A. G., (1960). Latitudinal variations in organic diversity. *Evolution, 14*(1), 64–81.

Flitcroft, R., et al., (2019). Theory and practice to conserve freshwater biodiversity in the Anthropocene. *Aquatic Conservation: Marine and Freshwater Ecosystems, 29*(7), 1013–1021.

Floyd, E. Y., Geist, J. P., & Werner, I., (2008). Acute, sublethal exposure to a pyrethroid insecticide alters behavior, growth, and predation risk in larvae of the fathead minnow (*Pimephales promelas*). *Environmental Toxicology and Chemistry: An International Journal, 27*(8), 1780–1787.

Garcia-Moreno, J., et al., (2014). Sustaining freshwater biodiversity in the Anthropocene. In: *The Global Water System in the Anthropocene* (pp. 247–270). Springer.

Geist, J., & Auerswald, K., (2007). Physicochemical stream bed characteristics and recruitment of the freshwater pearl mussel (*Margaritifera margaritifera*). *Freshwater Biology, 52*(12), 2299–2316.

Geist, J., (2011). Integrative freshwater ecology and biodiversity conservation. *Ecological Indicators, 11*(6), 1507–1516.

Geist, J., et al., (2007). Comparisons of tissue-specific transcription of stress response genes with whole animal endpoints of adverse effect in striped bass (*Morone saxatilis*) following treatment with copper and esfenvalerate. *Aquatic Toxicology, 85*(1), 28–39.

Geist, J., Porkka, M., & Kuehn, R., (2006). The status of host fish populations and fish species richness in European freshwater pearl mussel (*Margaritifera margaritifera*) streams. *Aquatic Conservation: Marine and Freshwater Ecosystems, 16*(3), 251–266.

Genskow, K. D., & Born, S. M., (2006). Organizational dynamics of watershed partnerships: a key to integrated water resources management. *Journal of Contemporary Water Research & Education, 135*(1), 56–64.

Gleick, P. H., (2003). Global freshwater resources: soft-path solutions for the 21st century. *Science, 302*(5650), 1524–1528.

Gleick, P. H., et al., (2009). *The World's Water 2008–2009: The Biennial Report on Freshwater Resources* (Vol. 6). Island Press.

Ioana-Toroimac, G., et al., (2020). Translating a river's ecological quality in ecosystem services: An example of public perception in Romania. *Ecohydrology & Hydrobiology, 20*(1), 31–37.

Ives, C. D., & Kendal, D., (2014). The role of social values in the management of ecological systems. *Journal of Environmental Management, 144*, 67–72.

Kingsford, R. T., (2011). Conservation management of rivers and wetlands under climate change–a synthesis. *Marine and Freshwater Research, 62*(3), 217–222.

Kondolf, G. M., & Yang, C. N., (2008). Planning river restoration projects: social and cultural dimensions. *River Restoration: Managing the Uncertainty in Restoring Physical Habitat*, 43–60.

Kozhov, M., (2013). *Lake Baikal, and its Life* (Vol. 11). Springer Science & Business Media.

Krebs, C. J., (2009). *Ecology: The Experimental Analysis of Distribution and Abundance*. San Francisco, CA: Pearson Benjamin Cummings.

Langhans, S. D., et al., (2019). Combining eight research areas to foster the uptake of ecosystem-based management in freshwaters. *Aquatic Conservation: Marine and Freshwater Ecosystems, 29*(7), 1161–1173.

Larned, S. T., et al., (2006). Macrophytes in urban stream rehabilitation: establishment, ecological effects, and public perception. *Restoration Ecology, 14*(3), 429–440.

MacArthur, R. H., (1965). Patterns of species diversity. *Biological Reviews of the Cambridge Philosophical Society*, (40), 510–533.

Marchetti, M. P., & Moyle, P. B., (2001). Effects of flow regime on fish assemblages in a regulated California stream. *Ecological Applications, 11*(2), 530–539.

Margules, C. R., & Pressey, R. L., (2000). Systematic conservation planning. N*ature, 405*(6783), 243–253.

May, R. M., (2000). The dimensions of life on earth. *Nature and Human Society*, 30–45.

Moyle, B. D., & Weiler, B., (2017). Revisiting the importance of visitation: Public perceptions of park benefits. *Tourism and Hospitality Research, 17*(1), 91–105.

Nel, J., et al., (2011). A new era in catchment management: Integration of environmental flow assessment and freshwater conservation planning. *Marine and Freshwater Research, 62*, 290–299.

Pimm, S. L., et al., (1995). The future of biodiversity. *Science, 269*(5222), 347–350.

Postel, S., & Richter, B., (2012). *Rivers for Life: Managing Water for People and Nature*. Island Press.

Rahel, F. J., (2002). Homogenization of freshwater faunas. *Annual Review of Ecology and Systematics, 33*(1), 291–315.

Reid, A. J., et al., (2019). Emerging threats and persistent conservation challenges for freshwater biodiversity. *Biological Reviews, 94*(3), 849–873.

Reis, V., et al., (2017). A global assessment of inland wetland conservation status. *Bioscience, 67*(6), 523–533.

Reis, V., et al., (2019). Characterizing seasonal dynamics of Amazonian wetlands for conservation and decision making. *Aquatic Conservation: Marine and Freshwater Ecosystems, 29*(7), 1073–1082.

Ricciardi, A., & Rasmussen, J. B., (1999). Extinction rates of North American freshwater fauna. *Conservation Biology, 13*(5), 1220–1222.

Rooney, R., et al., (2015). Replacing natural wetlands with stormwater management facilities: biophysical and perceived social values. *Water Research, 73*, 17–28.

Sanders, H. L., (1968). Marine benthic diversity: a comparative study. *The American Naturalist, 102*(925), 243–282.

Schneider, S. H., (1996). *Encyclopedia of climate and weather* (Vol. 2). Oxford University Press.

Stein, B. A., & Flack, S. R. (1997). species report card, The state of US plants and animals: Arlington, Va. *The Nature Conservancy*, 8.

Steinberg, C. E., Stürzenbaum, S. R., &. Menzel, R., (2008). Genes and environment—striking the fine balance between sophisticated biomonitoring and true functional environmental genomics. *Science of the Total Environment, 400*(1–3), 142–161.

Steinwender, A., Gundacker, C., & Wittmann, K. J., (2008). Objective versus subjective assessments of environmental quality of standing and running waters in a large city. *Landscape and Urban Planning, 84*(2), 116–126.

Strayer, D. L., & Dudgeon, D., (2010). Freshwater biodiversity conservation: recent progress and future challenges. *Journal of the North American Benthological Society, 29*(1), 344–358.

Strayer, D. L., (2006). Challenges for freshwater invertebrate conservation. *Journal of the North American Benthological Society, 25*(2), 271–287.

Strayer, D. L., et al., (2004). Changing perspectives on pearly mussels, North America's most imperiled animals. *BioScience, 54*(5), 429–439.

Tognelli, M. F., et al., (2019). Assessing conservation priorities of endemic freshwater fishes in the Tropical Andes region. *Aquatic Conservation: Marine and Freshwater Ecosystems, 29*(7), 1123–1132.

Townsend, C., Begon, M., & Harper, J., (2003). *Essentials of Ecology.* Blackwell Publishing. Malden, MA, USA.

Turner, J. R., (2004). Explaining the global biodiversity gradient: energy, area, history, and natural selection. *Basic and Applied Ecology, 5*(5), 435–448.

Van, S. A., et al., (2019). Potential impacts of agricultural development on freshwater biodiversity in the Lake Victoria basin. *Aquatic Conservation: Marine and Freshwater Ecosystems, 29*(7), 1052–1062.

Vaughn, C. C., & Hakenkamp, C. C., (2001). The functional role of burrowing bivalves in freshwater ecosystems. *Freshwater Biology, 46*(11), 1431–1446.

Ward, J., & Tockner, K., (2001). Biodiversity: towards a unifying theme for river ecology. *Freshwater Biology 46*(6), 807–819.

Waters, C. N., et al., (2016). The Anthropocene is functionally and stratigraphically distinct from the Holocene. *Science, 351*(6269).

Willig, M. R., Kaufman, D. M., & Stevens, R. D., (2003). Latitudinal gradients of biodiversity: pattern, process, scale, and synthesis. *Annual Review of Ecology, Evolution, and Systematics, 34*(1), 273–309.

Index

C

D

E